Fundamentals
of Environmental
Sampling and Analysis

BICENTENNIAL

1807

WILEY

2007

BICENTENNIAL

THE WILEY BICENTENNIAL–KNOWLEDGE FOR GENERATIONS

*E*ach generation has its unique needs and aspirations. When Charles Wiley first opened his small printing shop in lower Manhattan in 1807, it was a generation of boundless potential searching for an identity. And we were there, helping to define a new American literary tradition. Over half a century later, in the midst of the Second Industrial Revolution, it was a generation focused on building the future. Once again, we were there, supplying the critical scientific, technical, and engineering knowledge that helped frame the world. Throughout the 20th Century, and into the new millennium, nations began to reach out beyond their own borders and a new international community was born. Wiley was there, expanding its operations around the world to enable a global exchange of ideas, opinions, and know-how.

For 200 years, Wiley has been an integral part of each generation's journey, enabling the flow of information and understanding necessary to meet their needs and fulfill their aspirations. Today, bold new technologies are changing the way we live and learn. Wiley will be there, providing you the must-have knowledge you need to imagine new worlds, new possibilities, and new opportunities.

Generations come and go, but you can always count on Wiley to provide you the knowledge you need, when and where you need it!

WILLIAM J. PESCE
PRESIDENT AND CHIEF EXECUTIVE OFFICER

PETER BOOTH WILEY
CHAIRMAN OF THE BOARD

Fundamentals of Environmental Sampling and Analysis

Chunlong (Carl) Zhang

University of Houston-Clear Lake

BICENTENNIAL
1807
WILEY
2007
BICENTENNIAL

WILEY-INTERSCIENCE

A John Wiley & Sons, Inc., Publication

Library of Congress Cataloging-in-Publication Data:

Fundamentals of environmental sampling and analysis / Chunlong Carl Zhang.
    p. cm.
ISBN: 978-0-471-71097-4
1. Environmental sampling. 2. Environmental sciences–Statistical methods. I. Title.

GE45.S75Z43 2007

628–dc22   2006030899

To
My Parents

To My Wife Sue and
Two Sons
Richard and Arnold

Contents

10. Chromatographic Methods for Environmental Analysis   246

11. Electrochemical Methods for Environmental Analysis   289

Appendices   **402**

Index   423

Preface

The acquisition of reliable and defensible environmental data through proper sampling and analytical technique is often an essential part of the careers for many environmental professionals. However, there is currently a very diverse and diffuse source of literature in the field of environmental sampling and analysis. The nature of the literature often makes beginners and even skilled environmental professionals find it very difficult to comprehend the needed contents. The overall objective of this text is to introduce a comprehensive overview on the fundamentals of environmental sampling and analysis for students in environmental science and engineering as well as environmental professionals who are involved in various stages of sampling and analytical work.

Two unique features are evident in this book. First, this book presents a "know why" rather than a "know how" strategy. It is not intended to be a cookbook that presents the step-by-step details. Rather, fundamentals of sampling, selection of standard methods, chemical and instrumental principles, and method applications to particular contaminants are detailed. Second, the book gives an integrated introduction to sampling and analysis—both are essential to quality environmental data. For example, contrary to other books that introduce a specific area of sampling and analysis, this text provides a balanced mix of field sampling and laboratory analysis, essential knowledge in chemistry, statistics, hydrology, wet chemical methods for conventional chemicals, as well as various modern instrumental techniques for contaminants of emerging concerns.

Chapter 1 starts with an overview on the framework of environmental sampling and analysis and the importance for the acquisition of scientifically reliable and legally defensible data. Chapter 2 provides some background information necessary for the readers to better understand the subsequent chapters, such as review on analytical and organic chemistry, statistics for data analysis, hydrogeology, and environmental regulations relevant to sampling and analysis. The following two chapters introduce the fundamentals of environmental sampling—where and when to take samples, how many, how much, and how to take samples from air, liquid, and solid media.

Chapter 5 introduces the standard methodologies by the US EPA and other agencies. Their structures, method classifications, and cross references among various standards are presented to aid readers in selecting the proper methods. Quality assurance and quality control (QA/QC) for both sampling and analysis are also included in this chapter as a part of the standard methodology. Chapter 6 provides some typical operations in environmental laboratories and details the chemical principles of wet chemical methods most commonly used in environmental

xviii Preface

analysis. Prior to the introduction to instrumental analysis and applications in environmental analysis in Chapters 8–12, various sample preparation methods are discussed and compared in Chapter 7.

In Chapter 8, the theories of absorption spectroscopy for qualitative and quantitative analysis are presented. UV-visible spectroscopy is the main focus of this chapter because nowadays it is still the workhorse in many of the environmental laboratories. Chapter 9 is devoted to metal analysis using various atomic absorption and emission spectrometric methods. Chapter 10 focuses on the instrumental principles of the three most important chromatographic methods in environmental analysis, i.e., gas chromatography (GC), high performance liquid chromatography (HPLC), and ion chromatography (IC). Chapter 11 introduces the electrochemical principles and instrumentations for some common environmental analysis, such as pH, potential titrations, dissolved oxygen, ion selective electrodes, conductivity, and metal analysis using anodic stripping voltammetry. Chapter 12 introduces several analytical techniques that are becoming increasingly important to meet today's challenge in environmental analysis, such as various hyphenated mass spectrometries using ICP/MS, GS/MS and LC/MS. This last chapter concludes with a brief introduction to nuclear magnetic resonance spectroscopy (less commonly used in quantitative analysis but important to structural identifications in environmental research) and specific instrumentations including radiochemical analysis, electron scanning microscopes, and immunoassays.

The selections of the above topics are based on my own teaching and practical experience and the philosophy that sampling and analysis are equally important as both are an integral part of reliable data. An understanding of the principles of sampling, chemical analysis, and instrumentation is more important than knowing "specific how." It is not uncommon to witness how time and resources are wasted by many beginners in sampling and analysis. This occurs when samplers and analysts used improper sampling and analytical protocols. This occurs also when proper procedures from "step-by-step" cookbook were followed, but the samplers or analysts did not make proper modifications without understanding the fundamentals. Even skilled professionals are not immune to errors or unnecessary expenses during sampling, sample preparation and analytical processes if underlying fundamentals are either neglected or misinterpreted.

WHOM THIS BOOK IS WRITTEN FOR

This book is primarily targeted to beginners, such as students in environmental science and engineering as well as environmental professionals who are directly involved in sampling and analytical work or indirectly use and interpret environmental data for various purposes. It is, therefore, written as a textbook for senior undergraduates and graduate students as well as a reference book for general audiences.

Several approaches are used to enhance the book for such a wide usage. (1) Theories and principles will be introduced first in place of specific protocols followed by examples to promote logical thinking by orienting these principles to specific project applications. Questions and exercise problems are included in each chapter to help to understand these concepts. (2) Suggested readings are given at the end of each chapter for those who need further information or specific details from standard handbooks (EPA, ASTM, OSHA, etc.) or journal articles. This list of references is intended to provide an up-to-date single source information describing the details that readers will find for their particular needs. (3) Practical tips are given in most chapters for those who want to advance in this field. (4) A total of 15 experiments covering data analysis, sampling, sample preparation, and chemical/instrumental analysis are provided for use as a supplemental lab manual.

TO THE INSTRUCTOR

This 12-chapter book contains chapters of sampling (Chapters 2–4), standard methods and QA/QC (Chapter 5), wet analysis (Chapter 6), sample preparation (Chapter 7), and instrumental analysis (Chapters 8–12). It is designed to have more materials than needed for a one-semester course. Depending on your course focus (e.g., sampling vs. analysis) and the level of students, you can select topics most relevant to your course. These 12-chapters are better served in a lecture course but it can be used as a supplemental textbook for a lab-based course. Even though some chapter may not be covered in details for your particular course, the book can certainly be used as a valuable reference.

EXPERIMENTS

There are 15 experiments included in the book covering various sampling and analysis techniques, such as computer-based data analysis, field sampling, laboratory wet chemical techniques, and instrumental analysis. These 15 experiments alone should be suited to an instruction manual for a lab-based course. Several experiments require more advanced instruments such as IR, HPLC, GC, and GC-MS. If such resources are limited, the Instructor can select proper labs from the list, which can be based entirely on gravimetric, volumetric, and UV-visible spectroscopic analysis using analytical balance, burette, visible spectrometer, and pH meters. Each experiment can be completed during a 2–3 h lab session. Many of these lab procedures have been tested in author's lab with various TAs during the last several years. They have been debugged and are shown to be workable.

ACKNOWLEDGMENTS

Materials of this book have been used in several lecture and lab-based courses at the University of Houston. I first would like to thank my students for their comments, suggestions, and encouragement. These feedbacks are typically not technically detailed, but help me immeasurably to improve its readability. Certainly I would like to thank several technical reviewers (including several anonymous) from the review of the book proposal at the beginning to the review of the final draft. Thanks are due to Professors Todd Anderson (Texas Tech University), Joceline Boucher (Main Maritime Academy), and Kalliat Valsaraj (Louisiana State University). I have also invited several individuals to proofread all chapters in their expert areas, including Dr. Dennis Casserly (Associate Professor, University of Houston – Clear Lake), Dr. Dean Muirhead (Senior Project Engineer, NASA-JSC), Dr. Xiaodong Song (Principle Scientist, Pfizer Corporation), Mr. Jay Gandhi (Senior Development Chemist, Metrohm – Peak Inc.), and my thanks to them all.

Special thanks to Wiley's Executive Editor Bob Esposito for his vision of this project. Senior Editorial Assistant Jonathan Rose has been extremely helpful in insuring me the right format of this writing even from the beginning of this project. My sincere thanks to several other members of Wiley's editorial and production team, including Brendan Sullivan, Danielle Lacourciere, and Ekta Handa. It has been a pleasant experience in working with this editorial team of high professional standards and experience.

This book would not be accomplished without the support and love of my wife Sue and the joys I have shared with my two sons, Richard and Arnold. Even during many hours of my absence in the past two years for this project, I felt the drive and inspiration. This book is written as a return for their love and devotion. With that, I felt at some points the obligation of fulfilling and delivering what is beyond my capability.

ABOUT THE AUTHOR

Dr. Zhang is currently an associate professor in environmental science and environmental chemistry at the University of Houston-Clear Lake. He lectures extensively in the area of environmental sampling and analysis. With over two decades of various experience in academia, industries, and consulting, his expertise in environmental sampling and analysis covers a variety of first-hand practical experience both in the field and in the lab. His field and analytical work has included multimedia sampling (air, water, soil, sediment, plant, and waste materials) and analysis of various environmental chemicals with an array of classical chemical methods as well as modern instrumental techniques.

The author will be happy to receive comments and suggestions about this book at his e-mail address: zhang@uhcl.edu.

CHUNLONG (CARL) ZHANG
Houston, Texas
August 2006

Chapter 1

Introduction to Environmental Data Acquisition

This introductory chapter will give readers a brief overview of the purposes and scopes of environmental sampling and analysis. Sampling and analysis, apparently the independent steps during data acquisition, are in fact the integrated parts to obtain quality data—the type of data that are expected to sustain scientific and legal challenges. The importance of environmental sampling and the uniqueness of environmental analysis as opposed to traditional analytical chemistry are discussed. A brief history of classical to modern instrumental analysis is also introduced in this chapter.

1.1 INTRODUCTION

The *objectives of environmental sampling and analysis* may vary depending on the specific project (task), including regulatory enforcement, regulatory compliance, routine monitoring, emergency response, and scientific research. The examples are as follows:

1. To determine how much pollutant enters into environment through stack emission, wastewater discharge, and so forth in order to comply with a regulatory requirement.

Fundamentals of Environmental Sampling and Analysis, by Chunlong Zhang
Copyright © 2007 John Wiley & Sons, Inc.

2. To measure ambient background concentration and assess the degree of pollution and to identify the short- and long-term trends.

3. To detect accidental releases and evaluate the risk and toxicity to human and biota.

4. To study the fate and transport of contaminants and evaluate the efficiency of remediation systems.

This introductory chapter briefly discusses the basic process of environmental data acquisition and errors associated with field sampling and laboratory analysis. A unique feature of this text is to treat sampling and analysis as an entity. This is to say that sampling and analysis are closely related and dependent on each other. The data quality depends on the good work of both sampler and analyst.

The *importance of sampling* is obvious. If a sample is not collected properly, if it does not represent the system we are trying to analyze, then all our careful lab work is useless! A bad sampler will by no means generate good reliable data. In some cases, even if sampling protocols are properly followed, the design of sampling is critical, particularly when the analytical work is so costly.

Then what will be the data quality after a right sample is submitted for a lab analysis? The results now depend on the chemist who further performs the lab analysis. The *importance of sample analysis* is also evident. If the analyst is unable to define an inherent level of analytical error (precision, accuracy, recovery, and so forth), such data are also useless. The analyst must also know the complex nature of a sample matrix for better results. The analyst needs to communicate well with the field sampler for proper sample preservation and storage protocols.

1.1.1 Importance of Scientifically Reliable and Legally Defensible Data

All environmental data should be scientifically reliable. *Scientific reliability* means that proper procedures for sampling and analysis are followed so that the results accurately reflect the content of the sample. If the result does not reflect the sample, there is no claim of validity. Scientifically defective data may be a result of unintentional or deliberate efforts. The examples include the following:

- An incorrect sampling protocol (bad sampler)
- An incorrect analytical protocol (bad analyst)
- The lack of a good laboratory practice (GLP)
- The falsification of test results.

Good laboratory practice (GLP) is a quality system concerned with the organizational process and the conditions under which studies are planned, performed, monitored, recorded, archived, and reported. The term "defensible" means "the ability to withstand any reasonable challenge related to the veracity, integrity, or quality of the logical, technical, or scientific approach taken in a decision-making process." As scientific reliability must be established for all environmental data, legal defensibility may not be needed in all cases such as the one in most academic research projects. Legal defensibility is critical in many other circumstances such as in most of the industrial and governmental settings. Components of legally defensible data include, but are not limited to:

- Custody or Control
- Documentation
- Traceability.

Custody or Control: To be defensible in court, sample integrity must be maintained to remove any doubts of sample tampering/alteration. A *chain-of-custody* form can be used to prove evidence purity. The chain-of-custody form is designed to identify all persons who had possession of the sample for all periods of time, as it is moved from the point of collection to the point of final analytical results. "Control" over the sample is established by the following situations: (1) It is placed in a designated secure area. (2) It is in the field investigator's or the transferee's actual possession. (3) It is in the field investigator's or the transferee's physical possession and then he/she secures it to prevent tampering. (4) It is in the field investigator's or the transferee's view, after being in his/her physical possession (Berger et al., 1996).

Documentation: Documentation is something used to certify, prove, substantiate, or support something else. In a civil proceeding, documentation is anything that helps to establish the foundation, authenticity, or relevance leading to the truth of a matter. It may become evidence itself. Photos, notes, reports, computer printouts, and analyst records are all examples of documentation. The chain-of-custody form is one such piece of very important type of document. Documentary evidence is the written material that "speaks for itself."

Traceability: Traceability, otherwise known as a "paper trail," is used to describe the ability to exactly determine from the documentation that which reagents and standards were used in the analysis and where they came from. Traceability is particularly important with regard to the standards that are used to calibrate the analytical instruments. The accuracy of the standards is a determining factor on the accuracy of the sample results. Thus, each set of standards used in the lab should be traceable to the specific certificate of analysis (Berger et al., 1996).

Similar to the components of scientifically defective data, legally weak data may be a result of unintentional or deliberate efforts. In the case of misconduct, the person involved will be subjected to the same punishment as those who commit criminal acts, such as those frequently reported (Margasak, 2003). *Misconducts* in environmental sampling and analysis can be a result of the following:

- Outside labs oftentimes work for the people who hired them
- Poor training of employees in nongovernmental or private labs
- Ineffective ethics programs
- Shrinking markets and efforts to cut costs.

This text will not discuss the nontechnical or legal aspects of sampling and analysis, but the reader should be cautious about its importance in environmental data acquisition. The legal objective can influence the sampling and analytical effort by specifying where to sample, defining the method of sampling and analysis, adding additional requirements to a valid technical sampling design for evidentiary reasons, and determining whether the data are confidential (Keith, 1996). Sampling and analytical protocols must meet legal requirements for the introduction of evidence in a court, and the results of a technically valid sampling and analytical scheme might not be admissible evidence in a courtroom if the legal goal is not recognized early on in the presampling phase. A brief introduction to important environmental regulations will be presented in Chapter 2.

Practical tips

- In governmental and industrial settings, lab notebooks are the legal documents. In most of the research institutions, the rules about notebooks are loosely defined. In any case, date and signature are part of the GLP.

- Do not remove any pages and erase previous writings. Write contact information on the cover page in case of loss. A typical life of a laboratory notebook ranges from 10 to 25 years (Dunnivant, 2004).

1.1.2 Sampling Error vs. Analytical Error During Data Acquisition

During data acquisition, errors can occur anytime throughout the sampling and analytical processes—from sampling, sample preservation, sample transportation, sample preparation, sample analysis, or data analysis (Fig. 1.1). Errors of environmental data can be approximately divided into sampling error and analytical error. In general, these errors are of two types: (1) *Determinate errors* (systematic errors) are the errors that can be traced to their sources, such as improper sampling and analytical protocols, faulty instrumentation, or mistakes by operators. Measurements resulting from determinate error can be theoretically discarded. (2) *Indeterminate errors* (random errors) are random fluctuations and cannot be identified or corrected for. Random errors are dealt with by applying statistics to the data.

The quality of data depends on the integrity of each step shown in Figure 1.1. Although errors are sometimes unpredictable, a general consensus is that most errors come from the sampling process rather than sample analysis. As estimated, 90% or more is due to sampling variability as a direct consequence of the heterogeneity of environmental matrices. It is therefore of utmost importance that right samples are collected to be representative of the feature(s) of the parent material being investigated. A misrepresentative sample produces misleading information. Critical elements of a sample's representativeness may include the sample's physical dimensions, its location, and the timing of collection. If representativeness cannot be established, the quality of the chemical analysis is irrelevant (Crumbling et al., 2001).

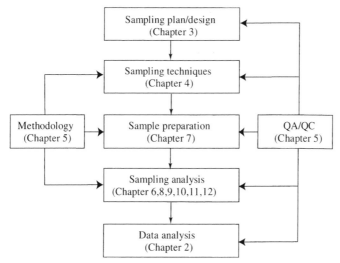

Figure 1.1.   Environmental data acquisition process

Unfortunately, there is a general misconception among many environmental professionals that the quality of data pertaining to a contaminated site is primarily determined by the nature of analytical methods used to acquire data. This assumption, which underestimates the importance of sampling uncertainties, can lead to a pronounced, negative impact on the cost-effective remediation of a contaminated site. In fact, it is of little use by placing an emphasis only on analytical uncertainty when sampling uncertainty is large and not addressed (Crumbling et al., 2001).

Errors in environmental data acquisition can be minimized through the proper design and implementation of a quality program. Two main parts of a quality program are quality control (QC) and quality assurance (QA). QC is generally a system of *technical* activities aimed at controlling the quality of data so that it meets the need of data users. QC procedures should be specified to measure the precisions and bias of the data. A QA program is a *management* system that ensures the QC is working as intended. QA/QC programs are implemented not only to minimize errors from both sampling and analysis, but many are designed to quantify the errors in the measurement. Details on QA/QC will be presented in Chapter 5.

1.2   ENVIRONMENTAL SAMPLING

1.2.1   Scope of Environmental Sampling

The scope of environmental sampling can be illustrated by a sample's life with the following seven consecutive steps (Popek, 2003). Since these steps are irreversible, a mistake can be detrimental. These seven steps of a sample's life are as follows: (1) a sample is planned ("conceived"); (2) a sampling point is identified; (3) the sample is collected ("born"); (4) the sample is transferred to the laboratory; (5) the sample is analyzed; (6) the sample expires and is discarded; and (7) the sample reincarnates as a chemical data point. Simply, the *scope of environmental sampling* addressed in this book will include the following aspects related to sampling:

- Where to take samples
- When to take samples
- How to take samples
- How many samples to take
- How often samples will be taken
- How much sample is needed
- How to preserve samples
- How long the sample will be stable
- What to take (air, soil, water)
- What to analyze (physical, chemical, biological)
- Who will take samples (sample custody)?

1.2.2 Where, When, What, How, and How Many

Many think of field sampling as simply going out to the field and getting some material, then bringing it back to a lab for analysis. Although such "random" sampling is often suggested as the basis of a sampling design, it is rarely the most appropriate approach, particularly when there is already some knowledge of the nature of the sample source and the characteristics of the variable being monitored. The choice of where (spatially) and when (temporally) to take samples generally should be based on sound statistics (simple random sampling, stratified random sampling, systematic sampling, composite sampling, as given in Chapter 3).

Although guidelines exists (detailed in Chapters 3 and 4), there is no set rule regarding the number, the amount, and the frequency of samples/sampling. For instance, the optimum number of samples to collect is nearly always limited by the amount of resources available. However, it is possible to calculate the number of samples required to estimate population size with a particular degree of accuracy. The best sample number is the largest sample number possible. But one should keep in mind that no sample number will compensate for a poor sampling design. In other words, quantity should not be increased at the expense of quality. Data in a poor quality will have more inherent error and, therefore, make the statistics less powerful.

1.3 ENVIRONMENTAL ANALYSIS

Whereas some of the environmental analyses are conducted in the field, the majority of the work is conducted in the laboratory. Depending on the data objectives, standard analytical methods should be consulted. This in turn depends on the analyte concentration, available instruments, and many other factors. Method selections are discussed in Chapter 5. This is followed by the common wet chemical analysis (Chapter 6) and instrumental methods (Chapters 8–12). Chapter 7 is devoted to sample preparation that is very critical to most of the complicated environmental samples.

1.3.1 Uniqueness of Modern Environmental Analysis

Environmental analyses are very different from the traditional chemical analysis entailed in analytical chemistry. Regardless, the majority of environmental analytical work has been traditionally and currently, are still, performed by the majority of analytical chemists. In the early days, this presented challenges to analytical chemists largely due to the complex nature of environmental samples and the analyses of trace concentrations of a wide variety of compounds in a very complex matrix. As Dunnivant (2004) stated, "my most vivid memory of my first professional job is the sheer horror and ineptitude that I felt when I was asked to analyze a hazardous waste sample for an analyte that had no standard protocol. Such was a life in the early days of environmental monitoring, when chemists trained in the isolated walls of a laboratory were thrown into the real world of sediment, soil, and industrial waste samples."

Today's analytical chemists, however, are better prepared for environmental analyses because a wealth of information is available to help them to conduct sampling and analysis. Nevertheless, professionals who are not specifically trained in this area need to be aware of the uniqueness of modern environmental analysis listed below (Fifield and Haines, 2000):

- There are numerous environmental chemicals, and the costs for analysis are high.
- There are numerous samples that require instrument automation.
- Sample matrices (water, air, soil, waste, living organisms) are complex, and matrix interferences are variable and not always predictable.
- Chemical concentrations are usually very low, requiring reliable instruments able to detect contaminants at ppm, ppb, ppt, or even lower levels.
- Some analyses have to be done on-site (field) on a continuous basis.
- Analysts need not only the technical competency but also the knowledge of regulations for regulatory compliance and enforcement purposes.

1.3.2 Classical and Modern Analytical and Monitoring Techniques

Environmental analyses are achieved by various "classical" and "modern" techniques. The difference between "classical" and "modern" analytical and monitoring techniques is a little arbitrary and ever changing as technology advances and many instruments continue to be modernized. For example, the analytical balance was, for a long time, considered to be the sophisticated instrument. With today's standards, however, those early balances were rather crude. Many of such yesterday's sophisticated instruments have become today's routine analytical tools. They are essential and continued to be improved and modernized (Rouessac and Rouessac, 1992).

Table 1.1 is a chronological listing of selected analytical instrumentations. In general, volumetric and gravimetric methods (wet chemicals) are the *classical methods*. Spectrometric, electrometric, and chromatographic methods are good examples of *modern analytical instrumentation*. Today's environmental analyses rely heavily on modern instrumentations. This, however, does not imply that classical methods will be vanished anytime soon. As can be seen, analytical instrumentations have become increasingly sophisticated to meet the analytical challenge. The advancement has made it possible to detect what would not had been detected in the past.

In the 1950s *gravimetric methods* were primarily used to estimate analyte's mass and concentration by precipitation, infiltration, drying, and/or combustion. Although gravimetric methods were sufficient, colorimetric and spectroscopic methods offered a greater precision. *Wet-chemistry-based methods* were developed that altered the spectroscopic properties of chemicals such as DDT and made these

Table 1.1 Selected milestones for analytical instrumentations

Year	Instrument
1870	First aluminum beam analytical balance by Florenz Sartorius
1935	First commercial pH meter invented by Arnold O. Beckman
1941	First UV-VIS spectrophotometer (Model DU) by Arnold O. Beckman
1944	First commercial IR instrument (Model 12) by Perkin-Elmer
1954	Bausch and Lomb introduced the Spectronic 20 (still used today in teaching)
1955	First commercial GC produced by the Burrell Corp (Kromo-Tog), Perkin-Elmer (Model 154), and Podbielniak (Chromagraphette)
1956	First spectrofluorometer by Robert Bowman
1956	First commercial GC/MS using time-of-flight (Model 12-101) by Bendix Corp.
1963	First commercially successful AA (Model 303 AA) by Perkin-Elmer
1963	First commercial NMR from a German company Bruker
1965	First true HPLC was built by Csaba Horváth at Yale University
1969	First commercially available FI-IR (FTS-14) introduced by Digilab
1970	First commercial graphite furnace AA by Perkin-Elmer
1974	First ICP-OES became commercially available
1977	First commercial LC–MS produced by Finnigan (now Thermo Finnigan)
1983	First commercial ICP–MS (Elan 250) by MDS Scix

AA = atomic absorption spectroscopy; ICP = inductively coupled plasma; OES = optical emission spectroscopy; IR = infrared spectroscopy; FT-IR = fourier transform infrared spectroscopy; NMR = nuclear magnetic resonance spectroscopy. See also Appendix A for a more detailed list of abbreviations and acronyms used in this text.

chemicals suitable for colorimetric determinations. These methods offered some advantages, but were still tedious and imprecise.

Soon, *chromatographic methods* made inroads into resolving separate components from a mixed solution. Chromatography is a physical method of separation that relies on the interaction of substances within a mixture when they are exposed to both a stationary and a mobile phase. Early thin-layer chromatography (TLC) and paper chromatography (PC) techniques used in the 1950s and 1960s separated compounds that were detected by measuring their intensity using ultraviolet and visible (UV-VIS) spectroscopy techniques.

Gas chromatography (GC) and high performance liquid chromatography (HPLC) were developed in the 1960s, becoming the methods of choice for residue analysis and replacing TLC and PC in the 1970s. GC and HPLC techniques efficiently "resolve" individual components from a complex mixture and can accurately quantify how much of an individual substance is present in the mixed component sample. The primary difference between GC and HPLC is that the former relies on resolution of substances being swept through a chromatography column in the gas phase at elevated temperatures, while the latter relies on the substance in a solution being chromatographically separated when in contact with a solid stationary phase.

Mass spectroscopy (MS) developments in the 1980s dramatically enhanced the scope of detection to include most semi- to nonpolar, and thermally-stable compounds. The first generation combined GC–MS relied on electron impact (EI) ionization to fragment the molecule into an array of positive mass ions. Continued refinement in GC–MS and maturation of HPLC-mass spectrometry has resulted in increasingly sensitive detections at even lower levels. Overall, the advances in instrumentation and technology have provided analysts with powerful tools to rapidly and accurately measure extremely low levels. Advanced analytical instrumentations tend to detect small quantities of almost anything.

REFERENCES

BERGER W, McCARTY H, SMITH R-K (1996), *Environmental Laboratory Data Evaluation*. Genium Publishing Corporation Amsterdaan, NY, pp. 2–1~2–62; pp. 3–1~3–12.

CLEAVES KS (2005), Ancillaries and Analyzers: Balances, pH meters, and more were critical to the rise of chemistry, *Enterprise of the Chemical Sciences*, 107–110 (http://pubs.acs.org/supplements/chemchronical2/107.pdf).

*CLEAVES KS, LESNEY MS (2005), Capitalizing on Chromatography: LC and GC have been key to the central science, *Enterprise of the Chemical Sciences*, 75–82 (http://pubs.acs.org/supplements/chemchronical2/075.pdf).

*CRUMBLING DM, GROENJES C, LESNIK B, LYNCH K, SHOCKLEY J, VAN EE J, HOWE R, KEITH L, McKENNA J (2001), Managing uncertainty in environmental decisions, *Environ. Sci. Technol.*, 35(19):404A–409A.

DUNNIVANT FM (2004), *Environmental Laboratory Excises for Instrumental Analysis and Environmental Chemistry*, John Wiley & Sons, Hoboken, NJ pp. xi–xii

---

ERICKSON B (1999), GC at a standstill.... Has the market peaked out at $1 billion? *Anal. Chem.*, 71(7):271A–276A.

ETTRE LS (2002), Fifty years of gas chromatography—The pioneers I knew, Part I. *LCGC North America*, 20(2):128–140 (http://www.chromatographyonline.com).

ETTRE LS (2002), Fifty years of gas chromatography—The pioneers I knew, Part II. *LCGC North America*, 20(5):452–462 (http://www.chromatographyonline.com).

FIFIELD FW, HAINES PJ (2000), *Environmental Analytical Chemistry*, 2nd Edition, Blackwell Science, Malden, MA, pp. 1–11.

*FILMORE D (2005), Seeing spectroscopy: Instrumental "eyes" give chemistry a window on the world, *Enterprise of the Chemical Sciences*, 87–91 (http://pubs.acs.org/supplements/chemchronical2/087.pdf).

KEITH LH (1996), *Principles of Environmental Sampling*, 2nd Edition, American Chemical Society Washington, DC.

MARGASAK L (2003), *Labs Falsifying Environmental Tests*, Houston Chronicle, January 22, 2003.

POPEK EP (2003), *Sampling and Analysis of Environmental Chemical Pollutants: A Complete Guide*, Academic Press San Diego, CA, pp. 1–10.

ROUESSAC F, ROUESSAC A (2000), *Chemical Analysis: Modern Instrumentation Methods and Techniques*, John Wiley & Sons West Sussex, England, p. xiii.

SKOOG DA, HOLLER FJ, NIEMAN TA (1992), *Principles of Instrumental Analysis*, 5th Edition, Saunders College Publishing Orlando, FL, pp. 1–15.

QUESTIONS AND PROBLEMS

1. Give examples of practice that will cause data to be scientifically defective or legally nondefensible.

2. Define and give examples of determinate errors and random errors.

3. Describe scopes of environmental sampling.

4. Why sampling and analysis are an integral part of data quality? Between sampling and analysis, which one often generates more errors? Why?

5. Describe how errors in environmental data acquisition can be minimized and quantified?

6. How does environmental analysis differs from traditional analytical chemistry?

7. Describe the difference between "classical" and "modern" analysis.

8. A chemist is arguing that sampling is *not* as important as analysis. His concern is whether there is a need for a sampling course in an environmental curriculum. His main rationale is that most employers and governmental agencies already have their own training courses and very specific and detailed procedures. Another consultant, on the contrary, argues that sampling should be given more weight than analysis. His main rationale is that a company always sends samples to commercial laboratories for analyses, and you do not become an analytical chemist by taking one course. For each of these two arguments, specify whether you agree or disagree and clearly state your supporting argument why you agree or disagree.

Chapter 2

Basics of Environmental Sampling and Analysis

In the real world, environmental sampling and analysis are performed by a group of people with different areas of expertise. This might be very different from small projects or research work in academic settings, where the sampler and analyst may be the same person. In any case, a basic knowledge of environmental sampling and analysis is essential for all individuals. As illustrated in this chapter, environmental sampling and analysis is an interdisciplinary field including chemistry, statistics, geology, hydrology, and law to name just a few. It is the purpose of this chapter to introduce readers to some of the relevant environmental sampling and analysis basics. An in-depth and detailed knowledge on particular topics is beyond the scope of this text and readers are referred to the suggested readings for further details.

2.1 ESSENTIAL ANALYTICAL AND ORGANIC CHEMISTRY

2.1.1 Concentration Units

Proper concentration units are important in environmental reporting. Although chemists prefer to use units such as percentage (% in m/v or v/v) or molarity (M) for chemical concentrations, these units are too large for common environmental

Fundamentals of Environmental Sampling and Analysis,   by Chunlong Zhang
Copyright © 2007 John Wiley & Sons, Inc.

contaminants that have low concentrations. Concentration units also vary with the types of environmental media (air, liquid, or solid) as described below.

Chemicals in Liquid Samples

For chemicals in liquid samples (water, blood, or urine), the mass/volume (m/v) unit is the most common. Depending on the numerical value, the concentration is expressed as mg/L, μg/L, or ng/L. For freshwater or liquids with density equal to 1.0 g/mL, the following units are equivalent. The units will not be equivalent if sea water, and denser or lighter liquids are concerned.

$$1 \text{ mg/L} = 1 \text{ ppm} \qquad 1 \text{ μg/L} = 1 \text{ ppb} \qquad 1 \text{ ng/L} = 1 \text{ ppt} \qquad (2.1)$$

Chemicals in Solid Samples

For chemicals in solid samples (soil, sediment, sludge, or biological tissue), the concentration unit is mass/mass (m/m) rather than mass/volume. Units such as mg/L or μg/L should not be used to express contaminant concentration in solid samples. The following two sets of units are equivalent:

$$1 \text{ mg/kg} = 1 \text{ ppm} \qquad 1 \text{ μg/kg} = 1 \text{ ppb} \qquad 1 \text{ ng/kg} = 1 \text{ ppt} \qquad (2.2)$$

In reporting such mass/mass units in solid samples, it should specify whether the mass is on a wet basis or on a dry basis. A dry basis is commonly adopted for comparison purposes when such samples have a large variation in moisture contents. A subsample should always be collected for the determination of the moisture content in addition to concentration measurement. Use the following to convert from wet basis to dry basis:

$$\text{mg/kg on dry basis} = \text{mg/kg on wet basis}/(1 - \% \text{ moisture}) \qquad (2.3)$$

Chemicals in Gaseous Samples

For chemicals in air, both sets of mass/volume (mg/m^3, $μg/m^3$, and ng/m^3) and volume/volume (ppm_v, ppb_v, ppt_v) are used, but they are not equivalent:

$$1 \text{ mg/m}^3 \neq 1 \text{ ppm} \qquad 1 \text{ μg/m}^3 \neq 1 \text{ ppb} \qquad 1 \text{ ng/m}^3 \neq 1 \text{ ppt}$$

To convert between these two sets of units at standard temperature and pressure (25°C, 1 atm), the following formula can be used.

$$\text{mg/m}^3 = \text{ppm} \times MW/24.5 \qquad \text{ppm} = (\text{mg/m}^3) \times (24.5/MW) \qquad (2.4)$$

Note that at other temperature and pressure conditions, the conversion factor (24.5) will be slightly different.

Practical tips

- Report concentrations with the right unit and right significant figures. Use common sense in reporting when to choose mg/L, µg/L, ng/L; or mg/kg, µg/kg, ng/kg; or even % if the concentration is high. Avoid very large or very small numbers.

- It is a good practice to use mass/volume (mg/L), mass/mass (mg/kg), and mass/volume (mg/m^3) for contaminants in water, soil, and air, respectively. Avoid using ppm, ppb, and ppt because they can be ambiguous. If chemicals in air are concerned, use ppm_v, ppb_v, and ppt_v to denote that the chemical concentration in air is based on the volume ratio.

- There are many other contaminant specific units, such as mg/m^3 for atmospheric particulate matter ($PM_{2.5}$ or PM_{10}) and lead (here ppm is an invalid unit because PM and lead cannot be expressed in volume); mg/L as $CaCO_3$ for water hardness, acidity, and alkalinity; nephelometric turbidity units (NTU) for turbidity; µS/m for conductivity; and pCi/L for radionuclides. The salinity unit is parts per thousands (ppt or ‰), which should not be confused with parts per trillion (ppt).

EXAMPLE 2.1. Maximum contaminant level (MCL) according to the U.S. EPA for 2,3, 7,8-TCDD (dioxin) in drinking water is 0.00000003 mg/L. Convert this concentration to ppt and molarity (M). What is the equivalent number of dioxin molecules per liter of water? The molecular weight of dioxin is 322 g/mol.

SOLUTION:

$$0.00000003 \, \text{mg/L} = 0.00000003 \, \text{mg/L} \times \frac{1 \, \text{ppm}}{1 \, \text{mg/L}} \times \frac{10^6 \, \text{ppt}}{1 \, \text{ppm}} = 0.03 \, \text{ppt}$$

$$0.00000003 \, \text{mg/L} = 0.00000003 \, \frac{\text{mg}}{\text{L}} \times \frac{1 \, \text{g}}{10^3 \, \text{mg}} \times \frac{1 \, \text{mol}}{322 \, \text{g}} = 9.32 \times 10^{-14} \, \frac{\text{mol}}{\text{L}}$$

$$= 9.32 \times 10^{-14} \, \text{M}$$

Using the Avogadro's constant of 6.022×10^{23} mol^{-1}, we can obtain the number of dioxin molecules per liter of water:

$$9.32 \times 10^{-14} \, \frac{\text{mol}}{\text{L}} \times 6.022 \times 10^{23} \, \frac{1}{\text{mol}} = 5.61 \times 10^{10} \, \frac{\text{dioxin molecule}}{\text{L}}$$

Dioxin is probably the most toxic contaminant on earth. The above calculations indicate that the unit of molarity (M) is a very large unit for such a low concentration. Even ppm (mg/L) is still not an appropriate unit. The above calculation also reveals that even though the MCL is so low at the ppt level, the number of molecules is significant, that is, 56.1 billion toxic dioxin molecules per liter of drinking water!

2.1.2 Common Organic Pollutants and Their Properties

There are an estimated total of 7 million chemicals with approximately 100,000 present in the environment. The number of chemicals commonly considered as important environmental pollutants, however, is likely a few hundred. Among them, few are routinely measured in environmental laboratories. A smaller list of "priority" chemicals has been established by various organizations based on selected factors such as quantity, persistence, bioaccumulation, potential for transport to distant locations, toxicity, and other adverse effects (Mackay, 2001). Such chemicals receive intense scrutiny and analytical protocols are always available.

The Stockholm Convention, signed by more than 90 countries in Sweden in May 2001, listed 12 key *persistent organic pollutants (POPs).* Commonly known as "the dirty dozen", the POPs are aldrin, chlordane, DDT, dieldrin, endrin, heptachlor, hexachlorobenzene, mirex, toxaphene, PCBs, dioxins, and furans. Most of the 12 key POPs are no longer produced in the United States, and their uses in developing countries have also declined. Of the 12 chemicals, 10 were intentionally produced by industries and 9 were produced as insecticides or fungicides. Only two of the 12 chemicals, dioxins and furans, are unintentionally produced in combustion processes (Girard, 2005).

The U.S. EPA published 129 *priority pollutants* (114 organic and 15 inorganic) in water. Many other countries have similar priority pollutant lists reflecting their pollution sources and monitoring priorities, such as the black and grey list by the European Union in 1975 and the black list by China (58 organic and 10 inorganic). The lists should be regarded as dynamic. For instance, the number of drinking water contaminants regulated by the U.S. EPA has increased from about five in 1940 to more than 150 in 1999 (Weiner, 2000). Appendix B is a list of 114 organic pollutants on the U.S. EPA priority list. Their chemical structures, molecular weights, and aqueous solubilities are also provided in this appendix.

Important chemical pollutants can be divided into nine categories based primarily on their chemical characteristics. Figure 2.1 lists the structure of several important organic compounds by their functional groups.

1. Element: Metals (Cu, Zn, Pb, Cd, Ni, Hg, Cr) and metalloids (As, Se)

2. Inorganic compounds: Cyanide, CO, NOx, asbestos

3. Organo-metallic and metalloid compounds: Tetraethyl lead and tributyl tin

4. Hydrocarbons: Saturated and unsaturated aliphatic and aromatic hydrocarbons including BTEX compounds (benzene, toluene, ethylbenzene, and xylene) and polycyclic aromatic hydrocarbons (PAHs)

5. Oxygenated compounds: Alcohol, aldehyde, ether, organic acid, ester, ketone, and phenol

6. Nitrogen compounds: Amine, amide, nitroaromatic hydrocarbons, and nitrosamines

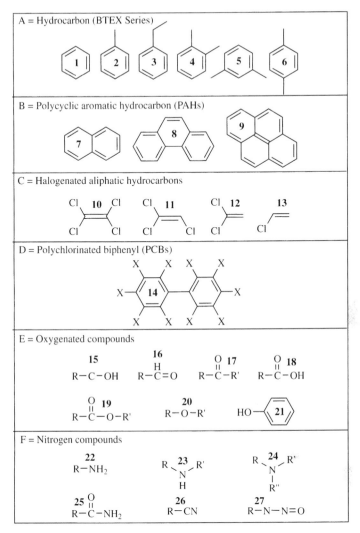

Figure 2.1   Structure of several important organic compounds by their functional groups (where R, R', R'' are alkyl groups and X can be either a chlorine or a hydrogen). (1) benzene, (2) toluene, (3) ethylbenzene, (4) o-xylene, (5) m-xylene, (6) p-xylene, (7) naphthalene, (8) phenanthrene, (9) pyrene, (10) tetrachloroethylene, (11) trichloroethylene, (12) dichloroethylene, (13) vinyl chloride, (14) polychlorinated biphenyl, (15) alcohol, (16) aldehyde, (17) ketone, (18) acid, (19) ester, (20) ether, (21) phenol, (22) primary amine, (23) secondary amine, (24) tertiary amine, (25) amide, (26) nitrile, (27) nitrosamine

7. Halogenated hydrocarbons: Aliphatic and aromatic halogenated hydrocarbon, polychlorinated biphenyls (PCBs), and dioxins

8. Organosulfur compounds: Thiols, thiophenes, mercaptans, and many pesticides

9. Phosphorus compounds: Many pesticides to replace organo-chlorine pesticides

Chemicals are sometimes categorized by one of their properties or by their analytical procedures (the so-called analytical definition). Examples include those loosely defined compound categories based on the following:

1. Density: *Heavy metals* are elements of generally higher atomic weight with specific gravities greater than five and especially those toxic to organisms such as Cu, Zn, Pb, Cd, Ni, Hg, Cr and the metalloid (As)

2. Volatility: *Volatile organic compounds (VOCs)* are group of compounds that vaporize at a relatively low (room) temperature. VOCs have boiling points below 200°C. Examples of VOCs include trichloroethane, trichloethylene, and BTEXs. *Semi-volatile organic compounds (SVOCs)* evaporate slowly at normal room temperature and are operationally defined as a group of organic compounds that are solvent-extractable and can be determined by chromatography including phenols, phthalates, PAHs, and PCBs. Volatility as defined by Henry's law constant will be introduced in Chapter 7.

3. Extractability: Extraction with methylene chloride initially under basic conditions to isolate the *base/neutral fraction* (B/N) and then acid condition to obtain the *acidic fraction* (mostly phenolic compounds).

2.1.3   Analytical Precision, Accuracy, and Recovery

Accuracy is the degree of agreement of a measured value with the true or expected value. Accuracy is measured and expressed as *% recovery* and calculated according to:

$$\% \text{ Recovery} = \text{Analytical value} \times 100/\text{True value} \qquad (2.5)$$

The true value (concentration) is rarely known for environmental samples. Thus, accuracy is typically determined by spiking a sample with a known quantity of a standard:

$$\% \text{ Recovery on spike} = \frac{\text{Spiked sample value} - \text{Sample value}}{\text{Spiked Value}} \times 100 \qquad (2.6)$$

An accurate test method should achieve a percentage recovery close to 100%. To determine percentage recovery, the amount of spiking chemical should be close to (0.5 ~ 2 times) the analytical concentration. In no case should the amount spiked exceed three times the concentration of the analyte.

Precision is the degree of mutual agreement among individual measurements $(x_1, x_2, \ldots x_n)$ as the result of repeated applications under the same condition. Precision measures the variation among measurements and may be expressed in different terms. The first term is *standard deviation* (s), which is defined as follows for a finite set of analytical data (generally $n < 30$):

$$s = \sqrt{\frac{\Sigma(x_i - \bar{x})^2}{n - 1}} \qquad (2.7)$$

Two other terms are *relative standard deviation* (RSD) and *relative percent difference* (RPD):

$$RSD = (s/\bar{x}) \times 100 = CV \times 100 \tag{2.8}$$
$$RPD = [(A - B)/(A + B)/2] \times 100 \tag{2.9}$$

where CV denotes coefficient of variation ($CV = s/\bar{x}$) and RPD is the difference between the duplicate values (A and B) divided by the average of the duplicate values and multiplied by 100. RSD is used for the evaluation of multiple replicate measurements, whereas RPD is used for measuring precision between two duplicate measurements.

Precision and Accuracy along with three other qualitative descriptors (Representativeness, Comparability, and Completeness), or PARCC, are termed *data quality indicators* (DQIs). The PARCC parameters enable us to determine the validity of environmental data.

Both accuracy and precision are needed to determine the data quality in a quantitative way. In an analogy, accuracy is how close you get to the bull's-eye, whereas precision is how close the repetitive shots are to one another. Hence, a good precision does not guarantee an accurate analysis. On the contrary, it is nearly impossible to have a good accuracy without a good precision.

2.1.4 Detection Limit and Quantitation Limit

The *method detection limit* is one of the secondary DQIs. The U.S. EPA defines method detection limit (MDL) as "the minimum concentration that can be measured and reported with 99% confidence that the analyte concentration is greater than zero" (EPA, 1984). To determine MDL, an analyte-free matrix (reagent water or laboratory-grade sand) is spiked with the target analyte at a concentration that is 3–5 times the estimated MDLs. This sample is then measured at a minimum of seven times. From these measurements, a standard deviation (s) is calculated (Eq. 2.7), and the MDL is calculated according to the formula:

$$MDL = s \times t \tag{2.10}$$

where t is obtained from "Student's t value Table" (Appendix C2), corresponding to $t_{0.98}$ and degree of freedom $df = n - 1$, where n is the number of measurements. For example if $n = 7$ (i.e., $df = 6$), then $t = 3.143$ at a 99% confidence level.

The MDLs are specific to a given matrix, method, instrument, and analytical technique. The MDL, however, is not the lowest concentration we can accurately measure during a routine laboratory analysis. EPA, therefore, uses another term, *practical quantitation limit* (PQL), and defines it as "the lowest concentration that can be reliably achieved within specified limits of precision and accuracy during routine operating condition" (EPA, 1996). Typically, laboratories choose PQL value at 2–10 times its MDLs. Therefore, the PQL may be also defined as "a concentration that is 2–10 times greater than the MDL and that represents the lowest point on the calibration curve during routine laboratory operation" (Popek, 2003). Sample PQLs are highly matrix dependent. For example, in measuring polynuclear aromatic hydrocarbons (EPA Method 8310), the matrix factor ranges from 10 (groundwater) to 10,000 (high level soil/sludge by sonication).

2.1.5   Standard Calibration Curve

The calibration curve or standard curve is a plot of instrumental response (absorbance, electrical signal, peak area, etc.) vs. the concentrations of the chemical of interest. Five or more standard solutions of known concentrations are first prepared to obtain the calibration curve – normally a linear regression equation:

$$y = ax + b \qquad (2.11)$$

where y is the instrument response and x is the concentration of the chemical. The slope factor (a) is the instrument response per unit change in the concentration and it is termed as the *calibration sensitivity*. Since intercept b equals the instrumental response when analyte is absent ($x = 0$), ideally b is a small number close to zero (i.e., the calibration curve ideally will run through the origin ($x = 0, y = 0$)). The concentration of the unknown sample is calculated using the standard curve and the measured instrument response. The calibration curve is essential for almost all quantitative analysis using spectrometric and chromatographic methods.

To obtain the regression equation ($y = ax + b$), one should be familiar with a spreadsheet program such as Excel. Procedures for the regression analysis using Excel are as follows:

- Select Tools|Data Analysis
- Select Regression from Analysis Tool list box in the Data Analysis dialog box. Click the OK button
- In the Regression dialog box

  ○ Enter data cell numbers in Input Y Range and Input X Range
  ○ Select Label if the first data cells in both ranges contain labels
  ○ Enter Confidence Level for regression coefficient (default is 95%)
  ○ Select Output Range and enter the cell number for regression output
  ○ Select Linear Fit Plots and/or others if needed
  ○ Click the OK button

- In the regression output, r^2, slope (a) and intercept (b) can be readily identified.

Figure 2.2 illustrates several common outcomes when the calibration curve is plotted. The calibration curve (a) is ideal due to its good linearity and the sufficient number of data points. Curve (b) is nonlinear, which is less favorable and

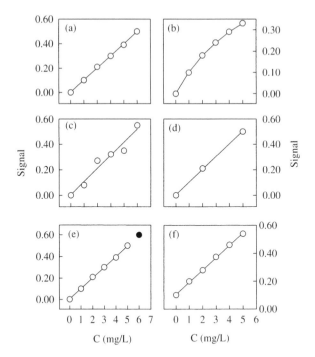

Figure 2.2 Correct and incorrect calibration curves: (a) Ideal, (b) Poor linearity, (c) Data too scattered, (d) Insufficient data points, (e) Incorrect concentration range, and (f) Systematic error (Meyer, VR, 1997, Reproduced with Permission, John Wiley & Sons Limited)

effort should be made to improve it prior to use. Data in (c) are too scattered, and you should redo to correct the errors. The curve in (d) does not have enough data points (five is normally the minimum). The concentration range for curve (e) is not correct, either the sample has to be diluted or a wider range of concentrations should be used. Calibration curve (f) does not run through the origin, implying the presence of some systematic errors such as positive or negative matrix interference (Meyer, 1997).

Practical tips

- It is always a good laboratory practice to run the calibration curve along with samples in the same batch, particularly when there is a considerable instrument drift. It takes more time to prepare and run, but it will improve the data validity and you can probably save time at the end.
- When a dilution is made and sample (injection) volume is different from the standard solution, corrections are needed to calculate sample concentration. In this case, a plot of instrument response vs. analyte mass (rather than concentration) is more appropriate.

2.2   ESSENTIAL ENVIRONMENTAL STATISTICS

2.2.1   Measurements of Central Tendency and Dispersion

Data from populations or samples can be characterized by two important *descriptive statistics*: the center (central tendency) and the variation (dispersion) of the data. Here the *population* refers to an entire subject body to be investigated (all fishes in a lake, all contaminated soils in a property), and a *sample* is the portion of the body taken in order to represent the true value of the population.

The central tendency is measured by three general methods: the mean, the median, and the mode. Population mean (μ) is the true value, but in most case, it is an unknown. Sample mean ($\bar{x}$), or the commonly used *arithmetic mean*, is calculated by adding all values and dividing by the total number of observations:

$$\bar{x} = \frac{1}{n} \sum_{i=1}^{n} x_i \qquad (2.12)$$

The *geometric mean* (GM) is used when the data is not very symmetrical (skewed), which is common in environmental data. To use the geometric mean, all values must be non-zero. Geometric mean is defined as:

$$GM = \sqrt[n]{x_1 \times x_2 \times x_3 \ldots \times x_n} \qquad \text{or} \qquad GM = \exp\left(\sum \ln(x_i)/n\right) \qquad (2.13)$$

Put another way, geometric mean is calculated by adding the logarithms of the original data, summing these, and dividing by the number of samples (n).

The *median* ($\tilde{x}$) is defined as the middle value of the data set when the set has been arranged in a numerical order. If the number of measurements (n) is odd, the median is the number located in the exact middle of the list (Eq. 2.14). If the number (n) is even, the median is found by computing the mean of the two middle numbers (Eq. 2.15).

$$\tilde{x} = x_{(n+1)/2} \qquad \text{(if } n \text{ is odd)} \qquad (2.14)$$
$$\tilde{x} = 1/2(x_{n/2} + x_{n/2+1}) \qquad \text{(if } n \text{ is even)} \qquad (2.15)$$

Median is sometimes a preferred choice if there are some extreme values. Unlike the mean, it is less affected by extreme data points (i.e., more "robust") and it also is not affected by data transformations. The median is, therefore, recommended for data distributions that are skewed as a result of the presence of outliers. Median is also advantageous as compared to mean in environmental data when non-numerical data exist, such as "not detected" or "below detection limit".

The *mode* (M) is the value in the data set that occurs most frequently. When two values occur with the same greatest frequency, each one is a mode and the data set is said to be bimodal. For the data set 1,1,2,2,2,9,10,11,11, the mode is 2. Mode does not always exist, that is, when no value is repeated, the data set is described as having no mode. The main use of mode is limited to a quick measure of the central value since little or no computation is needed.

Data variation or dispersion is a characteristic of how spread out the data points are from each other. Three common methods of describing the dispersion are variance, standard deviation, and range. The *range* of a set of data is the difference between the maximum and minimum values. Like mean, the range is influenced by extreme (low probability) observations, and its use as a measure of variation is limited. The *population variance*, denoted as σ^2, is defined as:

$$\sigma^2 = 1/N\Sigma(x_i - \mu)^2 \tag{2.16}$$

Sample variance (s^2) is the square of the standard deviation s (Eq. 2.7). It is calculated:

$$s^2 = 1/(n-1)\Sigma(x_i - \bar{x})^2 \tag{2.17}$$

In Eqs 2.16 and 2.17, N is the total population size and n the sample size. The term $(x_i - \bar{x})$ in Eq. 2.17 describes the difference between each value of a sample and the mean of that sample. This difference is then squared so that the differences of points above and below the mean do not cancel out each other. The sample variance, s^2, is an unbiased estimator of population variance, σ^2, meaning that the values of s^2 tend to target in on the true value of σ^2 in a population. This concept is important for inferential statistics. However, the unit of variance (units^2) is different from the original data set and must be used with care. *Standard deviation* is the square root of variance, that is, $\sigma = (\sigma^2)^{1/2}$ for a population and $s = (s^2)^{1/2}$ for a sample.

Variance is an important quantity, particularly, because variances are additive and the overall variance for a process may be estimated by summing the individual variances for its consistent parts, as expressed in equations. For example:

$$\sigma^2(\text{overall}) = \sigma^2(\text{field sampling}) + \sigma^2(\text{lab analysis})$$

This is an important relationship when considering the sources of the overall variability in a sampling and analytical process.

One should note that manual calculation of the above descriptive statistics using calculator are seldom a practice nowadays, for large data sets. Readers should get familiar with the Descriptive Statistics using Excel spreadsheet. This can be done by first selecting Tools|Data Analysis and then selecting Descriptive Statistics from the Analysis Tool list and click OK.

2.2.2 Understanding Probability Distributions

Normal (Gaussian) Distribution

This is a symmetrical and bell-shaped distribution of a given data set. Many of the environmental data sets are generally skewed, and a normal distribution is acquired after log-transformation of the original data. These data sets are termed as *log-normal distribution*. Mathematically, the normal distribution can be described as the following function:

$$y = \frac{e^{-\frac{1}{2}\left(\frac{x-\mu}{\sigma}\right)^2}}{\sigma\sqrt{2\pi}} \qquad (-\infty < x + \infty) \qquad (2.18)$$

This function can be defined by two parameters, the mean (μ) and the standard deviation (σ). Figure 2.3 shows the normal distribution of three different data sets (a, b, and c) and a normalized "standard normal distribution" (d). The *standard normal distribution* is similar in shape to the normal distribution with the exception that the mean equals zero and the standard deviation (σ) equals one. The normalization is done by converting x to z according to:

$$z = \frac{x - \mu}{\sigma} \qquad (2.19)$$

In the standard normal distribution, about 68% of all values fall within 1 σ, about 95% of all values fall within 2 σ, and about 99.7% of all values fall within 3 σ. These percentages are the probabilities at these particular ranges. With the use of standard normal distribution table (Appendix C1), we can find the probability at any given x value. For instance, we want to know the probability when x has a value from 10 to 20, $P(10 \leq x \leq 20)$, or, the probability when x is greater than 50, $P(x \geq 50)$. To use the standard normal table, the x value is first normalized into z by the given μ and σ, and Appendix C1 is then used to obtain the probability. An example of application using environmental data is given below.

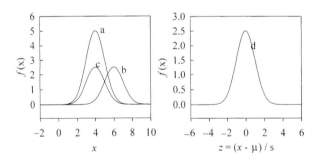

Figure 2.3 Normal (Gaussian) distribution (left) and standard normal distribution (right). Examples shown are: (a) $\mu = 4$, $\sigma = 0.5$, (b) $\mu = 6$, $\sigma = 1$, (c) $\mu = 4$, $\sigma = 1$, and (d) $\mu = 0$, $\sigma = 1$

EXAMPLE 2.2.   The background concentration of Zn in soils of Houston area is normally distributed with a mean of 66 mg/kg and a standard deviation of 5 mg/kg.

(a) What percentage of the soil samples will have a concentration <72 mg/kg?

(b) What percentage of the soil samples will have a concentration >72 mg/kg?

(c) What percentage of the soil samples will have a concentration between 61 and 72 mg/kg?

SOLUTION: For $x = 72, z = (72 - 66)/5 = 1.2; x = 61, z = (61 - 66)/5 = -1$

(a) $P(x < 72) = P(z < 1.2) = P(-\infty < z < 0) + P(0 < z < 1.2)$
$= 0.5 + 0.3849$ (Appendix C1) $= 0.8849(88.49\%)$

(b) $P(x > 72) = P(z > 1.2) = 1 - P(z < 1.2) = 1 - 0.8849 = 0.1151(11.51\%)$

(c) $P(61 < x < 72) = P(-1 < z < 1.2) = P(-1 < z < 0) + P(0 < z < 1.2)$
$= P(0 < z < 1) + P(0 < z < 1.2) = 0.3413 + 0.3849 = 0.7262(72.62\%)$

Student's *t* Distribution

The Student's t distribution was first described by William Gosset in 1908 for use with small sample sizes ($n \leq 30$). It is similar to normal distribution in that it is symmetric at about zero, and it becomes identical to normal distribution when n is large ($n \to \infty$). The t distribution is used to describe the distribution of the mean ($\bar{x}$) rather than x. The mathematical equation for the t distribution is:

$$t = \frac{\bar{x} - \mu}{s/\sqrt{n}} \tag{2.20}$$

where $s/\sqrt{n}$ is the *standard error of the mean* (SEM). It is often used to calculate the *confidence interval* (CI):

$$CI = \bar{x} \pm t_{n-1,1-\frac{\alpha}{2}}\left(\frac{s}{\sqrt{n}}\right) \tag{2.21}$$

where the degree of freedom (df) is $n - 1$, and $1 - \alpha/2$ is the one-sided confidence level. Note that a one-sided confidence level of 95% is equivalent to 90% of the two-sided confidence level. The t value can be obtained from Appendix C2.

EXAMPLE 2.3. If the same data shown in Example 2.2 were obtained from a sample size of $n = 8$ rather than a population (i.e., $\bar{x} = 66$ mg/kg, $s = 5$ mg/kg), what are the confidence intervals for two-sided confidence levels of (a) 90% or (b) 99%?

SOLUTION:

(a) $90\%(2\text{-sided}) = 95\%(1\text{-sided}) : 1 - \alpha/2 = 0.95, t_{7,0.95} = 1.895$ (Appendix C2)

$CI = 66 \pm 1.895 \times 5/\sqrt{8} = 66 \pm 3.3$ mg/kg (i.e., $62.7 \sim 69.3$ mg/kg)

(b) $99\% (2\text{-sided}) = 99.5\% (1\text{-sided}): 1 - \alpha/2 = 0.995, t_{7,0.995} = 3.499$ (Appendix C2)

$CI = 66 \pm 3.499 \times 5/\sqrt{8} = 66 \pm 6.2$ mg/kg (i.e., $59.8 \sim 72.2$ mg/kg)

F-distribution

Unlike the normal distribution and the *t*-distribution, *F-distribution* is skewed to the right. It is an appropriate model for the probability distribution of the ratio of two independent estimates of two population variances.

$$\text{Population } 1 = N(\mu_1, \sigma_1) \rightarrow \text{Sample } 1 : s_1^2, n_1$$
$$\text{Population } 2 = N(\mu_2, \sigma_2) \rightarrow \text{Sample } 2 : s_2^2, n_2$$

The F-value is defined as the ratio of the dispersion of two distributions:

$$F = s_1^2 / s_2^2 \tag{2.22}$$

where $s_1 > s_2$, otherwise switch s_1 and s_2. A comparison of F-value with tabulated $F_{\text{critical value}}$ (Appendix C3) will show whether the difference in variances is significant or not. The difference is significant when calculated F-value is greater than $F_{\text{critical value}}$. F distribution is the basis for the *analysis of variance* (ANOVA) and is useful, for example, (1) to compare one method of analysis to another method, (2) to compare the performance of one lab to that of another lab, and (3) to compare contaminated levels at two locations (are they statistically significantly different?). An alternative of F-test is to use Excel with One-Way ANOVA as described below:

- Select Tools|Data Analysis
- Select ANOVA: Single Factor from Analysis Tool list box in the Data Analysis dialog box. Click the OK button
- In the ANOVA: Single Factor dialog box. Enter the Input Range edit box
- Select the Grouped by Column option button and select the Labels in First Row check box
- Select Output Range and enter 0.05 in the Alpha edit box
- Click the OK button

EXAMPLE 2.4.   A groundwater sample was colleted from a benzene-contaminated site. This sample was split into two subsamples, which were then sent to two labs for independent analysis. Lab A results were as follows: 1.13, 1.14, 1.17, 1.19, 1.29 mg/L ($s_2 = 0.0639$ mg/L, $n_2 = 5$); Lab B results were as follows: 1.11, 1.12, 1.14, 1.19, 1.25, 1.34 mg/L ($s_1 = 0.0893$ mg/L, $n_1 = 6$). Are these results significantly different between two labs at a 5% level?

SOLUTION:

(a) Calculate F-value: $F = s_1^2 / s_2^2 = (0.0893)^2 / (0.0639)^2 = 1.95$

(b) $df_1 = n_1 - 1 = 5$ and $df_2 = n_2 - 1 = 4$, use Appendix C3, $F_{5,4,0.05} = 6.26$

(c) Because 1.95 is less than 6.26, the difference in the test results between the two labs is not significantly different.

For this example, one should also try the alternative method with one-way ANOVA using Excel. This method is more conservative because it differentiates the variance between the test groups (labs) and within the group (Details are omitted here, but the answer is: $F = 0.0257 < F_{\text{critical value}} = 5.117$, or p-value $= 0.876 > 0.05$, not significant).

2.2.3 Type I and II Errors: False Positive and False Negative

The previous discussion refers to a confidence level such as 95%. This is to say that the probability for correctly accepting a null hypothesis is 95%, and there is 5% chance of making an error. To illustrate two types of errors, let us first use an easy-to-understand example by setting:

Null hypothesis H_0: Defendant is innocent

Alternative hypothesis H_a: Defendant is not innocent

A *Type I error* occurs if we reject a null hypothesis when it is true, and a *Type II error* occurs if we accept a null hypothesis when it is false. This is equivalent to saying a Type I error is to put innocent people in jail whereas a Type II error is to set criminals free (Table 2.1).

Table 2.1 Type I and Type II errors in a court decision

Court decision	True state of nature	
	Defendant is innocent (H_0 is true)	Defendant is guilty (H_a is true)
Defendant is innocent	Correct decision	Type II error
Defendant is guilty	Type I error	Correct decision

If we denote the probability of committing a Type I error (reject H_0 when H_0 is true) and Type II error (accept H_0 when H_0 is false) is α and β, respectively, then we have:

$$\alpha + \beta = 1 \qquad (2.23)$$

The above equation tells us that reducing one error will certainly increase the other type of error. Let us now look at the following example to demonstrate why this concept is important in environmental analysis and decision-making (Table 2.2).

Table 2.2 Type I and Type II errors in environmental analysis

Possible decision based on analytical results of samples	True state of nature	
	Contaminant not present (H_0 is true)	Contaminant present (H_a is true)
Contaminant not detected	Correct decision	Type II error (False negative)
Contaminant detected	Type I error (False positive)	Correct decision

Reference: Keith (1991)

A Type I error in environmental analysis is to falsely report that an area is contaminated whereas in fact it is not (*false positive*), and a Type II error is to falsely report that an area is clean whereas it is in fact contaminated (*false negative*). Apparently, false positive is more conservative, but it will cost us unnecessary resources to clean-up the site. On the contrary, false negative will create health and environmental hazard/risk because of the failure in detecting the contaminant.

The next question is how to set up the hypothesis. An accepted convention in hypothesis testing is to always write H_0 with an equality sign ($H_0 = \#$). If the hypothesis is directional, then the one-tailed test is appropriate (i.e., $H_a > \#$ or $H_a < \#$). Otherwise, use a two-tailed test: $H_a \neq \#$ in which H_a does not specify a departure from H_0 in a particular direction.

2.2.4 Detection of Outliers

Outliers are observations that appear to be inconsistent with the remainder of the collected data. A rule of thumb is that one should never just throw data away without an explanation or reason. One should first examine the following cause of possible outliers: (1) The outlier may be the true outlier because of mistakes such as sampling error, analytical error (instrumental breakdowns, calibration problems), transcription, keypunch, or data-coding error. (2) The outlier may be because of inherent spatial/temporal variation of data or unexpected factors of practical importance such as malfunctioning pollutant effluent controls, spills, plant shutdown, or hot spots. In these cases, the suspected outlier is actually not the true outlier.

Possible remedies for suspected outliers are: (1) to *replace* the incorrect data by re-doing the sampling and analysis. Correct the mistake and insert the correct value. (2) To *remove* the outlier using a statistical test. However, no data should be discarded solely on the basis of a statistical test. (3) To *retain* the outliers and use a more robust statistical method that is not seriously affected by the presence of a few outliers. Discussed below are several major statistical tests to identify outliers of small data sets. For large data sets ($n > 25$) or data that are not normally distributed, the reader should consult other references for details.

Z-test

The mean and standard deviation of the entire data set are used to obtain the z-score of each data point according to the following formula. The data point is rejected if $z > 3$.

$$z = \frac{x - \bar{x}}{s} \tag{2.24}$$

This test does not require any statistical table, but is not very reliable because both mean and standard deviation themselves are affected by the outlier. Also the z-test requires the normality of the data set.

Grubbs' test

This test is recommended by the U.S. EPA (1992). First arrange the data in an increasing order $(x_1, x_2, \ldots x_n)$, the lowest (x_1) and the highest (x_n) data points can be tested for outlier by the following t-statistic:

$$T = \frac{\bar{x} - x_1}{s} \qquad \text{or} \qquad T = \frac{x_n - \bar{x}}{s} \tag{2.25}$$

The T value is then compared with a critical T value (Table 2.3) at the specified sample size and selected α (normally 5%). The data point is deemed to be an outlier if $T < T_{\text{critical}}$. The test can be continued to test for further possible outlier(s) (i.e., $x_2, x_{n-1}, \ldots$). EPA also recommends the use of Grubbs' test for log-normal data after the log transformation.

Table 2.3 Critical values used for the Grubbs' test

Degree of freedom $(n - 1)$	Critical T $T_{0.05}$	$T_{0.01}$	Degree of freedom $(n - 1)$	Critical T $T_{0.05}$	$T_{0.01}$
3	1.153	1.155	13	2.331	2.607
4	1.463	1.492	14	2.371	2.659
5	1.672	1.749	15	2.409	2.705
6	1.822	1.944	16	2.443	2.747
7	1.938	2.097	17	2.475	2.785
8	2.032	2.221	18	2.504	2.821
9	2.110	2.323	19	2.532	2.854
10	2.176	2.410	20	2.557	2.884
11	2.234	2.485	30	3.103	2.745
12	2.285	2.550	50	2.956	3.336

Dixon's test

This test is used for a small number of outliers with a sample size between 3 and 25 (Gibbons, 1994). First arrange data in an increasing order, $x_1 < x_2 < x_3 < \ldots < x_n$, then calculate the D value based on the number of observations (Table 2.4). An

Table 2.4 Formula to calculate D values for the Dixon's test

n	If x_n is suspected	If x_1 is suspected
$3 \sim 7$	$D_{10} = \frac{x_n - x_{n-1}}{x_n - x_1}$	$D_{10} = \frac{x_2 - x_1}{x_n - x_1}$
$8 \sim 10$	$D_{11} = \frac{x_n - x_{n-1}}{x_n - x_2}$	$D_{11} = \frac{x_2 - x_1}{x_{n-1} - x_1}$
$11 \sim 13$	$D_{21} = \frac{x_n - x_{n-2}}{x_n - x_2}$	$D_{21} = \frac{x_3 - x_1}{x_{n-1} - x_1}$

outlier is identified by comparing D with $D_{critical}$ shown in Table 2.5 (i.e., reject the data point suspected of an outlier if $D > D_{critical}$).

Table 2.5   Critical values for the Dixon's test[*]

Critical value	n	Risk of false rejection			
		0.5%	1%	5%	10%
	3	0.994	0.988	0.941	0.886
	4	0.926	0.899	0.765	0.679
D_{10}	5	0.821	0.780	0.642	0.557
	6	0.740	0.698	0.560	0.482
	7	0.680	0.637	0.507	0.434
	8	0.727	0.683	0.554	0.479
D_{11}	9	0.677	0.635	0.512	0.441
	10	0.639	0.597	0.477	0.409
	11	0.713	0.679	0.576	0.517
D_{21}	12	0.675	0.642	0.546	0.490
	13	0.649	0.615	0.521	0.467

*Reference: Robracher (1991)

The Dixon's test is not as efficient as the *Rosner test* (EPA recommended) for detecting multiple outliers of large data sets. If multiple outliers are suspected, the least extreme value should be tested first and then the test repeated. However, the power of the test decreases as the number of repetitions increases.

2.2.5 Analysis of Censored Data

Suppose that the following mercury concentrations were obtained from several representative samples of a drinking water supply: 2.5, <1.0, 1.9, 2.6 µg/L. The analyst reported the limit of quantitation is 1.0 µg/L. What are the mean and standard deviation? Does the quality of drinking water meet the regulatory standard of 2.0 µg/L (the maximum contaminant level allowable in drinking water)?

The non-numerical data such as "not detected," "less than" in the above example are the so-called *censored data*. The mean and standard deviation from such measurements cannot be computed, hence a comparison with the legal standard cannot be made. The presence of such censored data makes it difficult or impossible to apply typical statistical analyses, specifically parametric tests typically used for hypothesis testing, (i.e., comparisons of means, variances, or regression analyses). Tests that can be applied will also have a decreased reliability as the amount of censoring increases. Censoring can be a formidable problem in environmental analysis, particularly for trace contaminants in waters where the amount of data censoring can be as high as 80–95% of the data set.

How should we handle censoring data? First, deletion of censored data is probably the worst procedure and should never be used because it causes a large and variable bias in the parameter estimates (Helsel, 1990). After deletion, comparisons made are between the upper $X\%$ of one group with the upper $Y\%$ of another group. Further statistical tests would have little or no meaning. Another common mistake is to substitute the "less than" values with zero. This produces estimates of the mean and median that are biased low.

There are many approaches available but not readily used by practitioners in dealing with censored data. Some of these approaches require an in-depth statistical theory, and interested readers should refer to the suggested readings (Cohen, 1991; Ginevan and Splitstone, 2004; Helsel, 2004; Manly, 2001). Fortunately the U.S. EPA, in a manual on practical methods of data analysis, recommended a very simple approach depending on the degree of data censoring (EPA, 1998):

- With less than 15% of values censored, replace these values with DL (detection limit), DL/2, or a very small value.
- With 15–50% of censored values, use maximum likelihood estimate (MLE), or estimate the mean excluding the same number of large values as small values.
- With more than 50% of the values censored, just base an analysis on the proportion of data values above a certain level.

The simple substitution method, although widely used, has no theoretical basis. The MLE method assumes a distribution of the data and the likelihood function (which depends on both the observed and the censored values) is maximized to estimate population parameters. To use these various methods, a computer program called UNCENSOR can be downloaded for censored data analysis (Newman et al., 1995). These data analysis tools are seldom found in standard statistical packages.

2.2.6 Analysis of Spatial and Time Series Data

Often, environmental samples are taken for a period of time at a specific location or a snapshot concentration pattern related to samples' physical locations in a two-dimensional space around a source are taken. Data acquired through these sampling plans are either concentrations vs. time (*temporal data*) or concentrations vs. x and y (*spatial data*). The researcher may just want to know the average temporal concentration (daily, weekly, monthly, or yearly average), so that this temporal average can be compared to background concentrations or to regulatory standard concentrations. In most cases, the researcher may further want to know whether the concentrations present any temporal or spatial patterns.

One should be cautious that averaging all temporal or spatial data to obtain the mean concentration and standard deviation is not a statistically sound approach, if these data show serious temporal or spatial patterns. In applying the standard approach of calculating mean and standard deviation, we have assumed the randomness of the data. The randomness is apparently violated for data with temporal or spatial patterns. In either case, such patterns should be identified and particular statistical tools should be used.

The analysis of specific spatial or time series data is a highly specialized area in statistics. Its theories are not even covered in standard statistics texts and are beyond the reach of most environmental professionals. Fortunately, there are particular computer programs that are manageable for the general reader. One example of such useful software is the Geostatistical Environmental Assessment Software (Geo-EAS) that was developed for performing two-dimensional analyses of spatially distributed data (downloadable from www.epa.gov/ada/csmos/models/geoeas.html). One of its unique features is the ability to plot contour maps of pollutant concentration or other variables. Examples of applications include Pb and Cd concentrations in soils surrounding a smelter site, outdoor atmospheric NO_2 concentrations in metropolitan areas, and regional sulfate deposition in rainfall (EPA, 1991). Another software package, called ANNIE, was developed by the USGS for the analysis of time-series data (USGS, 1995).

2.3 ESSENTIAL HYDROLOGY AND GEOLOGY

2.3.1 Stream Water Flow and Measurement

Measurement of stream flow (i.e., discharge in cubic feet per second, or cfs) is important for surface water quality monitoring. In many large waterways in the US, flow data, gauge height, and discharge, can be obtained in almost real time from the USGS Web site (http://water.usgs.gov/index.html). If the sampling site is close to the USGS gauging station, field personnel can refer to the gauge height during sampling. The gauge height (ft) is converted to stream flow (cfs) by a rating curve that is calibrated by the USGS for that particular station.

When a stream flow needs to be determined, it is often conducted prior to the collection of chemical and biological water samples. There are several acceptable methods for flow measurement, but a flow meter is most commonly used. The flow meter measures the water flow velocity (V), in feet per second, which is then converted to the discharge by using the cross-sectional area of the stream (A). A variety of flow meters are available; some are based on electromagnetism (e.g., Marsh-McBirney, Montedoro-Whitney) and others are propeller type (Price Pigmy).

Since velocities vary both horizontally and vertically across a stream cross-section (velocity is slower toward the bottom and near the bank), multiple measurements must be made to determine the average flow:

$$Q = (W_1 \times D_1 \times V_1) + (W_2 \times D_2 \times V_2) + \ldots (W_n \times D_n \times V_n) \qquad (2.25)$$

where W and D denote width and depth, respectively. The number (n) of flow cross sections depends on the stream width (use 0.5 ft if total width is less than 5 ft; use minimum of $n = 10$ and $n = 20–30$ for a wider stream). For the same cross section, measure the velocity twice, once at 20% and once at 80% of the water depth if water is deeper than 2.5 ft, or measure once only at 60% if water depth is less than 2.5 ft (TNRCC, 1999). The average velocity is at 6/10th of the total water depth from the water surface.

2.3.2 Groundwater Flow in Aquifers

Groundwater flow measurement is extremely important. Imagine if your measurements result in the groundwater flowing north when it is actually flowing south, sampling expenses will be wasted and the whole remediation effort will be in vain. Unlike surface water, groundwater flows very slowly (several feet to several hundred feet per year) and the flow direction cannot be visualized but has to be determined by other means, such as using wells (Chapter 3). To understand the groundwater flow, which is critical to groundwater sampling and monitoring, some basic concept and terminologies are introduced below.

Groundwater is stored in an *aquifer*, which typically is soil or rock that has a high porosity and permeability. Here a *rock* refers to two or more minerals with defined chemical composition (e.g., SiO_2 for quartz), and *soil* is the product of the physical breakdown of rock. Soils that produce pumpable groundwater include gravel, sand, and silt. Groundwater in clay can be abundant but it is often not possible to pump water from a clay formation. In bedrock, abundant water can generally be found in highly fractured rock. A few unfractured rock types can also support abundant water, for example, sandstone, limestone, and dolomite.

The grain size dictates where and how fast water and contaminant will flow. In large-sized gravels, the permeability is high, so water and contaminants will flow quickly through gravel. Of the four sizes of soil, gravel is the largest and clay is the smallest. For example, gravel measures less than 76.2 mm (3-inches) and greater than 4.74 mm (0.19 inches). Sand measures between 4.74 mm (0.19 inches) and 75 μm (0.003 inches). Silt is between 0.075 mm (75 μm) and 0.002 mm (2 μm), whereas clay particles measure less than 2 μm (Bodger, 2003). In soil sample preparation, these sizes are measured by sieves for gravel and sand, and by a hydrometer for silt and clay. Table 2.6 is a representation of common sieves used in soil sample preparation.

Most groundwater is present within 300 ft of the surface, but it can be as deep as 2000 ft (Moore, 2002). The vertical distribution of subsurface water can be illustrated in Fig. 2.4. The groundwater table divides the subsurface into an unsaturated zone (vadose zone) and a saturated zone. The *unsaturated zone* extends from the surface to the water table through *soil water zone* to the *capillary fringe*. The soil water zone is usually a few feet in thickness, and the capillary rise ranges

Table 2.6 Soil sieving (nominal dimensions of standard sieves)

Tyler Equivalent (mesh/in)	USA standard sieve size (ASTM 11-61)	British standard (mesh/in.)	Sieve opening (mm)
250	230	240	0.063
100	100	-	0.149
60	60	60	0.250
20	20	-	0.841
10	10	12	1.68

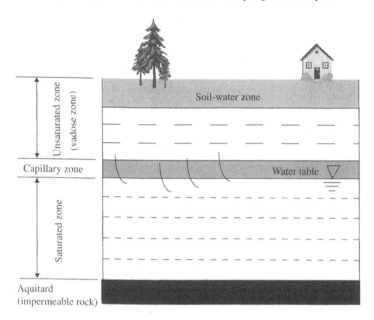

Figure 2.4   Vertical zones of subsurface water

from 2.5 cm for fine gravel to more than 2 m for silt (Todd, 1980). The entire vadose zone range can vary from a few feet for high water table conditions to hundreds of feet (meters) in arid regions. In the vadose zone, water (moisture) content generally decreases with the vertical distance from the water table with an exception of the soil water (root) zone where the amount of water depends primarily on recent exposure to rainfall, infiltration, and vegetation. Water in the vadose zone is held in place by surface forces, only the infiltrating water passes downward to the water table by a gravity flow.

In the *saturated zone* beneath the water table, water is held in the pores and moves through pores that are connected. *Porosity*, the ratio of the volume of voids to the total volume, determines the amount of groundwater or the storage capacity of an aquifer. The *permeability* or *hydraulic conductivity* is the measure of an aquifer's ability to transmit water, hence the flow rate of groundwater. Hydraulic conductivity is defined as the change in head (decrease in water level) per unit distance. The greater the hydraulic conductivity, the lesser the resistance to the groundwater flow.

2.3.3 Groundwater Wells

A *well* is a vertical hole that is extended into an aquifer to a certain depth. Wells are very important because they serve various purposes, such as pumping of water for domestic, agricultural, or industrial uses (supply wells), pumping of contaminated water for site clean-up (recovery well), injection of water or disposal of chemicals (injection wells), or water level measurement and water quality sampling (observation and monitoring wells). A *piezometer* is a special small diameter, nonpumping well used to measure the elevation of water table or potentiometric surface.

Figure 2.5 shows a typical *monitoring well* and several terminologies one should be familiar with for groundwater sampling. The steel or PVC plastic pipe that extends from the surface to the screened zone is called the *casing*. The *well screen*, a portion of the well that is open to the aquifer, is typically a section of slotted pipe allowing water to flow into the well whereas coarse soil particles are screened out. The annular space around the screen, often filled with sand or gravel to limit the influx of fine aquifer materials, is termed the *filter pack*. Above the filter pack is an *annular seal*, which is the seal around the annulus of the well casing. It is usually cemented up to the surface with a low permeability material, such as bentonite clay,

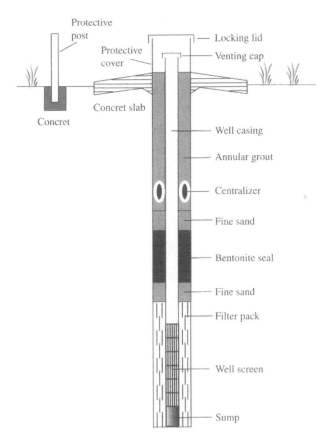

Figure 2.5 Schematic of a monitoring well (adapted from the Louisiana Department of Environmental Quality, 2000)

to protect leakage from above the screened interval or percolating rainwater. All materials used to install a monitoring well are critical for the quality of monitoring data. Many states environmental regulatory agencies therefore require that wells be installed by licensed drillers. Common monitoring wells are 2 and 4 inch in diameter and are installed in 6 and 10 inch diameter *boreholes*.

Wells, in general, come in various types depending on whether the well is located in an unconfined or confined aquifer (Fig. 2.6). A *water table well* is situated in an *unconfined aquifer*; its water table rises or declines in response to rainfall and changes in the stage of surface water to which the aquifer discharges. An *artesian well* draws water from a *confined aquifer*, which is overlain by a relatively impermeable (confining) unit (*aquitard*). The water level in artesian wells stands at some height above the water table because of the pressure (artesian pressure) of the aquifer. The level at which it stands is the *potentiometric (or pressure) surface* of the aquifer. If the potentiometric surface is above the land surface, a flowing well or spring results and is referred to as a *flowing artesian well*.

Lastly, a measurement of the water level of a well is typically conducted prior to sampling. A Lufkin steel tape can be used for this purpose. Before the tape is lowered down the well, the bottom 1–2 ft is coated with carpenter's chalk. The tape is then lowered into the well until the lower part is submerged. This method is now mostly replaced by electrical tape water level indicators. When an electrical probe is lowered into the water, an electrical circuit is completed causing a buzzer or light to be activated. Using a pressure transducer is another method for testing an aquifer where the water level changes rapidly. Water level data can be automatically transferred to data logger for a continuous measurement.

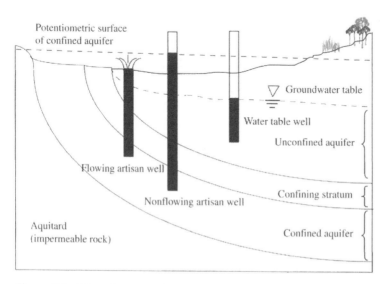

Figure 2.6 Schematic cross-section of unconfined and confined aquifers showing various types of wells

2.4 ESSENTIAL KNOWLEDGE OF ENVIRONMENTAL REGULATIONS

Environmental laws and regulations are the driving forces behind almost all of the environmental sampling and analytical work (academic research excluded). This statement, perhaps, applies to all sampling and analysis for regulatory compliance in industries, for regulatory enforcement by federal, state, and local environmental agencies, and for services performed by third parties such as environmental consulting firms and commercial labs. A statement by Smith (1997) well reflects this point from an industry perspective: "The only reason that the plant manager takes a sample is because he is directed to do so by a federal or state government regulation. The environmental industry... exists solely because it services government requirements for monitoring, remediation and pollution prevention. Persons in the industry who refuse to recognize this fundamental motivating force and fail to keep themselves informed of changes in the regulations... will eventually run into the reality of what it means to be ignorant of government directives."

It is, therefore, important for these entities to be well informed of the current regulations – perhaps keep a copy of Title 40 "Protection of Environment" of the Code of Federal Regulations (40 CFR). Recent years' CFR can be retrieved electronically from the U.S. Government Printing Office's Web site (http://www.gpoacess.gov/cfr). Unfortunately, the specific requirements for sampling, analysis, and monitoring are dispersively distributed in the entire 40 CFR from Part 1 to Part 1517, which makes it very inefficient, if not impossible, for general use. The contents given below summarize the framework of these regulations relevant to sampling and analysis.

2.4.1 Major Regulations Administrated by the U.S. EPA

Five major environmental regulations are highlighted in Table 2.7 with a focus on the relevancy to sampling and analysis. Although the U.S. EPA administrates all these regulations, in many cases, an environmental agency of a state has its own regulatory counterparts and should be consulted as well.

2.4.2 Other Important Environmental Regulations

There are several other environmental regulations governing the protocols of sampling and analysis that one may encounter occasionally. These include but are not limited to the following:

- Federal Insecticide, Fungicide, and Rodenticide Act (FIFRA)
- Toxic Substances Control Act (TSCA)
- Food, Drug, and Cosmetic Act (FDCA)
- Occupational Safety and Health Act (OSHA)

Table 2.7   Major EPA regulations and their relevancies to sampling and analysis

1. Clean Water Act (CWA): 40 CFR 100–140

- Established standards for *surface water* (40 CFR 131) and municipal and industrial *wastewater* (40 CFR 125). The latter is referred to as National Pollutant Discharge Elimination System (NPDES) permits.
- Approved methods of analysis along with approved sampling containers, preservatives, and holding times (40 CFR 136.3 (e)).
- Lists three classes of monitoring pollutants: Conventional pollutants (BOD, COD, pH, TSS, bacteria, oil and grease fecal coliform), nonconventional pollutants (N, P, NH_3, etc.), and toxic pollutants (129 EPA *priority pollutants*).

2. Safe Drinking Water Act (SDWA): 40 CFR 141–149

- Established two tiers of analytes in *drinking water* for The National Primary Drinking Water Standards (40 CFR 141) and The National Secondary Drinking Water Standards (40 CFR 143).
- Includes chemical, microbiological and radiological parameters that directly affect human health to meet the Primary Standard requirement. Includes materials that affect taste, odor, color, and other nonhealth related qualities to meet the Secondary Standard requirement.
- Approved proper sampling and analytical methodologies, and the Manual for Certification of Laboratories Analyzing Drinking Water (EPA, 1997).

3. Clean Air Act (CAA): 40 CFR 50–99

- Established National Primary and Secondary Ambient Air Quality Standards for *ambient air* and required EPA to establish *emission* standards for various industries for air permit issuance (similar to NPDES permit for wastewater).
- Approved analytical methods for six atmospheric *criteria pollutants*, CO, SO_2, NO_2, Pb, PM, O_3.
- Developed a number of methods for the analysis of 189 compounds called *hazardous air pollutants* (HAP), including volatiles (vp > 0.1 mm Hg, BP < 300°C), semivolatiles (vp 10^{-1} to 10^{-7} mm Hg, BP 300–600°C), and particulates.
- The applicable methods encompass sampling and analysis in the 00xx series methods in SW-846, the draft Air CLP-SOW, 40 CFR 60 & 61 and the TO-1 to −14 manual.

4. Resource Conservation and Recovery Act (RCRA): 40 CFR 240–299

- Published sampling and analytical methods for *solid wastes*: Test Methods for Evaluating Solid Wastes, Physical/Chemical Methods commonly referred to as *SW-846* (http://www.epa.gov/epaoswer/hazwaste/test/main.htm)
- Includes methods for (1) identification of a waste as hazardous on the basis of toxicity, ignitability, reactivity, and corrosivity, (2) determining physical properties of wastes, and (3) determining chemical constituents in wastes.

(*continued*)

Table 2.7   (*Continued*)

---

5. Comprehensive Environmental Response, Compensation, and Liability Act (CERCLA or Superfund Act): 40 CFR 300–399

- Authorized EPA to investigate the origins of waste found in hazardous sites (i.e., the *Superfund site* on the National Priority List) and force the generators and other responsible parties to pay for the remediation.
- Provided analytical support for investigation and remediation under CERCLA through the Contract Laboratory Program (CLP), with the detailed methods contained in the Statement of Work (SOW) (http://www.epa.gov/superfund/programs/clp/methods.html)

Both the FIFRA and TSCA are administered by the U.S. EPA whereas the FDCA and OSHA are administered by the FDA and OSHA, respectively. The FIFRA requires the EPA to oversee the manufacture and use of "-icides" in the US, but most commercial labs have little to no contact with the provisions in the FIFRA. Under the TSCA, the EPA requires chemical producers to supply information dealing with risk assessment of proposed products 90 days before proposed manufacture or import. Such risk assessment includes tests for chemical fate, environmental, and health effect, which are only tested in very specialized labs. Labs performing the TSCA studies are required to comply with Good Laboratory Practice Standards (40 CFR 792). Although the FDCA is administered by the FDA, the wastes and byproducts generated from the manufacture of these substances are controlled by the EPA.

OSHA is probably the most important for industrial hygienists and safety professionals in environmental health and safety. This act promulgates the standards, sampling, and analytical methods for *workplace environments* such as asbestos, dusts, and toxic vapors. One important aspect related to environmental sampling is OSHA's requirement for a sampler to obtain *HAZWOPER training certification* (29 CFR 1910.120) prior to actual sampling work on hazardous wastes and materials.

REFERENCES

ANDERSON TW (1958), *The Statistical Analysis of Time Series*, John Wiley & Sons New York, NY.

*BEDIENT PB, RIFAI HS, NEWELL CJ (1999), *Ground Water Contamination: Transport and Remediation*, 2nd Edition, Prentice Hall PTR, Upper Saddle River, NJ.

BODGER K (2003), Fundamentals of Environmental Sampling, Government Institute Rockville, MD.

BRANKOV E, RAO ST, PORTER PS (1999), Identifying pollution source regions using multiply censored data, *Environ. Sci. Technol.*, 33(13):2273–2277.

BRILLINGER DR (1981), *Time Series: Data Analysis and Theory*, Holden Day, Inc. San Francisco, CA.

*Suggested Readings

*Bruice PY (2001), *Organic Chemistry*, 3rd Edition, Prentice Hall, Upper Saddle River, NJ.

Clarke JU (1998), Evaluation of censored data methods to allow statistical comparisons among very small samples with below detection limit observations, *Environ. Sci. Technol.*, 32(1):177–183.

Cohen AC (1991), Truncated and Censored Samples: Theory and Applications. Marcel Dekker, NY.

Fifield FW, Haines PJ (2000), *Environmental Analytical Chemistry*, 2nd Edition, Blackwell Science, Malden, MA pp. 1–11.

Gilbert RO (1987), Statistical Methods for Environmental Pollution Monitoring, Van Nostrand Reinhold, NY.

Gibbons RD (1994), Statistical Methods for Groundwater Monitoring, John Wiley & Sons New York, NY.

Gilliom RJ, Hirsch RM, Gilroy EJ (1984), Effect of censoring trace-level water-quality data on trend-detection capability, *Environ. Sci. Technol.*, 18(7):530–535.

Ginevan ME, Splistone DE (2004), *Statistical Tools for Environmental Quality Measurement*, Chapman & Hall/CRC, Boca Raton, FL.

*Helsel DR (1990), Less than obvious: Statistical treatment of data below the detection limit. *Environ. Sci. Technol.*, 24(12):1766–1774.

Helsel DR (2004), *Nondetects and Data Analysis: Statistics for Censored Environmental Data*. John Wiley & Sons Hoboken, NJ.

Kebbekus BB, Mitra S (1998), *Environmental Chemical Analysis*, Blackie Academic & Professional London, UK.

Keith LH (1991), *Environmental Sampling and Analysis: A Practical Guide*, Lewis Publishers, Boca Raton, FL.

Mackay D (2001), Multimedia Enviromental Models: The Fugacity Approach, 2nd Edition, Lewis Publishers, Boca Raton.

Manly BFJ (2001), *Statistics for Environmental Science and Management*, Chapman & Hall/CRC, Boca Raton, FL.

Meyer VR (1997), *Pitfalls and Errors of HPLC in Pictures*, Hüthig Verlag Heidelberg.

Moore JE (2002), *Field Hydrogeology: A Guide for Site Investigations and Report Preparation*, Lewis Publishers, Boca Raton, FL.

Newman MC, Greene KD, Dixon's PM (1995), UnCensor, Version 5.1. Savanah River Ecology Laboratory, Aiken, SC. (Software available from http://www.vims.edu/env/research/software/vims_software.html#uncensor).

Popek EP (2003), Sampling and Analysis of Environmental chemical pollutants: A complete Guide, Academic Press, San Diego, CA.

Rantz SE et al. (1982), Measurement and Computation of Streamflow, Volume 1. Measurement of Stage and Discharge, U.S. Geological Survey Water Supply Paper 2175.

Rorabacher, DP (1991), statistical treatment for rejection of deviant values: critical values of Dixon's "Q" parameter and related subrange ratios at the 95% confidence level *Anal. Chem.*, 63(2):139–146.

*Sincich TL, Levine DM, Stephan D (2002), *Practical Statistics by Example using Microsoft Excel and Minitab*, Prentice Hall, 2nd Edition Upper Saddle River, NJ.

*Smith RK (1997), *Handbook of Environmental Analysis*, 3rd Edition, Genium Publishing Corporation Amsterdam, NY.

*Sawyer CN, McCarty PL, Parkin GF (1994), *Chemistry for Environmental Engineering*, 4th Edition, McGraw-Hill, Inc. NY.

Sullivan TF (1999), *Environmental Law Handbook*, 15th Edition, Government Institutes, Rockville, MD.

Texas Natural Resource Conservation Commission (1999), *Surface Water Quality Monitoring Procedures Manual*, Austin, TX.

Todd DK (1980), *Ground Water Hydrology*, 2nd Edition, John Wiley & Sons, New York, NY.

USEPA (1984), Definition and Procedure for the Determination of the Method Detection Limit, Code of Federal Regulations Title 40, Part 136, Appendix B, Federal Register Vol. 49, No. 209, US Government Printing Office.

USEPA (1991), GEO-EAS 1.2.1 User's Guide, EPA/600/9-91/008.

USEPA (1992), Statistical Training Course for Groundwater Monitoring Data Analysis, EPA/530-R-93-003, Office of Solid Waste, Washington, D.C.

USEPA (1994), Guidance for the Data Quality Objectives Process, EPA QA/G-4. (http://www.epa.gov/quality/qs-docs/g4-final.pdf).

USEPA (1996), Test Methods for Evaluating Solid Waste, Physical/Chemical Methods, SW-846.

USEPA (1997), Manual for the Certification of Laboratories Analyzing Drinking Water: Criteria and Procedures Quality Assurance, Office of Ground Water and Drinking, Cincinnati, OH. (http://www.michgan.gov/deq).

USEPA (1998), Guidance for Data Quality Assessment: Practical Methods for Data Analysis. Report EPA/600/R-96/084, Office of Research and Development, Washington, D.C. (http://www.epa.gov/quality/qs-doc/g9-final.pdf).

USGS (1995), User's Manual for ANNIE, version 2, A Computer Program for Interactive Hydrologic Data Management, USGS Water-Resources Investigations Report 95-4085, Reston, Virginia.(http://water.-usgs.gov/software).

WILSON N (1995), *Soil Water and Groundwater Sampling*, 1st Edition, CRC Press, Inc. Lewis Publishers, Boca Raton, FL.

QUESTIONS AND PROBLEMS

1. Describe the difference between (a) accuracy and precision, and (b) method detection limits and practical quantitation limits.

2. A water sample has $10\,\mu g/L$ Hg^{2+} and a density of $1.0\,g/mL$ (atomic weight of $Hg = 200.59$). Calculate the following: (a) Hg^{2+} in ppb, (b) Hg^{2+} in μM, and (c) the number of Hg^{2+} ion in $1\,L$ of this sample containing $10\,\mu g/L$ Hg^{2+} (Avogadro's number $= 6.022 \times 10^{23}$ /mol).

3. The concentrations of arsenic (As) and selenium (Se) in a drinking water well were 2.0 and 3.8 ppb. (a) Convert arsenic concentration into ppm and mg/L, (b) Convert selenium concentration into molarity (M) and micro-molarity (μM), (c) Have the concentrations exceeded the maximum contaminant level (MCL) of 50 mg/L for both elements? The atomic weight of Se is 79.

4. Carbon monoxide (CO) and hydrocarbons (HC) are the two main exhaust gases produced by the combustion of gasoline-powered vehicles. Their concentrations are regulated by required car inspections in some states. (a) Convert the regulatory standard of 220 ppm HC into $\mu g/m^3$ (assuming a nominal molecular weight $MW = 16$). (b) If the actual exhaust concentration of CO ($MW = 28$) at a low speed emission test is 0.16%, what is the CO concentration in ppm and mg/m^3?

5. The exhaust gas from an automobile contains 0.002% by volume of nitrogen dioxide ($MW = 46$). What is the concentration of NO_2 in $\mu g/m^3$ at standard temperature and atmosphere pressure ($25°C$ and 1 atm pressure)?

6. Indicate the category of the following organic compounds from this list: (1) aliphatic acid, (2) aliphatic ether, (3) saturated aliphatic hydrocarbon, (4) polycyclic aromatic hydrocarbon, (5) aliphatic alcohol, (6) aliphatic amine, (7) unsaturated aliphatic hydrocarbon, (8) chlorinated hydrocarbon.

(a) $CH_3CH_2NH_2$

(b) CH_3CH_2COOH

(c) $CH_3CH_2CH_2OH$

(d) $CH_3CH_2OCH_3$

(e)

(f) $CH_2{=}\!\!=\!\!CH_2$

(g)

(h)

7. An analytical method is being developed to measure trichloracetinitrile in water. The analyst has performed a number of spiked and recovery experiments in different water matrices to determine the method detection limits (MDLs). The data are reported in the following table.

				Relative	
	Number	True	Mean	standard	Standard
	of sample	conc.	conc., $\bar{x}$	deviation	deviation
Matrix	(n)	(μg/L)	(μg/L)	(%RSD)	(s)
Distilled Water	7	4.0	2.8	6.8	0.190
River Water	7	4.0	4.55	6.9	0.314
Ground Water	7	4.0	5.58	2.6	0.145

Trichloroacetonitrile Recovery and Precision Data

(a) Which water matrix yields the most precise results? Why?

(b) Which water matrix yields the most accurate results? Why?

(c) Calculate the MDLs with a 98% confidence level for each water matrix.

8. A Cr^{6+} solution of 0.175 mg/L was prepared and analyzed eight times over the course of several days. The results were 0.195, 0.167, 0.178, 0.151, 0.176, 0.155, 0.154, 0.164 mg/L. The standard deviation of this data set was calculated to be 0.0149 mg/L. (a) Calculate the method detection limit (MDL), (b) What is the estimated practical quantitation limit (PQL)? (c) Is it accurate to measure samples containing Cr^{6+} around 0.1 mg/L?

9. Describe in what cases the use of "median" can be preferred over that of "mean" in environmental data reporting?

10. Define Type I error and Type II error. Explain why both "false positive" and "false negative" should be avoided in the analysis and monitoring of environmental contaminants?

11. Define "outliers" and also describe how the "outlier" data should generally be dealt with and be removed?

12. Evaluate the pros and cons when using mean or median to evaluate the following two data sets (unit in μg/L) – Data A: 1, 1, 1,1, ….1, 10^6 ($n = 1,000$). Data point 10^6 could be an outlier. Data B: 0.11, ND, 0.13, 0.15. ND denotes "not detected."

13. For the following set of analyses of an urban air sample for carbon monoxide: 325, 320, 334, 331, 280, 331, 338 μg/m^3, (a) determine the mean, standard deviation, median, mode, and range using Excel's Descriptive Statistics. (b) calculate the coefficient of variation and RSD. (c) is there any value in the above data set that can be discarded as an outlier?

14. The following absorbance data at wavelength 543 nm were obtained for a series of standard solutions containing nitrite (NO_2^-) using a colorimetric method: A = 0 (blank), A = 0.220 (5 μM), A = 0.41 (10 μM), A = 0.59 (15 μM), A = 0.80 (20 μM). Using Excel (a) to plot the calibration curve, and (b) to determine the calibration equation and the regression coefficient (R^2).

15. For the raw data set $x_i = 1, 100, 1000$ and its logarithmic (base 10) transformation data set $y_i = 0, 2, 3$, what are the means and medians for these two data sets (*Hint:* For

comparison, you need to anti-log the mean and median calculated after log transformation)? What conclusion can be drawn regarding the effect of logarithmic transformation on the values of mean and median?

16. Soil samples were collected at two areas surrounding an abandoned mine and analyzed for lead. At each area several samples were taken. The soil was extracted with acid, and the extract was analyzed using flame atomic absorption spectrometry. In Area A, Pb concentrations were 1.2, 1.0, 0.9, 1.4 mg/kg. In Area B, Pb concentrations were 0.7, 1.0, 0.5, 0.6, 0.4 mg/kg. (a) Are these two areas significantly different from each other with Pb concentrations at 90% confidence level? (b) Perform a Single Factor ANOVA test (i.e., One-Way ANOVA) using Excel.

17. Samples of bird eggs were analyzed for DDT residues. The samples were collected from two different habitats. The data reported are:

Sample collection area	Number of samples	Mean conc. DDT (ppb)	Standard deviation (s)
Area 1	4	1.2	0.33
Area 2	6	1.8	0.12

(a) Are the two habitats significantly different ($\alpha = 5\%$) from each other in the amount of DDT to which these birds are exposed?

(b) Calculate the 90% Confidence Intervals for DDT concentration in both areas.

(c) If DDT concentration in Area 2 can be assumed to have a normal distribution with the same mean and standard deviation as the sample measurement listed in the above table (i.e., $\mu = 1.8$ ppb, $\sigma = 0.12$ ppb), what is the probability of a bird egg having residual DDT of greater than 2.2 ppb? less than 2.2 ppb? or between 1.3 and 2.2 ppb?

18. PCB concentrations in fish tissues in two rivers have been determined (Data source: Kebbekus and Mitra 1998). In River A the concentrations (μg/kg) were as follows: 2.34, 2.66, 1.99, 1.91 (Mean $= 2.225$, $s = 0.3449$); in River B (μg/kg): 1.55, 1.82, 1.34, 1,88 (Mean $= 1.648$, $s = 0.2502$). Determine the statistical significance for the concentration difference between the two rivers (a 90% confidence level assumed).

19. The U.S. EPA wishes to test whether the mean amount of Radium-226 in a soil exceeds the maximum allowable amount, 4 pCi/kg. (a) Formulate the appropriate null and alternative hypotheses, (b) Describe a Type I error and a Type II error.

20. The following calibration data were obtained for a total organic carbon analyzer (TOC) measuring TOC in water:

Concentration, μg/L	Number of replicates	Mean signal (mv)	Standard deviation
0.0	20	0.03	0.008
6.0	10	0.45	0.0084
10.0	7	0.71	0.0072
19.0	5	1.28	0.015

(a) Do the regression analysis (y = mean signal, x = concentration) using Excel. Include at least the Summary Output, ANOVA, Line Plot, Residual Plot, and Normal Probability Plot.

(b) What is the best straight line by least squares fit? Indicate the slope and intercept of this linear fit. Indicate whether or not this linear regression is significant and whether or not the intercept and slope are significant? Use significant level of 0.05.

(c) What is the calibration sensitivity in mv/(μg/L)?

21. A survey on the background concentration of lead in soil was conducted in an Environmental Impact Assessment of a former battery plant. Preliminary results of six soil samples in the protected area give results in mg/kg: 28, 32, 21, 29, 25, 22. (a) Is there any value in this data set that can be discarded as an outlier at 90% confidence level? (b) Calculate CV, RSD, and 95% Confidence Interval (CI)? (c) If sufficient numbers of soil samples were taken and the population was known to have a normal distribution with a mean of 25.5 mg/kg and standard deviation of 1.2 mg/kg, what would be the probability of a soil sample having Pb concentration of greater than 29.8 mg/kg; less than 21 mg/kg; and in the range of 23.1 to 27.4 mg/kg?

22. A new method is being developed for the analysis of a pesticide in soils. A spiked sample with a known concentration of 17.00 mg/kg was measured five times using the new method (data: 15.3, 17.1, 16.7, 15.5, 17.3 mg/kg) and the established EPA method (data: 15.4, 15.9, 16.7, 16.1, 16.2 mg/kg). The Excel outputs of descriptive statistics and one-way ANOVA are given below ($\alpha = 0.05$). On the basis of these output, (a) are these two methods significantly different at a 95% confidence level? (b) Which method is more accurate, why? (c) Which method is more precise, why?

(1) Excel output of descriptive statistics

EPA method		New method	
Mean	16.06	Mean	16.38
Standard error	0.211187121	Standard error	0.412795349
Median	16.1	Median	16.7
Mode	#N/A	Mode	#N/A
Standard deviation	0.472228758	Standard deviation	0.923038461
Sample variance	0.223	Sample variance	0.852
Kurtosis	1.012889863	Kurtosis	−2.910136878
Skewness	−0.105406085	Skewness	−0.400799173
Range	1.3	Range	2
Minimum	15.4	Minimum	15.3
Maximum	16.7	Maximum	17.3
Sum	80.3	Sum	81.9
Count	5	Count	5
Largest(1)	16.7	Largest(1)	17.3
Smallest(1)	15.4	Smallest(1)	15.3
Confidence Level (95.0%)	0.586350662	Confidence Level (95.0%)	1.146106

(2) Excel output of one-way ANOVA

Summary

Groups	Count	Sum	Average	Variance
EPA method	5	80.3	16.06	0.223
New method	5	81.9	16.38	0.852

ANOVA

Source of variation	SS	df	MS	F	p-value	F crit
Between Groups	0.256	1	0.256	0.4762791	0.5096343	5.317644991
Within Groups	4.3	8	0.5375			
Total	4.556	9				

23. Pyrene was analyzed using GC-MS at the selected ion mode. The calibration curve of peak area (GC-MS response) vs. concentration is shown below along with the regression output. A blank sample was analyzed seven times, giving a mean peak area of 125 and standard deviation of 15. A check standard at 0.5 ppm was also measured seven times, giving a standard deviation of 300. The raw data, Excel's printouts of linear regression along with the line plot are attached. (a) Perform your own Excel analysis and compare with what is given. (b) What are the linear regression equation and the R^2? (c) What is the calibration sensitivity? (d) Is the linear regression statistically significant, if yes then why? (e) If an unknown liquid sample (after a dilution of 10 times) was injected into GC-MS and gave a peak area of 35720, what is the concentration of this sample?

Concentration of pyrene (ppm)	GC-MS response
0	0
0.1	7456
0.2	14929
0.3	21219
0.4	28783
0.6	43496

SUMMARY OUTPUT

Regression Statistics	
Multiple R	0.999793298
R Square	0.999586638
Adjusted R Square	0.999483298
Standard Error	353.7486093
Observations	6

ANOVA

	df	SS	MS	F	Significance F
Regression	1	1210430543	1210430543	9672.75953	6.408E-08
Residual	4	500552.3143	125138.0786		
Total	5	1210931095			

	Coefficients	Standard error	t Stat	p-value	Lower 95%	Upper 95%	Lower 95.0%	Upper 95.0%
Intercept	107.2428571	242.8861165	0.441535559	0.68164403	−567.11851	781.6042	−567.12	781.60422
Concentration	72024.71429	732.3291958	98.35018827	6.4084E-08	69991.438	74057.99	69991.4	74057.99

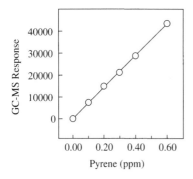

24. Two analysts were given a standard reference material (SRM) and were asked to determine its copper concentration (in mg/kg) using the EPA standard method. Each analyst was given sufficient time so they could produce as many accurate results as possible. Their results (mg/kg) were as follows:

    Analyst A   45.2, 47.3, 51.2, 50.4, 52.2, 48.7 (Standard error $s = 2.62$)

    Analyst B   49.4, 50.3, 51.6, 52.1, 50.9 (Standard error $s = 1.06$)

    (a) For Analyst A, is 52.2 mg/kg a possible outlier?

    (b) Calculate the RSD for each student; which analyst is more precise?

    (c) If the known concentration of copper in SRM is 49.5 mg/kg, which analyst is more accurate (without considering the deletion of outliers if any)?

25. Draw a schematic diagram of a monitoring well showing as a minimum the casing, filter pack, annular seal, well screen, and the above ground well protective cover and post.

26. Define the following geological and hydrological terms: clay, vadose zone, piezometer, water table well, confined aquifer, and hydraulic conductivity.

27. For the environmental standard related to following: (a) ambient air, (b) workplace air, (c) drinking water, and (d) wastewater effluent discharged into receiving water, identify the particular environmental regulation (e.g., CAA, CWA) and search the CFR Web site for the numerical standards for (a) to (d).

28. Which CFR would you find methodologies for drinking water: (a) 29 CFR 1900, (b) 40 CFR 141, (c) 40 CFR 401, (d) 21 CFR 189?

Chapter 3

Environmental Sampling Design

Before going into the field for the actual sampling (Chapter 4), a planning stage is executed. A key component of this planning process is to develop a sampling design, which indicates *how many* samples to take, from *where* the samples will be taken, and *when* the samples will be taken. The purpose of this chapter is to address the basic principles of environmental sampling design. The "representativeness," determined by several statistical and nonstatistical approaches, is the focal point of this design process. These sampling design approaches are introduced in plain language for beginners with a "how to" approach rather than technical details and jargon. Examples are then provided to help readers to understand these principles so one can relate them to sampling of various environmental matrices (air, soil, groundwater, and biological). For field professionals, standard EPA sampling protocols are available and are listed at the end of this chapter.

3.1   PLANNING AND SAMPLING PROTOCOLS

The planning process is critical for the overall data quality and the successful completion of the project. Outcomes of this planning include the development of data quality objectives (DQOs) and a sampling and analysis work plan. To develop

Fundamentals of Environmental Sampling and Analysis,   by Chunlong Zhang

such a plan, a team with a broad range of expertise (or a well-informed Principal Investigator for a small research project) is needed (Fig. 3.1). This multidisciplinary undertaking is common for most environmental projects, such as remediation investigation/feasibility study, site assessment, waste management, remediation action, and risk assessment.

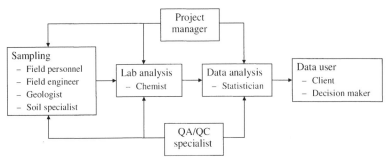

Figure 3.1 Personnel involved in project planning: From sampling to data user

As shown in Figure 3.1, experienced field personnel, who are experts in collecting samples and solving field problems, are called to interact with an analytical chemist who knows how to preserve, store, and analyze samples. The analytical chemist must also understand sampling theory and practice in addition to measurement methods (Kratochvil et al., 1984). An engineer helps in a complex manufacturing process to optimize sampling location and safety. A statistician will then verify that the resulting data is suitable for any required statistical calculation or decision. A QA/QC representative will review the applicability of standard operating procedure (SOP), determine the numbers of QA/QC samples (blanks, spikes, and so forth), and document the accuracy and precision of the resulting database. A data end-user ensures that data objectives are understood and incorporated into the sampling and analytical plan. Depending on the project, other individuals might include a geologist, facility manager, local citizen, and an EPA representative. For beginners in this field, exposure to a team of various expertises could be an intimidating but a rewarding experience.

3.1.1 Data Quality Objectives

The planning process is critical in the development of DQOs. The DQOs are defined as "qualitative and quantitative statements that define the appropriate type of data, and specify the tolerable levels of potential decision errors that will be used as basis for establishing the quality and quantity of data needed to support decision" (EPA, 2000). The DQOs were first developed by the U.S. EPA specifically for projects under EPA's oversight (EPA, 1993). However, its planning principles can serve as a checklist and apply to any projects that require environmental data collection. The main idea of the DQO process is to have the least expensive data collection scheme but not at the price of providing answers that have too much uncertainty. The DQOs are a written document developed before data collection and preferably agreed to by all stakeholders shown in Figure 3.1. This seven-step DQO process is shown in Figure 3.2.

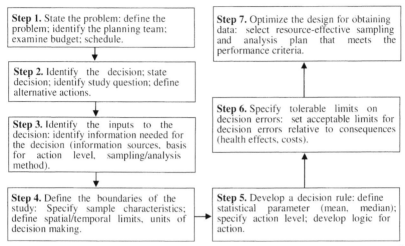

Step 1. State the problem: define the problem; identify the planning team; examine budget; schedule.

Step 2. Identify the decision; state decision; identify study question; define alternative actions.

Step 3. Identify the inputs to the decision: identify information needed for the decision (information sources, basis for action level, sampling/analysis method).

Step 4. Define the boundaries of the study: Specify sample characteristics; define spatial/temporal limits, units of decision making.

Step 5. Develop a decision rule: define statistical parameter (mean, median); specify action level; develop logic for action.

Step 6. Specify tolerable limits on decision errors: set acceptable limits for decision errors relative to consequences (health effects, costs).

Step 7. Optimize the design for obtaining data: select resource-effective sampling and analysis plan that meets the performance criteria.

Figure 3.2 The data quality objective (DQO) process (U.S. EPA, 2000)

As it is shown in Figure 3.2, the DQO process is not a straightforward task for beginners as well as many practitioners. Steps that are directly relevant to sampling and analysis include Step 3 in confirming that appropriate sampling techniques and analytical methods exist to provide the necessary data, Step 4 in defining spatial/temporal boundaries that data must represent or where/when the sampling will be conducted, and Step 7 in developing general sampling design and analytical plans. To illustrate DQOs in simple terms, Popek (2003) described DQOs as a process of asking questions and finding answers with regard to the following *what, who, why, when, where, and how*:

- What is the project's purpose?
- What is the problem that requires data collection?
- What types of data are relevant for the project?
- What is the intended use of data?
- What are the budget, schedule, and available resources?
- What decisions and actions will be based on the collected data?
- What are the consequences of a wrong decision?
- What are the action levels?
- What are the contaminants of concern and target analytes?
- What are the acceptance criteria for the PARCC parameters?
- Who are the decision-makers?
- Who will collect data?
- Why do we need to collect the particular kind of data and not the other?

- When will we collect the data?
- Where will we collect the data?
- How will we collect the data?
- How will we determine whether we have collected a sufficient volume of data?
- How will we determine whether the collected data are valid?
- How will we determine whether the collected data are relevant?

3.1.2 Basic Considerations of Sampling Plan

In developing a sampling plan to formulate the number and location of samples (how many, where, and when), four primary factors need to be considered (Gilbert, 1987, Fig. 3.3). Of these four factors (objectives, variability, cost factors, nontechnical factors), a project objective is probably the determining factor in sampling design. For example, sampling efforts in water quality monitoring will be very different depending on whether the objective is for trend analysis or baseline (background) type investigation. The former needs a long-term but less frequent sampling scheme, while the latter requires more samples and perhaps a one-time sampling event. The project objective with regard to the required data quality also affects the number of samples to be collected. In other words, the sample number will increase significantly as the allowable margin of error is reduced. Environmental variability or the spatial/temporal patterns of contamination is another important consideration. The variations in different environmental matrices will be further discussed in Section 3.2, and the sampling approaches to address such variations in order to obtain "representative" samples will be discussed in Section 3.3.

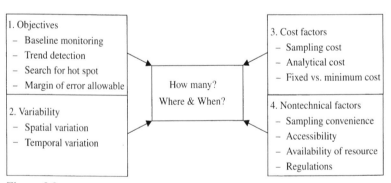

Figure 3.3   Criteria for selecting a sampling plan

Expenses associated with sampling and analysis should certainly be a consideration in all environmental projects. The cost-effectiveness of a sampling design should be evaluated in the design stage so that the chosen sampling design will achieve a specified level of data quality at a minimum cost or, an acceptable level of data quality at a prespecified cost. Last but not least important thing is that sampling design should consider other factors such as sampling convenience, site accessibility, availability of sampling equipment, and political aspects. On the top of the above considerations, regulation constraints in most cases always need to be consulted first. A good example is the sampling related to remediation projects under the Superfund Act. The U.S. EPA has specific guidelines for representative sampling of soil, air, biological tissue, waste, surface water, and sediment in Superfund investigations. Specific details can be found in the suggested readings. These nontechnical factors have a significant impact on the final sampling design, but they are beyond the scope of this text.

3.2 SAMPLING ENVIRONMENTAL POPULATION

3.2.1 Where (Space) and When (Time) to Sample

Environmental sampling can be viewed in a space-time domain (Gilbert, 1987). In the time domain, there is only one dimension (1-D). Specific time can be designated as $t_1, t_2, t_3, \ldots t_N$ in a time period such as days, weeks, months, or years. In the space domain, contaminant variations and hence sampling points can be designated by the coordinates in 1-D (x), 2-D (x, y), or 3-D (x, y, z). An example of 1-D problem is an outfall of an industrial wastewater discharge. We would only be concerned about the concentration as a function of the downstream distance from the discharge point. An example of 2-D problem is the radiochemical contents in surface soil due to atmospheric deposition from a nuclear weapon testing site or the lead content in surface soils downwind from a local smelter facility. In these two cases, the soil depth may not be the primary interest, so soil samples may be designated by longitude and latitude, or by measurements relative to an existing structure. A 3-D sampling site is common in a large body of water, or a solid/hazardous waste landfill site where depth is a variable for contaminant variations.

In many cases, contaminant variations in both space and time are of interest. That is, we are interested in both spatial and temporal patterns of contamination. For example, in a wetland area subjected to seasonal nonpoint source pollution from agricultural activities, wetland soil samples can be designated by longitude and latitude as well as by time. Note that in the following discussions on various sampling design approaches (Section 3.3), space and time are interchangeable from the design standpoint—meaning that the statistical design approach applicable to 1-D space will also be applicable to temporal variation in the 1-D time domain.

3.2.2 Obtain Representative Samples from Various Matrices

"Representativeness" is one of the five so-called "data quality indicators (DQIs)," or PARCC mentioned in the previous chapter (Section 2.1.3). Unlike precision and accuracy, representativeness is a qualitative parameter in defining data quality. The U.S. EPA defines *representativeness* as "a measure of the degree to which data accurately and precisely represent a characteristic of a population, a parameter variation at a sampling point, a process condition, or an environmental condition." It can be further defined in three different levels (Popek, 2003): (1) *Ultimate representativeness* is a measure of how well the overall data collection design (including the sampling design and analysis) represents a characteristic of a population and the environmental condition at the site; (2) *Sampling point representativeness* is a measure of how well a sample collected represents the characteristic properties of the sampling point; (3) *Collected sample representativeness* is a measure of how well subsampling techniques are during analysis and the use of proper preparation and analytical procedures.

Representative Solids Samples

The representativeness depends largely on the types of sample matrices. Contaminants accumulated in soils at a certain depth typically do not vary much temporarily in a short-term. However, significant vertical variations may occur at different soil depths. In special cases, heterogeneity is even more severe at the microscopic scale (among soil particles) than the large-scale spatial variations, which presents a challenge when subsamples are taken. This is the case when the contaminant of interest is present in a "nugget" form commonly found in contaminated soils, sludge, and other waste materials in solid samples. Uncertainties associated with the representativeness of these types of heterogeneous samples frequently far exceed those inherent in sample collection and analysis (Keith, 1990), making sample preparation process (such as subsampling, mixing, grounding, and sieving) particularly important for representative samples. Mixing is impossible for wastes with immiscible phases. In such cases, representative samples should be collected from each individual phase.

Representative Air Samples

Air is a unique matrix with potentially extreme variations and heterogeneity compared with soil and water. Atmospheric contaminant concentrations at the same location may have several orders of magnitude difference within minutes depending on the changes in local meteorological conditions (wind velocity and direction). It is thus important to determine whether the air sample collected is representative of the "typical" or "worst case" site condition both spatially and temporally. Meteorological and topographical factors must be incorporated into an air-sampling plan. These precautions also apply to atmospheric precipitation samples such as snow, rain, fog, and dew.

Representative Water Samples

Water and groundwater samples may have typical seasonal variations depending on the water balance due to recent precipitation and water usage. There are some special techniques for representative groundwater samples, which will be introduced in Chapter 4. Surface waters of various types can be very heterogeneous both spatially and temporally as a result of flow and stratification, making it difficult to collect representative samples (Keith, 1990). Stratification is a common problem in oceans, deep lakes during the stratified seasons (summer and winter), and slow-flowing streams (>5 m). Stratification also occurs when two streams of water merge, such as at the outfall of a wastewater discharge point. Estuarine water may present special challenge because strata move up rivers unevenly (Keith, 1990).

Representative Biological Samples

The heterogeneity of biological samples presents unique challenges because of the difference in species, size, sex, mobility, and the tissue variations. Consultation with a trained ecologist or biologist is often recommended. For example, the selection of an inappropriate species may introduce an error, which can be minimized by selecting a species that is representative of the habitat and whose life cycle is compatible with the timing of the study. Migratory or transient species should be avoided. Tissues for residual chemical analysis should be well homogenized. Ideally, tissue homogenates should consist of organisms of the same species, sex, developmental stage, and size, since these variables affect chemical uptake. Fur and shell should be removed from tissue, as they cannot be practically homogenized as a whole (EPA, 1997).

Keep this in mind that, in addition to the physical variations discussed above, "representativeness" always depends on the project objective. This means that samples are representative in one case, but the same samples will not be representative for another purpose. For illustration, consider Figure 3.4, a site map for a dry lagoon formerly fed by a pipe discharging wastewater. The analytical results of soil samples drawn from randomly located sites A, B, and C may be representative if the objective is to address whether the pipe released a particular contaminant. However, these data are not representative if the objective is to estimate the average concentration in the entire old lagoon. For that estimation,

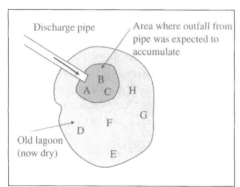

Figure 3.4 Sampling for an old lagoon to illustrate the dependence of sample representativeness on the project objective (U.S. EPA, 2002)

random sampling locations should be generated from the entire site of the lagoon (perhaps including samples at D, E, and F). The selection of the appropriate sampling design (Section 3.3) is necessary in order to have data representative of the problem being investigated.

3.3 ENVIRONMENTAL SAMPLING APPROACHES: WHERE AND WHEN

Discussed below are several sampling approaches commonly used in environmental data collection defining where and when to collect samples, including judgmental sampling, simple random sampling, stratified sampling, and systematic sampling. Judgmental sampling is a nonstatistically based approach, and the other three are termed as *probabilistic sampling*. Note that a combination of these approaches can be used for a specific project or at various stages of the same project. There are also many other design approaches including those innovative ones that have potential for improving the quality of more cost-effective data. These will be briefly defined in Section 3.3.5. Details on advantages, disadvantages, example applications, and statistical formulas are beyond the coverage of this text. Interested readers are referred to two excellent sources by Gilbert (1987) and EPA's *Guidance on Choosing a Sampling Design for Environmental Data Collection* (EPA, 2002).

3.3.1   Judgmental Sampling

Judgment sampling refers to the subjective selection of sampling locations based on professional judgment using prior information on the sampling site, visual inspection (e.g., leaks and discoloration), and/or personal knowledge and experience. It is the preferred sampling approach when schedule (such as for emergency spill response) and budget are tight. Judgment sampling is also preferred at the early stage of site investigation or when the project objective is to just screen an area for the presence or absence of contamination so a prompt decision can be made whether or not a follow-up statistical sampling is needed.

Judgment sampling is the primary representative sampling approach used for groundwater assessment in the selection of monitoring wells (EPA, 1995). This is because monitoring wells are complex, expensive, and time-consuming to install. In order to best determine the nature of a suspected contaminant plume, monitoring wells need to be placed in areas most likely to intercept the plume. Locations are selected based on the investigator's knowledge of the suspected contaminants, site geology, and hydrology. The alternative approaches using randomization would likely result in too many wells that will possibly miss the contaminant plume. Random sampling and systematic sampling discussed below are generally not used for groundwater sampling (EPA, 1995).

There are cases when judgment sampling cannot be applied. As judgment sampling includes no randomization, it does not support any statistical interpretation of the sampling results. This is equivalent to saying that when the level of confidence needs to be quantified, judgment sampling is not applicable. Keith (1990) further indicated that samples collected for research studies allow use of prior knowledge in order to obtain as much useful data as possible, but sampling for legal purposes often requires absolutely random samples.

3.3.2 Simple Random Sampling

Simple random sampling, also referred to as *random sampling*, is the arbitrary collection of samples by a process that gives each sample unit in the population (e.g., a lake) the same probability of being chosen (Fig. 3.5a). The word "arbitrary" is probably misleading because a random process (e.g., random number table) is used and each sampling point is selected independently from all other points in random sampling. This differs from the nonstatistical *haphazard sampling*, in which the sampling person may consciously or subconsciously favor the selection of certain units of the population. The haphazard sampling claims "any sampling location will do" and hence encourages taking samples at convenient locations or times. Simple random sampling is the simplest and most fundamental probability-based sampling design. It is also the benchmark against which the efficiency and cost of other sampling designs are often compared.

Simple random sampling assumes that variability of a sampled medium is insignificant, and hence it is appropriate for relatively uniform or homogeneous populations such as holding vessel, lagoon, and so forth. This random sampling approach also applies for sites with little background information or for site where obvious contaminated areas do not exist or are not evident. One advantage of using simple random sampling is that statistical analysis of the data is simple and straightforward. Explicit formulas as well as tables and charts are readily available to estimate the minimum sample size needed to support many statistical analyses. The formula for the calculation of (arithmetic) mean ($\bar{x}$) is the same as Eq. 2.12 and the standard deviation (s) can be obtained from the sample variance (s^2) in Eq. 2.17. Both were discussed in Section 2.2:

$$s = \sqrt{\frac{\sum_{i=1}^{n}(x_i - \bar{x})^2}{n-1}} = \sqrt{\frac{\sum_{i=1}^{n}x_i^2 - \left(\sum_{i=1}^{n}x_i\right)^2 / n}{n-1}} \tag{3.1}$$

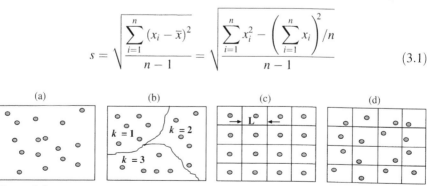

Figure 3.5 Three common probability-based sampling designs for sampling in two-dimensional space: (a) simple random sampling; (b) stratified sampling (three strata); (c) systematic grid sampling; and (d) systematic random sampling (L = spacing).

Simple random sampling is not applicable for heterogeneous populations. The higher the degree of the heterogeneity, the lesser the random sampling approach will adequately characterize true conditions. Furthermore, while the complete randomization approach defines the statistical uncertainty, it could also lead to several problems. First, sampling points by random chance may not be uniformly dispersed in space and/or time particularly when sample size is small. This drawback can be overcome by systematic sampling. Second, simple random sampling, by ignoring prior site information or professional knowledge, often leads to more samples, which is not as cost-effective as other sampling designs. Third, randomly selected sampling points could be harder to locate precisely. Because of this limitation in its implementation, simple random sampling is seldom recommended for use in practice except for relatively uniform populations. The U.S. EPA guideline does not recommend simple random sampling in flowing water bodies and states that it is only practicable for sediment bed sampling in nonflowing (static) water bodies (EPA, 1995).

3.3.3 Stratified Random Sampling

As shown in Figure 3.5b, *stratified random sampling* divides sampling population into several nonoverlapping (mutually exclusive) strata and within each stratum a simple random sampling is employed. Each stratum is relatively more homogeneous than the population as a whole. The selection of strata, however, requires some prior knowledge of the population to be sampled.

"Strata" could be "temporal" or "spatial." *Temporal strata* permit different samples to be selected for specified time periods, for example, day vs. evening, weekdays vs. weekend, four seasons of the year, and so forth. *Spatial strata* are more common and come with many varieties. Spatial strata can be based on sampling depth (stratified lakes, soil, or sediment cores), ages and sex of population (men, women and children), topography, geographical regions, land types and uses, zones of contamination, wind direction (downwind vs. upwind), political boundaries, and so forth.

A major advantage of stratified random sampling is that the sample size can be adjusted depending on the variations or the cost of sampling in various strata. Strata expected to be more variable or less expensive should be sampled more intensively. This provides greater precision and cost-saving than simple random sampling. It also implies that stratified random sampling results in smaller standard deviation than the simple random sampling, particularly if the strata are quite different from one another. Another advantage is the additional information it provides regarding the mean and standard deviation within each stratum, which may be of interest for a particular project.

Within each stratum, the formulas used to calculate mean and standard deviation are the same as those for simple random sampling (Eqs 2.12 and 3.1). Suppose we have r strata ($k = 1, 2, \ldots r$), the stratum mean ($\bar{x}_k$) and stratum standard deviation (s_k) are then combined to estimate the population mean ($\bar{x}$) and standard deviation (s)

by the following formulas:

$$\bar{x} = \sum_{k=1}^{r} w_k \bar{x}_k \tag{3.2}$$

$$s^2 = \sum_{k=1}^{r} \frac{w_k^2 \, s_k^2}{n_k} \tag{3.3}$$

where w_k = the fraction or the weight of the population represented by stratum k. Eqs 3.2 and 3.3 indicate that the mean and standard deviation for stratified random sampling are the weighted average of all strata. The weight, w_k, is assumed to be known before any sampling takes place.

The following discussions focus on various methods to allocate the number of samples into each stratum. Due to the complex nature of the stratified random sampling, interested readers are referred to Gilbert (1987) for a more thorough discussion on these mathematical equations.

Equal allocation: Each stratum is assigned the same number of samples. Since the total number of stratum is r, the number of sample in each stratum $(n_k) = n/r$, where n is the total number of samples from all strata.

Proportional allocation: The number of samples in each stratum is proportional to the size of the stratum. The larger the stratum, the more the sample can be collected. Assume N = total population units, N_k = total population units in stratum k, n = total sample units, and n_k = total sample units in stratum k, then:

$$\frac{n_k}{n} = \frac{N_k}{N} \equiv w_k \qquad \text{or} \qquad n_k = n \, w_k \tag{3.4}$$

Optimal allocation: The cost is considered through either the optimal precision for a fixed study cost or the optimal cost for a fixed level of precision. The number of samples in the kth stratum is related to the variation (s_k) and the cost per sampling unit in the kth stratum (C_k) by the following equation:

$$n_k = n \frac{w_k s_k / \sqrt{C_k}}{\sum_{k=1}^{r} (w_k s_k / \sqrt{C_k})} \tag{3.5}$$

A special case is when the sampling cost per sampling unit is the same for all strata, then the above equation reduces to Eq. 3.6, which is frequently called *Neyman allocation*.

$$n_k = n \frac{w_k s_k}{\sum_{k=1}^{r} w_k s_k} \tag{3.6}$$

3.3.4 **Systematic Sampling**

Systematic sampling involves selecting sample units according to a specified pattern in time or space, for example, at equal distance intervals along a line or a grid pattern. Some slight variations exist for systematic sampling as shown in Figure 3.5 for a two-dimensional space sampling, *systematic grid sampling* (Fig. 3.5c), and *systematic random sampling* (Fig. 3.5d). The systematic grid sampling subdivides the area of concern by using a square or triangular grids and then collects samples from the nodes (the intersections of the grid line) or a fixed location (e.g., center) of each grid. The first sample to be collected from a population is randomly selected (e.g., through a random number table), but all subsequent samples are taken at a fixed space or time interval. The systematic random sampling subdivides the area into grids and then collects a sample from within each grid cell using simple random sampling.

The systematic grid sampling is easier to implement and more convenient to field personnel than simple random sampling. An important application of this approach is groundwater sampling at a fixed well over time. There are several reasons for this preference: (a) extrapolation from the same period to future periods is easier with a systematic sample; (b) seasonal cycles can be easily identified and accounted for in the data analysis; (c) a systematic sample will be easier to administrate because of the fixed sampling schedule; (d) most groundwater samples have been traditionally collected using a systematic sample, making comparisons to background more straightforward (EPA, 2002).

The second advantage of grid sampling is its more uniform distribution over the space or time domain, which helps to delineate the extent of contamination and define contaminant concentration gradients. This has been proved to be efficient for a full characterization of soil contamination, because grid sampling insures that all areas are represented in the sample and provides confidence that a site has been fully characterized. The same is true for its application to geostatistical applications to delineate the temporal or spatial patterns for correlated data in time and space (see Section 2.2.6).

The calculation of mean and standard deviation for systematic sampling is straightforward. The simplest way to analyze the data from systematic sampling is to treat it as though it was collected using simple random sampling (Eqs 2.12 and 3.1). A critical part of the systematic sampling design is to choose the right grid spacing (L in Fig. 3.5). The grid spacing should be small enough to detect the spatial/temporal patterns or to search hot spots. Otherwise, grid sampling will likely either overestimate or underestimate the population. This also occurs when the grid spacing L coincides with the spatial or temporal pattern of the variable of interest. For example, temperature or other temperature-dependent parameters (e.g., dissolved oxygen in water) will have a 24-h cyclic pattern; if all samples are taken at the late afternoon of sunny days, it will overestimate the temperature but will underestimate the dissolved oxygen. These problems associated with the grid sampling, however, can be minimized by the use of systematic random sampling.

The grid spacing (L) in one-dimensional systematic sampling (such as sampling in time or along a line in a stream) is easy to calculate. First, calculate the spacing interval $k = N/n$, where N is the total population units and n is the predetermined number of samples. Round k to the nearest integer. Then randomly pick a starting position/time point, and select the second point k distance from the first point, the third k distance from the second, and so on until n samples have been defined. For example, to design a 5-day sample scheme in a month of 31 days, $k = 31/5 = 6.2 \approx 6$. If we randomly choose a random number 3 from 1 to 31 using a random number table, this would give us the sampling days of 3, 9, 15, 21, and 26.

The grid spacing (L) in a two-dimensional area can also be readily calculated once we know the total number of samples to be collected (n) and the area to be sampled (A). This results in the following equations:

$$L = \sqrt{A/n} \qquad \text{for square grid} \qquad (3.7)$$

$$L = \sqrt{A/0.866n} \qquad \text{for triangular grid} \qquad (3.8)$$

3.3.5   Other Sampling Designs

So far we have discussed three commonly used probabilistic sampling designs—simple random sampling, stratified random sampling, and systematic sampling. There are also a number of other design strategies that may be useful in environmental data collection. They are briefly described below.

If sampling cost is much less than analytical cost, which is typical for many trace analysis, *composite sampling* is a valuable means in cost-saving. If the goal is to estimate the average concentration rather than the variability or extreme concentrations for grab samples, this physical mixing will provide the same degree of precision and accuracy as the mathematically computed average from the analysis of all samples. Compositing is frequently done in the monitoring of wastewater discharge and contaminated soils. One exception is that it should not be used for volatile organic compounds, because mixing is not allowed for these compounds.

If the primary goal is to locate the hot spots (areas with elevated contamination that exceed applicable clean-up standards), then *search sampling* is an effective design approach. Search sampling utilizes either a systematic grid or systematic random sampling approach to define the minimum grid size in order to locate them. Generally, the smaller the hot spots are, the smaller the grid spacing must be. In other cases, an acceptable error of missing hot spots is given; then the smaller the allowable error is, the smaller the grid spacing must be. Mathematical equations are available for interested readers.

Transect sampling is another variation of systematic grid sampling. It involves establishing one or more transect lines across a surface. Samples are collected at regular intervals along the transect lines at the surface and/or at one or more given depths. Multiple transect lines may be parallel or nonparallel to one another. The primary benefit of transect sampling vs. systematic grid sampling is the ease of establishing and relocating individual transect lines. This approach is applicable to waste piles or impoundments. It is applicable to characterizing water flow and contaminant characteristics and contaminant depositional characteristics in sediments. This method is not most desirable in large lakes and ponds or areas of surface water that are accessible only by a boat.

In summary, Table 3.1 lists various sampling approaches and ranks that approaches from most to least suitable, based on the sampling objectives. It is intended to provide general guidelines rather than site-specific evaluation in choosing a sampling design approach (EPA, 1995). Sampling design is a complex business; interested readers can consult some software for assistance with this task. One such example is the visual sample plan downloadable at http://dqo.pnl.gov/vsp/vspsoft.htm Battelle Memorial Institute, 2006.

Table 3.1   Comparison of various sampling design approaches based on project objectives

Sampling objective	Judgmental sampling	Simple random	Stratified random	Systematic grid	Systematic random	Search sampling	Transect sampling
				Sampling approach			
Establish threat	1	4	3	2[a]	3	3	2
Identify sources	1	4	2	2[a]	3	2	3
Delineate extent of contamination	4	3	3	1[b]	1	1	1
Evaluate treatment and disposal options	3	3	1	2	2	4	2
Confirm clean-up	4	1[c]	3	1[b]	1	1	1[d]

1 = Preferred approach; 2 = Accepted approach; 3 = Moderately acceptable approach; 4 = Least acceptable approach. [a]Should be used with field analytical screening; [b]Preferred only where known trends are present; [c]Allows for statistical support of clean-up verification if sampling over entire site; [d]May be effective with compositing technique if site is presumed to be clean.

EXAMPLE 3.1.   Selection of sampling strategies. You are asked to estimate the weekly average concentration of SO_2 emitted from a stack in a coal power plant. Four alternative sampling plans were proposed to take exhaust gas SO_2 samples at the outlet of the stack. Discuss the advantages and disadvantages. (a) Randomly collect two samples everyday for seven days; (b) Collect two samples per day for seven days, sampling interval is 12 h (9:00 am and 9:00 pm); (c) Collect two samples everyday for seven days. One sample is randomly collected during the daytime, and the second sample is randomly collected during the night shift; (d) Same as (c), but four random samples are collected during the day shift and three random samples are collected during the night shift.

SOLUTION: All four sampling approaches are concerned with the sampling for temporal data. They are simple random sampling (a), systematic grid sampling (b), systematic random sampling (c), and stratified random sampling (d). For (b), samples are collected at a fixed schedule, it is easier to implement than (c). It might be a little inconvenient for operators to collect samples around midnight time for (c), but it will better represent the data than (b) if SO$_2$ discharge peaks at a certain time of the day. Option (d) would generate better precision if SO$_2$ discharge present a clear difference between day and evening. As the project goal is to estimate the weekly average mean concentration, this appears to be too costly. Overall, option (c) may least likely to give the unbiased estimate.

EXAMPLE 3.2.   Simple random sampling from two waste tanks. A manufacturing plant had been generating a liquid waste over a period of years and storing it in a large open-top tank. As this tank approached its capacity, some of the waste was allowed to overflow to a nearby small tank (Fig. 3.6). Prior to sampling the following information is available:

(a) The large tank had a diameter of 50 ft, a height of 20 ft, and an approximate volume of 295,000 gal. It was encircled and traversed by catwalks to allow access to the entire waste surface. Contaminant concentration appears to be heterogeneous due to the long time of operation, but it was determined that vertical composite samples (not the vertical strata) are sufficient.

(b) The small tank had a diameter of 10 ft, a height of 10 ft, and an approximate volume of 6000 gal. There is only a small inspection port on the top allowing a limited access. A decision was made to use two tank trucks and to sample the waste randomly over time as it drains from the tank into the tank trucks. It was estimated that it would take 300 min to drain the tank at a rate of 20 gal/min.

If financial constrains allow for a total of 15 samples to be collected from each tank, design a sampling strategy using simple random sampling approach.

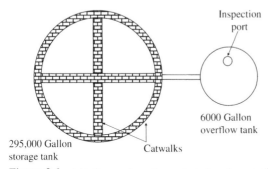

Figure 3.6   A bird's eye view of a waste tank and an overflow tank (U.S. EPA, 1986)

SOLUTION: This example was excerpted and slightly modified from a guideline by the U.S. EPA. Here a simple random sampling over a two-dimensional space is used for the large waste tank because the operators are interested in the average composition and variability of the waste, and not interested in determining if different vertical strata existed. For the small overflow tank, a random sampling over time was used since sampling over space is impossible due to the limited access from the only inspection port on the top. We are also given that the total number of samples is 15 for each tank, so the question remains is fairly easy, that is, how to generate the random numbers to represent those locations in space (large tank) and time points (small tank).

There are many ways to acquire random numbers. For instance, we can just simply use the page number of a book to obtain them while we are in the field. One can also use a random number generator at http://www.random.org. The random numbers shown below are obtained from the Tool I Data Analysis I Random Number Generation function using Excel. Readers should be aware of the fact that your answers will be different because of the randomization process each time you use it.

For the large tank, the two-dimensional coordinates can be denoted by a radius (25 ft) and a circumference ($2 \times \pi \times 25 = 157$ ft). Hence, 15 pairs of random numbers are generated for the circumference coordinate from 0 to 157, and radius coordinate in the range of 0–25:

(1, 1), (25, 5), (130, 21), (15, 3), (156, 25), (26, 5), (107, 17), (102, 16), (138, 22), (20, 4), (18, 4), (138, 22), (27, 5), (92, 15), (124, 20).

For the overflow small tank, 15 random numbers between 0 and 300 min are needed, and they are: 2, 155, 31, 126, 127, 168, 62, 28, 108, 18, 41, 261, 238, 18, 120.

EXAMPLE 3.3.   Calculating means and standard deviation for a stratified random sampling. Studies were conducted to test lead (Pb) concentrations in surface soils due to atmospheric deposition of emissions from a smelter. Prior information revealed that Pb is higher in the prevailing downwind direction (from East to West). The Pb concentration is also higher in clayed soils than sandy soils. It was determined that a stratified random sampling approach was appropriate. A total of 30 samples were collected, and the number of samples in each stratum was proportionally allocated based on the estimated percentage land area under the specified wind direction. The stratum weights and the resulting statistics of analytical results are shown in Table 3.2, and the raw data are shown in the footnotes. Estimate the overall means and standard deviation.

SOLUTION: Use Eqs 3.2 and 3.3 to calculate the overall mean and variance as follows.

$$\bar{x} = \sum_{k=1}^{r} w_k \bar{x}_k = 0.5 \times 79.0 + 0.2 \times 65.5 + 0.17 \times 56.8 + 0.13 \times 50.0 = 68.8 \, \text{mg/kg}$$

$$s^2 = \sum_{k=1}^{r} \frac{w_k^2 \, s_k^2}{n_k} = (0.5^2 \times 7.0^2/15 + 0.2^2 \times 3.4^2/6 + 0.17^2 \times 2.9^2/5 + 0.13^2 \times 2.6^2/4)$$

$$= 0.972 \, (\text{mg/kg})^2$$

$$s = (s^2)^{1/2} = (0.972)^{1/2} = 0.99 \, \text{mg/kg}$$

Table 3.2   Summary statistics for lead contents in four strata of soils in a nearby smelter[a]

Stratum	Sample number (n_k)	Mean($\bar{x}_k$)	Standard deviation (s_k)	Stratum weight (w_k)
Downwind clayed soil (S1)	15	79.0	7.0	0.5
Downwind sandy soil (S2)	6	65.5	3.4	0.2
Perpendicular wind clayed soil (S3)	5	56.8	2.9	0.17
Perpendicular wind sandy soil (S4)	4	50.0	2.6	0.13
Overall	$n = 30$	$\bar{x} = ?$	$s = ?$	$\Sigma W_k = 1.00$

[a]Raw data (unit in mg/kg): S1 = 80, 75, 89, 65, 73, 77, 74, 83, 82, 85, 76, 87, 77, 90, 72; S2 = 66, 68, 65, 60, 70, 64; S3 = 60, 55, 59, 57, 53; S4 = 53, 51, 49, 47.

The results using stratified random sampling can be expressed as $\bar{x} \pm s$ or 68.8 ± 0.99 mg/kg, which compares favorably with 68.7 ± 12.54 mg/kg if the same raw data were obtained using simple random sampling. Although the means are very close in this example, the standard deviation is much smaller using stratified random sampling.

3.4 ESTIMATING SAMPLE NUMBERS: HOW MANY SAMPLES ARE REQUIRED

The best sample number is the largest sample number possible. Unfortunately, it is very unlikely to take too many samples due to the limited time and budget resources for sample collection and analysis. Oftentimes, investigators should avoid taking too few samples that could make data scientifically unreliable, or even worse, lead to error conclusions.

In principle, the sample number (n) is a function of the project goal, type of sampling approaches, environmental variability (s), cost (C), tolerable error, and other factors. For example, a judgmental sampling intended to determine the presence or absence of a contaminant requires only a few samples. On the contrary, a grid sampling requires a much greater sample number to delineate the extent of contamination. Each of the above probability sampling methods discussed in the previous section has its own ways of calculating sample numbers. There is no universal formula to calculate the adequate sample size. Given below is an example of sample number (n) calculation for simple random sampling. This could serve as a conservative estimate since simple random sampling usually results in the maximal number of samples required. Readers are referred to the references listed at the end of this chapter for further details on other specific equations and assumptions.

Manly (2001) gave the following equation in relating sample size (n) to the variability (σ) and acceptable error (δ) in simple random sampling from a normally distributed population:

$$n = \frac{4\sigma^2}{\delta^2} \tag{3.9}$$

where σ is the population standard deviation and δ is half of the width of a 95% confidence interval on the mean ($\bar{x} \pm \delta$). To use this equation, an estimate or best guess of σ must be known. Sources of a preliminary estimate of population variance (σ^2) include: a pilot study of the same population, another study conducted with a similar population, or an estimate based on a variance model combined with separate estimate for the individual variance components. In the absence of prior information, the U.S. EPA guideline recommended the following equation to obtain the crude approximation of standard deviation ($\hat{\sigma}$) by dividing the expected range of the population by 6, that is

$$\hat{\sigma} = \frac{\text{Maximum (expected)} - \text{Minimum (expected)}}{6} \tag{3.10}$$

A particular equation used to estimate sample size for solid wastes is given by the U.S. EPA, which relate required sample number (n) to the sample variance (s^2) and the closeness of the estimate mean ($\bar{x}$) to the regulatory threshold (RT):

$$n = \frac{t^2 s^2}{e^2} \tag{3.11}$$

where e is the acceptable level of error, $e = \text{RT} - \bar{x}$ for hazardous materials. Eq. 3.11 indicates that the sample number is directly proportional to the contaminant variability and more samples should be collected if contaminant concentration is closer to its regulatory standard. Since the tabulated t-value in Eq. 3.11 depends on the number of sample ($df = n - 1$), a trial-and-error procedure is typically needed to estimate the optimal sample size (n).

It is common that the sample sizes calculated using simple random sampling formulae exceed the study budget. Then adjustment needs to be done, including redefining the study goal or the DQOs (a smaller study), and modifying the required precision. Most often, choosing other sampling designs may reduce a significant number of samples. For instance, stratified random sampling may result in a smaller sample size and composite sampling may reduce a significant cost if the goal is to determine the mean.

EXAMPLE 3.4.   Historical data from a contaminated site suggested Hg concentrations in the range of 2–20 µg/kg, and a standard deviation of 3.25 µg/kg. A thorough survey is needed for a planned remediation of this site. Estimate the number of samples required so that the sample mean would be within ±1.5 µg/kg of the population mean at a 95% confidence level.

SOLUTION: From Eq. 3.11, it is apparent that an initial guess of n is required. Let us assume $n = 10$ (i.e., $df = 9$), this gives a t-value of 2.262 (Appendix C2). Plug these known values ($s = 3.25$, $e = 1.5$, $t = 2.262$) into Eq. 3.11, we have:

$$n = \frac{t^2 s^2}{e^2} = (2.262 \times 3.25/1.5)^2 = 24.01.$$

Since the calculated $n = 24$ is much larger than the assumed $n = 10$, a further iteration is needed. The new t-value for $n = 24$ (95%) is 2.069 (Appendix C2), then calculate $n = (2.069 \times 3.25/1.5)^2 = 20.10 \approx 20$, which is closer to the assumed value of 24. A third iteration at n = 24 gives a new value of $n = (2.093 \times 3.25/1.5)^2 = 20.56 \approx 21$. Therefore a total of 21 samples should be tested.

EXAMPLE 3.5.   Calculation of required numbers of samples. An abandoned waste pile needs to be sampled and analyzed due to recent complaints from local residents. Historical data show that the wastes were mainly from a solvent recovery facility in 1970s with a PCB concentration of ~0.70 mg/kg (average of five samples) and standard deviation of 0.12 mg/kg. The regulatory soil screening level for PCB is 0.74 mg/kg. Use a 80% confidence level. (a) Estimate the number of samples required; (b) Estimate the number of sample required if the mean was 0.58 mg/kg.

SOLUTION: Apply Eq. 3.11 for both questions, we have:

(a) $t_{0.20} = 1.533$ for $df = 5 - 1 = 4$; two-tailed $= 0.20$ or one-tailed 0.10 (80% confidence), $n = [1.533 \times 0.12/(0.74 - 0.70)]^2 = 22$

(b) $n = [1.533 \times 0.12/(0.74 - 0.58)]^2 = 2$. Now only two samples are needed if the actual concentration is very different from the regulatory level.

REFERENCES

Battelle Memorial Institute (2006), Visual Sample Plan (VSP) Version 4.6D. Richland, WA (downloadable from http://dqo.pnl.gov/vsp/vspsoft.htm).

*GILBERT RO (1987). *Statistical Methods for Environmental Pollution Monitoring*, Van Nostrand Reinhold, New York, NY.

KEBBEKUS BB and MITRA S (1998), *Environmental Chemical Analysis*, Blackie Academic & Professional London, UK.

KEITH L.H. (1990). Environmental sampling: A summary. *Environ. Sci. Technol.*, 24(5):610–617.

KRATOCHVIL B, WALLACE D, TAYLOR JK 1984. Sampling for chemical analysis, *Anal. Chem.*, 56: 113–129.

MANLY BFJ (2001), *Statistics for Environmental Science and Management*, Chapman & Hall/CRC Boca Raton, FL.

POPEK EP (2003), *Sampling and Analysis of Environmental Chemical Pollutants: A Complete Guide*, Academic Press San Diego, CA.

US EPA (1986), Chapter 9. Sampling Plan. Testing Methods for Evaluating Solid Wastes: Physical/Chemical Methods (http://www.epa.gov/epaoswer/hazwaste).

US EPA (2000), Guidance for the Data Quality Objectives Process, EPA QA/G-4; EPA/600/R96/055, U.S. EPA: Washington, DC, August 2000 (http://www.epa.gov/quality/qs-doc/g4-final.pdf).

US EPA (1993), Data Quality Objectives for Superfund: Interim Final Guidance. EPA/540/G-93/071.

*US EPA (2002), Guidance on Choosing a Sampling Design for Environmental Data Collection for Use in Developing a Quality Assurance Project Plan, EPA QA/G-5S. Office of Environmental Information, EPA/240/R-02/005.

US EPA (1995), Superfund Program Representative Sampling Guidance: Volume 1, Soil, Interim final, EPA 540-R-95-141, OSWER Directive 9360.4-10, PB96-963207, December 1995.

US EPA (1995), Superfund Program Representative Sampling Guidance: Volume 2, Air (Short-Term Monitoring), Interim final, EPA 540-R-95-140, OSWER Directive 9360.4-09, PB96-963206, December 1995.

US EPA (1997), Superfund Program Representative Sampling Guidance: Volume 3, Biological, Interim final, EPA 540-R-97-028, OSWER Directive 9285.7-25A, PB97-963239, December 1995.

US EPA (1995), Superfund Program Representative Sampling Guidance: Volume 4, Waste, Interim final, EPA 540-R-95-141, OSWER Directive 9360.4-14, PB96-963207, December 1995.

US EPA (1995), Superfund Program Representative Sampling Guidance: Volume 5, Surface Water and Sediment, Interim final, OSWER Directive 9360.4-16, December 1995.

QUESTIONS AND PROBLEMS

1. Why are the data quality objectives (DQOs) important prior to the implementation of environmental sampling?

---

*Suggested Readings

2. Define "representativeness" and discuss how sample "representativeness" is dependent on both the project objective and sample matrices? Give examples to illustrate.

3. Make a list of important factors (criteria) that are important in developing a sampling design including where, when, and how many samples are collected?

4. Describe various reasons that can cause heterogeneity, that is, the difficulties in obtaining representative samples from (a) surface water (river, stream, lake) and (b) atmosphere.

5. Describe various reasons that can cause heterogeneity, that is, the difficulties in obtaining representative samples from (a) solids (e.g., soils) and (b) biological samples (e.g., fishes).

6. Describe the differences between: (a) haphazard sampling; (b) judgmental sampling; and (c) simple random sampling.

7. Define: (a) composite sampling; (b) transect sampling; and (c) search sampling.

8. Describe the advantages and disadvantages of (a) composite sampling and (b) systematic sampling.

9. Describe the differences of three allocation methods used in stratified random sampling: (a) equal allocation; (b) proportional allocation; and (c) optimal allocation.

10. Describe (a) the difference between systematic grid sampling and systematic random sampling and (b) the method to calculate the mean and standard deviation from systematic sampling.

11. Why the standard deviation is typically smaller for stratified random sampling than simple random sampling, particularly for a heterogeneous population with a geographical (spatial) or temporal pattern?

12. In a site assessment for the identification of contaminated source, which one of the following is likely the best and the least favorable: (a) judgmental sampling; (b) simple random sampling; and (c) systematic random sampling? Briefly explain.

13. To confirm whether a site has been cleaned up or not, which one of the following is likely the best and the least favorable: (a) judgmental sampling; (b) simple random sampling; and (c) systematic random sampling? Briefly explain.

14. A petroleum refinery plant discharges toxic chemicals into a river at unknown periodic intervals. (a) If water samples below the discharge pipe will be collected to estimate the weekly average concentration of the effluent at that point, is simple random, stratified random, or systematic sampling the best? Briefly discuss the advantages and disadvantages of these three designs for this situation. (b) If the data objective is to estimate the maximum concentration for each week of the year, is your choice of a sampling plan different than in (a) where the objective was to estimate the weekly mean? Briefly state why?

15. Given the following particular situation, identify which sampling design approach should be used and discuss why this is the most appropriate?

    (a) If you are performing a screening phase of an investigation of a relatively small-scale problem, and you have budget and/or a limited schedule. Your goal is to assess whether further investigation is warranted that should include a detailed follow-up sampling.

    (b) If you are estimating a population mean and you have known the existence of spatial or temporal patterns of the contaminant. Your goal is to increase the precision of the estimate with the same number of samples, or achieve the same precision with fewer samples and lower cost.

(c) If you are estimating a population mean and you have an adequate budget. In the meantime, you are also interesting in knowing the information of a spatial or temporal pattern.

(d) If you are estimating a population mean and you have budget constrains. The analytical costs are much higher compared with sampling costs. Your goal is to produce an equally precise or a more precise estimate of the mean with fewer analyses and lower cost.

(e) If you are developing an understanding of where contamination is present and you have an adequate budget for the number of samples needed, your goal is to acquire coverage of the area of concern with a given level of confidence that you would have detected a hot spot of a given size.

(f) If you are developing an understanding of where contamination is present and you have an adequate budget for the number of samples needed, your goal is to acquire coverage of the time periods of interest.

16. A former small pesticide manufacturing facility was surveyed for pesticide residues in surrounding soils. Historical data have shown that the pesticide is very stable in soil, concentration is in the range of 40–200 ppb with a standard deviation of 5 ppb.

(a) If an error level of ± 2 ppb is acceptable, how many samples are needed to be 95% confident that the requirement is met?

(b) If an area of 1 km^2 is to be surveyed (see the figure below), design the locations using the method of simple random sampling. Use Excel to generate random numbers and use the coordinate as shown below (i.e., $x = 0$, $y = 0$ for the manufacturing facility, $x = -500 \sim 500$; $y = -500 \sim 500$). Attach the random number from your Excel output and plot a x–y scatter plot showing the locations of all samples calculated from (a).

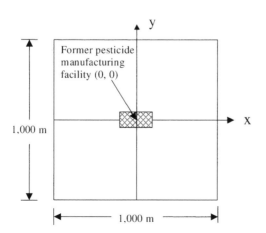

17. A pigment manufacturing process has been generating waste over a number of years. The wastes were discharged into a 40-acre lagoon. The pigment is generated in large batches that involve a 24-h cycle: 16 h of high percentage of large-sized black particulate matter and 8 h of smaller white pigment particles that are much smaller, settle more slowly and travel further toward the effluent pipe. The lagoon has a distinct color strata, with the black and the gray sludge each covering a quadrant measuring 1320 ft by 330 ft, and the white sludge covering the remaining area of the lagoon, which measured 1320 ft by 660 ft. The leachable contaminant (barium) was assumed to be associated with the black sludge, which was concentrated in the first quadrant. The sludge had settled to a uniform thickness throughout the lagoon and was covered with 2 ft of water. Design a sampling plan (locations) using stratified random sampling. Assume that we need to collect a total of 10 samples from the black sludge area (heavily contaminated), 10 samples from the gray area (mixed, less contaminated), and 20 samples from the large strata (least contaminated).

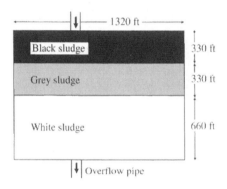

18. Thermal stratification is common in lakes located in climates with distinct warm and cold seasons. It divides lakes into three zones (top: epilimnion; middle: thermocline; bottom: hypolimnion). Because of the stratification, the vertical mixing of water is prohibited. A stratified random sampling is designed to collect water samples for nitrogen concentrations. The following data were obtained:

Stratum	Volume ($\times 10^6$ gal)	Number of samples taken	Nitrogen concentration (mg/L)
Epilimnon (0–8 ft)	5	8	8, 6, 11, 16, 9, 17, 4, 13
Thermocline (8–10 ft)	1	2	11, 16
Hypolimnion (10–25 ft)	14	10	25, 17, 18, 11, 12, 10, 22, 16, 10, 11

(a) One of the objectives was to estimate the mean, standard deviation, and confidence interval of the entire lake based on this stratified random sampling plan. Use 80% confidence level.

(b) If the above nitrogen concentrations were obtained from simple random sampling (i.e., total number of samples = 8 + 2 + 10 = 20), calculate the mean, standard deviation, and confidence interval at a 80% of confidence level.

(c) Compare the results from (a) and (b) and comment the difference.

19. A stratified random sampling plan was adopted for a contaminated site as a result of a recent oil spill in an open agricultural land. Three strata were chosen as shown in the diagram below: (1) the heavily contaminated surface soil (0–2 ft); (2) the unsaturated soil (2 ft to the groundwater level at 7 ft deep); and (3) the saturated zone (7–27 ft). The results of a probe chemical (benzene) are shown in the table below. If one of the objectives was to estimate the degree of contamination in the entire aquifer (0–27 ft):

(a) Estimate the overall mean, standard deviation, and confidence interval (at 80% confidence level) using proportional allocation method (i.e., based on the depth or volume ratio).

(b) Estimate the overall mean, standard deviation, and confidence interval (at 80% confidence level) using optimal allocation method (i.e., based on the variation of each stratum).

Stratum	Depth (ft)	Number of samples taken	Benzene concentration (mg/kg)
Surface soil (0–2 ft)	2	15	$\bar{x}_1 = 25; s_1 = 4.0$
Unsaturated soil (2–7 ft)	5	6	$\bar{x}_2 = 17; s_2 = 2.0$
Saturated soil (7–27 ft)	20	5	$\bar{x}_3 = 10; s_3 = 1.0$

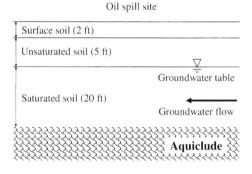

Oil spill site

Surface soil (2 ft)

Unsaturated soil (5 ft)

Groundwater table

Saturated soil (20 ft)

Groundwater flow

Aquiclude

20. A field scale remediation demonstration project was studied to test the efficiency of soil vapor extraction to remove BTEX compounds from contaminated soils in the unsaturated zone due to a rupture of chemical container. It was estimated that the contaminated area should be within the boundary of 100 m (north–south) × 150 m (east–west). After the demonstration project was completed, a detailed sampling plan needs to be executed to precisely map the spatial pattern of BTEX concentrations. Budgetary constrains allows for 50 samples to be collected for BTEX analysis of soil cores. (a) If systematic grid sampling was employed, determine the grid spacing (L) using square grid? (b) Draw a schematic diagram showing how these 50 sampling points are located?

21. A lagoon waste pit has the following historical data for the barium concentration based on a simple random sampling ($n = 4$): 86, 90, 98, 104 mg/kg (the lower two thirds of lagoon). The regulatory threshold for barium is 100 mg/kg. The waste on this site was categorized to be hazardous, and therefore a more thorough sampling plan is needed. Determine the number of samples required so that the reported mean has a 90% confidence level.

Chapter 4

Environmental Sampling Techniques

From Chapter 3 we learned the fundamentals of sampling design including issues regarding *when* and *where* to take samples and *how many* samples need to be taken. Materials covered in Chapter 4 will include *how much* sample should be collected, *how long* the sample can be preserved prior to analysis, *what* sampling tools and containers should be used, and *how to* collect representative samples for various analytes and matrices. The purpose of this chapter is to give readers an overview on the basics of environmental sampling techniques employed in the field. These contents are not intended to substitute for the detailed guidelines available from many governmental and nongovernmental agencies, but should help beginners understand the principles and get started from these generally exhaustive guidelines. A good knowledge of these sampling principles should also help you develop site and project-specific sampling and analysis plans and eliminate some of the common problems encountered during sampling.

4.1 GENERAL GUIDELINES OF ENVIRONMENTAL SAMPLING TECHNIQUES

We limit our discussions in this section to some guidelines common to all environmental sampling, including sequence of sampling matrices and analytes, sample amount, sample preservation and storage, and selection of sample containers and equipments.

Fundamentals of Environmental Sampling and Analysis, by Chunlong Zhang
Copyright © 2007 John Wiley & Sons, Inc.

There are other topics that are equally important, such as sample custody, sample labeling and tracking, sample packaging, and shipment (see Popek 2003 for details). Note that ignorance of any one of these steps will likely invalidate the data acquired.

4.1.1   Sequence of Sampling Matrices and Analytes

When a project deals with multimedia and/or multiple parameters, collect samples according to the following sequence. Remember to use a commonsense approach whenever in doubt. A list of tips is as follows:

- Collect from the least to the most contaminated sampling locations within the site whenever possible. This applies to sampling water, groundwater, soil, and biological samples. For example, if the goal is to investigate a ground-water contaminant plume as a result of known gasoline leaks from an underground storage tank, start from down gradient and work on to up gradient. This will reduce potential cross-contamination between samples and minimize decontamination efforts.

- If you collect both sediment and water at the same site, collect water first and then sediment to minimize effects from suspended bed materials.

- For shallow surface water in a stream, start sampling downstream and work upstream to minimize the effects of sediment due to sampling disturbance. If you collect a sediment sample while standing in the water, be sure to stand downstream of the collecting point. For deep waters, the collection sequence from downstream to upstream is less important.

- If sampling at different water depths is needed, collect surface water samples first and then proceed to a deeper interval.

- Always collect VOCs first, followed by SVOCs such as extractable organics (PCBs, pesticide), oil and grease, and total petroleum hydrocarbons (TPHs). Then proceed to other parameters in the order of total metals, dissolved metals, microbiological samples, and inorganic nonmetals.

4.1.2   Sample Amount

The minimally required sample amount depends primarily on the concentration of the analytes present in sample matrices. The sample volume should be sufficient to perform all required laboratory analyses with an additional amount remaining to provide for analysis of QA/QC samples including duplicates and spikes (Chapter 5). Another determining factor is the representativeness associated with the sample amount. For heterogeneous samples, a larger portion is generally required to be a representative of the actual sample variations such that possible biased results can be minimized. This larger portion of the collected sample is then homogenized thoroughly followed by subsampling. Taking too large or too many samples should be avoided because storage (cold room and freezer), transportation, and disposal costs can become a burden to a big project.

Water/wastewater samples

Minimal liquid sample volume varies considerably in the range of 5 mL for TPHs in liquid wastes, 100 mL for total metals, and 1 L for trace organics such as pesticides. This bulk estimate of sample size represents a volume sufficient to perform one analysis only, and as a general guide, the minimum volume collected should be three to four times the amount required for the analysis (EPA, 1995).

Soil/sediment/solid wastes

A full characterization of soil physicochemical properties including particle size and soil texture, as required for most environmental projects, usually needs a minimum of approximately 200 g of soil. For most of the contaminant analysis, however, a dry mass of approximately 5–100 g is sufficient. More soil samples are needed if the goal is to detect low solubility (hydrophobic) organic contaminants in the aquifer materials since these chemicals tend to accumulate in the first six inches of surface soils (EPA, 1995). The required sample size for the sediment is smaller than the water samples because contaminants tend to accumulate in the sediment as a sink. Because waste samples are generally of high concentrations, sample volumes are of less concern, but the volume should be kept at a minimum to reduce disposal costs.

Air samples

Similar to other matrices, the volume of air required depends on the minimum chemical concentration that can be detected and the sensitivity of the measurement. Because the concentration range may be unknown, the sample size will have to be determined by trial and error. Trial samples of more than $10 \, m^3$ may be required to determine ambient concentrations (Christian, 2004). This volume should be lower if smoke stack emission samples are collected for the target contaminant.

Water/sediment samples for toxicity testing

Unlike chemical analysis, the water and sediment used as a substrate for toxicity testing demand more samples. For instance, 20–40 L of water is needed to perform an effluent acute toxicity test. For the sediment, 15 L is required to conduct bioaccumulation tests (based on an average of 3 L sediment per test chamber and 5 replicates), and 8–16 L sediment is needed to conduct benthic macrointevertebrate assessments (EPA, 2001).

4.1.3   Sample Preservation and Storage

The purpose of sample preservation is to minimize any physical, chemical, and/or biological changes that may take place in a sample from the time of sample collection to the time of sample analysis. Three approaches (i.e., refrigeration, use of proper sample container, and addition of preserving chemicals) are generally used to minimize such changes (Fig. 4.1).

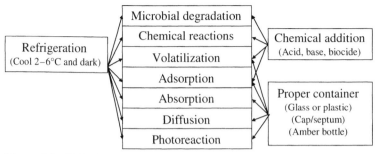

Figure 4.1   Methods of sample preservation to minimize potential changes of analytes during sample transportation and storage

As shown in Figure 4.1, refrigeration (including freezing) is a universally applicable method to slow down all loss processes. The only exception that refrigeration does not help is when acidified water samples are preserved for metal analysis. Cold storage will adversely reduce metal solubility and enhance precipitation in the solution. The proper selection of containers (material type and headspace) is critical to reduce losses through several physical processes, such as volatilization, adsorption, absorption, and diffusion. Colored (amber) bottles help preserve photosensitive chemicals such as PAHs. The addition of chemicals is essential to some parameters for their losses due to chemical reaction and bacterial degradation. Chemical addition or pH change can also be effective to reduce metal adsorption to glass container walls. Table 4.1 lists examples of chemical loss mechanisms and the preservation methods.

Table 4.1   Preservation methods of selected analytes and their physicochemical and biological changes during sample storage

Analytes	Change during storage	Preservation
Metals (M)	Adsorption to glass wall	Use plastic bottle
	Precipitation (MO, $M(OH)_2$)	Add HNO_3 to pH < 2
Phthalate ester	Diffusion from plastics	Use Teflon or glass bottle
Oil	Adsorption to plastics	Use glass bottle
VOCs	Volatilization	Avoid headspace
NH_3	Volatilization	Add H_2SO_4 to pH < 2
S^{2-}	Volatilization	Add zinc acetate and NaOH to pH > 9
CN^-	Volatilization	Add NaOH to pH > 12
	Chemical reaction with Cl_2	Add ascorbic acid to remove free Cl_2
Organic (drinking water)	Chemical reaction with Cl_2	Add $Na_2S_2O_3$ to remove free Cl_2
PAH	Photochemical degradation	Use amber glass container
Organic	Biodegradation	Low pH and temperature; add $HgCl_2$ to kill bacteria
Total phenolics	Bacterial degradation	Add H_2SO_4 to stop bacterial degradation

Adapted from Keith (1988) and Popek (2003).

Even with the proper preservation, no samples can be stored for an extended period of time without significant degradation of the analyte. The *maximum holding times* (MHTs) are the length of time a sample can be stored after collection and prior to analysis (or pretreatment) without significantly affecting the analytical results. MHTs start with the moment of sampling and end with the beginning of the analytical procedure. Samples that have exceeded their MHTs should be discarded in order to not jeopardize data quality.

MHTs vary with analyte, sample matrix, and analytical methodology used to quantify the analyte's concentration. Considering that multiple agencies designate MHTs, they can be confusing when several matrices or multiple projects are involved. The most frequently cited MHTs are those promulgated under the CWA (40 CFR 136) for the analyses of wastewater relating to NPDES permits (refer to Chapter 2). Other programs have established the MHTs as well, such as SW-846 used to comply RCRA regulations, and the Contract Lab Program for laboratories performing work under contract with the U.S. EPA in Superfund site investigations. Other agencies that set holding time requirements are the ASTM, USGS, APHA, AWWA, and WEF. For the most part, the holding times established in these various programs are consistent. Some minor inconsistencies do exist that can create confusion as to the selection of appropriate holding times. For instance, holding times for VOCs range from 5 days from sample receipt for Superfund work, to 7 days from sampling for NPDES permits, to 14 days from sampling for RCRA groundwater analyses (Keith 1988; Popek, 2003).

Figure 4.2 depicts the general MHTs requirements recommended by APHA (1998) for common water quality parameters. Several parameters must be measured immediately in the field, such as pH, temperature, salinity, and DO. Many other parameters must be measured within 1–2 days after sample collection. For these parameters, a careful pre-planning is needed to avoid sampling on Friday, Saturday, or near Holiday weekends. Many of the organic compounds (purgeable

	ASAP	6–48 h	7–28 days	6 months
MHTs	pH Salinity Cl_2, ClO_2 CO_2, I_2, O_3 DO (by electrode) Temperature	Color (48 h) PO_4^{3-}, NO_3^- (48 h) Surfactant MBAs (48 h) Chlorophyll (24–48 h) Acidity/alkalinity (24 h) CN^-, Cr^{6+} (24 h) Turbidity (24 h) DO (Winkler) (8 h) BOD (6 h) Odor (6 h)	Oil and grease (28 d) Total P (28 d) F^-, S^{2-}, SO_4^{2-} (28 d) B, Si, Hg (28 d) Conductance (28 d) Solids (7 d) Base/neutral/acid organics (7 d) Pesticide (7 d) Purgeable hydrocarbons (7 d) NH_3, TKN, COD, TOC (7 d)	Metals Hardness

MHTs

Figure 4.2   Maximum holding times (MHTs) for common water quality parameters. DO: dissolved oxygen; BOD: biochemical oxygen demand; TKN: total Kjeldahl nitrogen; COD: chemical oxygen demand; TOC: total organic carbon

hydrocarbons, pesticides, and base/neutral/acid extractable organics) have only 7 days of permissible storage time until sample pretreatment. Only samples for hardness and general metals can be stored for up to 6 months after the addition of HNO_3 to pH < 2.

Practical tips

- The guidelines on preservation appear to be tedious, but understanding the physicochemical and biological mechanisms behind various preservation and storage requirements will help you make more sense.

- Prior to sample transport, packaged samples must be kept in ice at 2–6 °C in insulated coolers. "Wet" ice (frozen water) should be double-bagged to prevent water damage from melting ice. The "blue ice," a synthetic glycol packaged in plastic bags and frozen, is less effective. Dry ice is not recommended either, because it freezes samples and causes the containers to break. When samples are kept frozen in a freezer, fill containers only 90% full to avoid container breakage.

- The MHTs are the "rules" of regulatory importance. Do not take chances and violate the rules if the data will have to sustain legal challenges! In other cases, while exceeding the holding time does not necessarily negate the veracity of analytical results, it causes "flagging" of any data not meeting all of the specified acceptance criteria (EPA, 1997).

- MHTs, for the most part, are not based on rigorous scientific evaluation. In research labs with limited resources, if recovery can be justified in a spiked sample conducted parallel to the study, results could be acceptable in the scientific community. The general consensus among environmental professionals is that the holding times for high molecular weight analytes, such as PCBs, PAHs, organochlorine pesticides, and dioxins, could be reasonably extended without adversely affecting the data quality. Organic and inorganic parameters susceptible to degradation in storage, such as VOCs, COD, BOD, TOC and alkalinity, should be analyzed as soon as possible after sampling, especially at low-level concentrations (Popek, 2003).

4.1.4   Selection of Sample Containers

In selecting appropriate sample containers, the factors to be considered are costs, ease of use, and cleanliness. More important to the data quality, the containers should be compatible to the analytes in a particular matrix. For water samples, containers are selected as follows:

- *Glass vs. plastics*: Glass containers may leach certain amount of boron and silica, and significant sorption of metal ions may take place on the container wall. Glass containers are generally used for organic compounds, and plastic containers are used for inorganic metals. For trace organics, the cap and liner should be made of inert materials so that sorption and diffusion will not be a potential problem. For example, use 1 L narrow necked glass bottles with Teflon-lined caps for extractable organics, including phthalates, nitrosamine, organochlorine pesticides, PCBs aromatics, isophorone, PAHs, haloethers, chlorinated hydrocarbons, and dioxin. Glass containers with polytetrafluoroethylene (PTFE)-lined caps, which are less expensive than Teflon-lined, can be used for purgeable halocarbons, aromatics or pesticide. In cases when either plastic or glass can be used, plastic is preferred because it is easier to transport and less likely to break. Plastic containers can be used for physical properties, inorganic minerals, and metals.

- *Headspace vs. no headspace*: No headspace is allowed for the storage of samples used for the analysis of volatile organic compounds (VOCs). Even a very small bubble will invalidate the analytical results. Generally, a 40-mL glass vial with Teflon-lined septum is the choice for VOCs samples. On the contrary, samples for oil and grease should be half-filled in wide-mouth glass bottles. Do not overfill the sample container and do not subdivide the sample in the laboratory for oil and grease analysis.

- *Special containers*: Use specific container such as DO/BOD bottles and VOC vials. The BOD bottle is a narrow-mouth glass-stopped container with 300-mL capacity, which has a tapered and pointed ground-glass stopper and flared mouth.

We have not thus far discussed the preservation and container requirements for other sample matrices such as soil, biological, and air samples. For soil, sediment, and sludge samples, low temperature storage is most likely all that is needed to preserve the sample integrity (EPA, 1989). Chemical preservatives are not needed, as they are for water samples, except for the addition of methanol or sodium bisulfate for VOC analysis. This is the only common chemical preservation method for soil samples (Popek, 2003). Wide-mouth containers are used for soil samples. If soils or sediments are anaerobic, they should not be exposed to air.

When biological samples (e.g., fish) are collected for chemical analysis, aluminum foil and closed glass containers with inert seals or cap liners can be used. Aluminum foils should not be used if mercury is the target analytes, and the shiny side should not contact the sample because it is coated with a slip agent (Keith, 1991). Individually wrapped sterile plastic cups can be used for microbiological samples.

Preservation, containers, handling, and storage for air samples can be found on the U.S. EPA and OSHA methods. Generally after air sample collection, the sampling media such as filter cassettes or adsorbent tubes (discussed below) are immediately sealed. The samples are then placed into suitable containers (e.g., resealable bags or culture tubes) that are then placed into a shipping container.

For detailed requirements on sample amount, container, preservation, and maximum holding time, the readers should consult relevant agency's guidelines. Appendix D is an excerpt of these specific details.

4.1.5 Selection of Sampling Equipment

Several factors need to be considered when selecting sampling equipment. The ease of use and decontamination of the device, physical location of the sampling point, and the type of sample matrix are some of the factors. While matrix type is a determining factor, the materials of sampling equipment are of general importance to all sampling medium. The compatibility of contaminants being sampled with the composition of sampling device is a priority because contaminants can be contributed to or removed from the sample. Sampling equipment can be made of plastic, glass, Teflon, stainless steel, or other materials. Like sample containers, plastic is generally used when analyzing metals. Teflon or stainless steel is used when analyzing for organic compounds. Most of the sampling tools are commercially available or can be custom-made. This section presents the sampling equipment most commonly used in surface water, wastewater, groundwater, soil and sediment, wastes, biological samples, and air matrices.

Surface Water and Wastewater Sampling

In Figure 4.3, three common sampling tools are depicted: the pond sampler, the weighted bottle sampler, and the Kemmerer bottle. The *pond sampler* (grab sampler) is used for near shore sampling where cross-sectional sampling is not appropriate. It is also used for sampling from an outfall pipe along a disposal pond, lagoon, and pit bank where direct access is limited. The *weighted bottle sampler* is used to collect samples in a water body or impoundment at a predetermined depth. The *Kemmerer bottle* is a Teflon, acrylic, or stainless steel tube attached to a rope, and it has stoppers at each end. It is best used when access is from a boat or structure such as a bridge or pier, and where discrete samples at specific depths are required. When the Kemmerer bottle lowers vertically into the water at the desired depth, a "messenger" is sent down to the rope and hits the trigger to close the stoppers. The *van Dorn sampler* is a slight

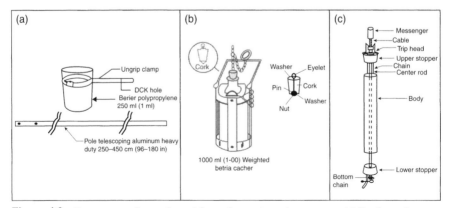

Figure 4.3   Common sampling tools used for surface water and wastewater: (a) Pond sampler, (b) Weighted bottle sampler, and (c) Kemmerer bottle sampler (U.S. Army Corps of Engineers, 2001)

variation of the Kemmerer that lowers horizontally and is more appropriate for estuary sampling.

Groundwater Sampling

Groundwater samples are collected from a well by a bailer or pumps of various types (Fig. 4.4). A *bailer* is a pipe with an open top and a check valve at the bottom (3 ft long with a 1.5 inch internal diameter and 1 L capacity). A line is used to mechanically lower the bailer into the well to retrieve a volume of water. A bailer is easy to use and transport but it may generate turbulence when the bailer is dropped down to the well and it is exposed to atmospheric O_2 when the sampled water is poured into a container. Bottom-filled bailers, which are more commonly used, are suitable provided that care is taken to preserve VOCs.

A *peristaltic pump* consists of a rotor with ball-bearing rollers. Dedicated tubings are attached to both ends of the rotor. One end is inserted into the well and the other end is a discharge tube. A peristaltic pump is suitable for sampling wells of small diameter (e.g., 2 inches) and has a depth limitation of 25 ft. Cross-contamination is not of concern because dedicated tubing is used and the sample does not come in contact with the pump or other equipment. It can result in a potential loss of VOCs due to sample aeration.

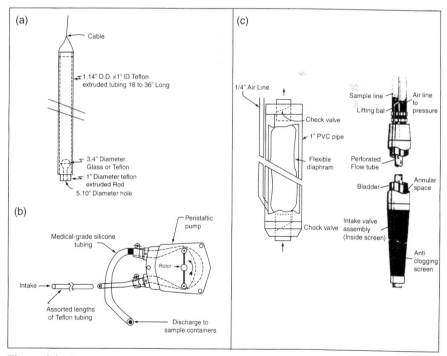

Figure 4.4   Common sampling tools used for groundwater: (a) Bailer, (b) Peristaltic pump, and (c) Bladder pump (U.S. Army Corps of Engineers, 2001)

A *bladder pump* consists of a stainless steel or Teflon housing that encloses a Teflon bladder. It is operated using a compressed gas source (bottled gas or an air compressor). Groundwater enters the bladder through a lower check valve, compressed gas moves the water through an upper check valve and into a discharge line. A bladder pump can be used to sample a depth of approximately 100 ft. It is recommended for VOC sampling because it causes a minimal alteration of sample integrity. The pump is somewhat difficult to decontaminate and should be dedicated to a well.

Not shown in Figure 4.4 are other pumps used for sampling or well development. A *submersible pump* can be used for high flow rates (gals/min), but it is hard to decontaminate. A *suction pump* is good for well development, when the proposed well area has a lot of sediment. The disadvantage is that it is gasoline-driven and the pump needs to be primed, meaning water needs to be added to the well in order to do this. It is effective only to a depth of 20 ft. An *air-lifter pump* operates by releasing compressed air. The air mixes with water in the well to reduce the specific gravity of the water column and lift the water to the surface. It is therefore used in well development rather than sampling.

Soil Sampling

Soil depth and whether or not each soil horizon is needed are the major factors to consider in the proper selection of soil sampling equipment. Scoops or trowels (Fig. 4.5)

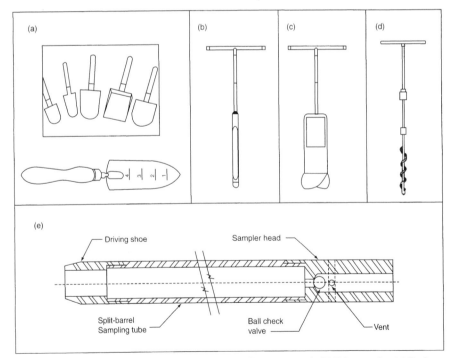

Figure 4.5 Common sampling tools used for soil: (a) Scoop or trowel, (b) Tube sampler, (c) Bucket auger, (d) Hand auger, and (e) Standard split spoon sampler (U.S. Army Corps of Engineers, 2001)

are used for soft surficial soil samples. A tube sampler can be used at depths of 1–10 ft soft soil samples. By inserting an acetate sleeve, a tube sampler can also preserve soil core. Auger samplers (*bucket auger* and *hand auger*) can be used to take deeper soils (3 inches – 10 ft) but will disrupt and mix soil horizons, making the precise horizon interface difficult to determine. The *split-spoon sampler* has excellent depth range and is useful for hard soils. It may be used in conjunction with drilling rigs for obtaining deep core soil profiles.

Sediment Sampling

Scoops and trowels, which are used in soil sampling, can be used for surface sediments around shoreline for shallow and slow-moving waters. Most sediment samples, however, are collected using either one of the devices shown in Figure 4.6. The *Ekman dredge* is a small and lightweight (10 lbs) device to collect bottom materials that are usually soft. If the sediment is compacted, or if much gravel, rocks, vegetation-covered bottom or large debris cover the bottom, then heavier Petersen or

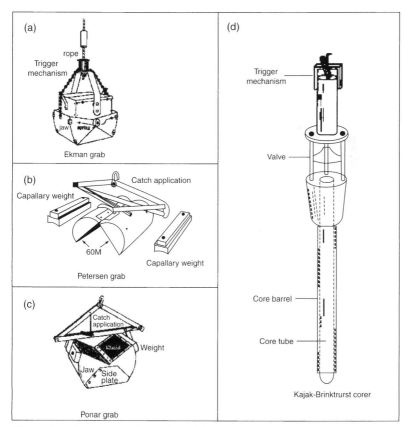

Figure 4.6   Common sampling tools used for sediment: (a) Ekman dredge, (b) Petersen dredge, (c) Ponar dredge, and (d) Sediment core sampler (U.S. Army Corps of Engineers, 2001)

Ponar dredges (30–70 lbs) must be used. Due to the weight of the equipment, a cable is required to winch from a boat, bridge or pier for raising or lowering the sampler. The *ponar dredge* is used to sample sediments that range in size from silt to granular materials. The "petit" size Ponar can be operated by one person without the use of a winch or crane. The *Petersen dredge* is used when a bottom is rocky, in deep water or in a stream with high velocity. Because of the dredge's large size, a large volume of water is displaced when the jaws close and fine sediment can be swept away. A *sediment core sampler* such as the one shown in Figure 4.6d is needed (1) to compare recent surficial vs. historical deeper sediments, or (2) to collect undisturbed sediment samples with minimal loss of fine-grained sediment fraction and exposure to oxygen.

Hazardous Waste Sampling

Sampling tools for hazardous wastes depend on the nature of waste and the type of sources. If waste sludges are sampled, the Ponar or Ekman dredge can be used to collect waste sludge samples from the bottom of impoundments, lakes, or other standing water bodies. If solid waste samples are taken (such as from waste piles), tools used for soil sampling can be used, such as scoop, trowel, and bucket auger.

The composite liquid waste sampler (*coliwasa*) (Fig. 4.7) is a tool typically used for sampling stratified liquid in drums and other similar containers. It is a transparent or opaque glass, PVC, or Teflon tube approximately 60 inches in length and one inch in diameter. A neoprene stopper at the bottom of the tube can be opened and closed

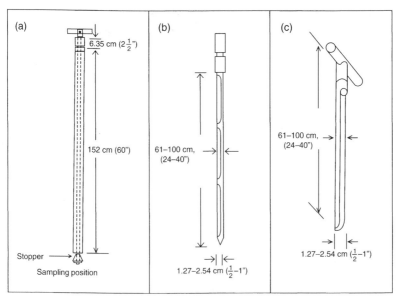

Figure 4.7   Common sampling tools used for wastes: (a) Composite liquid waste samplers (coliwasa), (b) Thief (grainer sampler), and (c) Trier (U.S. EPA, 1986)

via a rod that passes through the length of the sampler. A *thief* (grain sampler), consisting of two slotted concentric tubes, is another commonly used drum sampling device particularly useful for grain-like material. A sampling *trier* is used to collect sticky solids and loosened soils. The trier is usually a stainless steel tube, cut in half lengthwise with a sharpened tip, which allows the sampler to cut the solid sample. A trier is relatively easy to use and decontaminate.

Biological Sampling

Biological sampling for chemical residual analysis has very unique and diverse equipment compared with other environmental sampling processes. Consultation with an experienced ecologist or biologist is strongly recommended. For instance, in collecting mammals, trapping is the most common method. Both live and kill trap methods may be used through commonly used traps such as Museum Special, Havahart, Longworth, and Sherman traps (EPA, 1997). To collect fish, electrofishing, gill nets, trawl nets, sein nets, and minnow traps are the common methods. In collecting vegetation samples, samples should be harvested during the growing season for the herbaceous plants or during the growing or dormant season for woody plants. For benthic macroinvertebrate samples, both the Petersen and Ekman dredges can be used as discussed above.

Air and Stack Emission Sampling

There are many direct-reading instruments and types of monitoring equipment that provide instantaneous (real-time) data on the contaminant levels. Despite this, air sampling followed by laboratory analysis (off- or on-site) is still required for the most part of trace level analysis of atmospheric contaminants. Air sampling equipment is expensive and techniques are complex. Therefore, most of the compliance (such as source stack emission monitoring) are contracted to professional stack-testing firms. Since air sampling equipment and sampling device/media are so diverse, we can only give general discussions herein. Readers in this field can consult an excellent introductory book on air sampling by Wight (1994).

Common air sampling equipments includes (Fig. 4.8) (1) high volume, total suspended particle (TSP) samplers, (2) PM-10 samplers, (3) High volume PS-1 samplers, (4) personal sampling pumps, and (5) canister samplers. The *TSP sampler* collects all suspended particles by drawing in air at 40 cubic feet per minute (cfm) across an 8 by 10 inch glass-quartz filter. The mass of TSPs is determined by weighing the filter before and after sampling. The *PM-10 sampler* collects particulates with a diameter of 10 μm or less from ambient air. PM-10 samplers can be high volume (40 cfm) or low volume (36 cfm). The high volume sampler uses 8 by 10 inch glass-fiber filter, and the low volume PM-10 sampler collects its sample on 37-mm Teflon filters.

The *PS-1 sampler* draws air through polyurethane foam (PUF), or a combination of foam or adsorbing materials, and a glass-quartz filter at a rate of

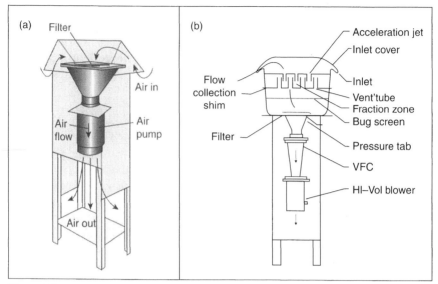

Figure 4.8   Common sampling equipments used for air. (a) Total suspended particulate matter (TSP) sampling system (Girard, JE, principles of Environmental Chemistry, 2005, Jones and Bartlett publishers, sudbury, MA. WWW. jbpub. com. Reprinted with permission.), and (b) PM-10 sampling system (Norwegian Institute for Air Research, 2001)

5-10 cfm. This system is excellent for measuring low concentrations of SVOCs, PCBs, pesticides, or chlorinated dioxins in ambient air. The *personal sampling pumps* are portable sampling devices that draw air through a number of sampling media including adsorbent tubes, impingers, and filters (discussed below).

Air sampling collection media and device shown in Figure 4.9, in principle, can be divided into four major categories: (1) by the collection of an air sample in a container such as a canister or Tedlar bag, (2) by filtration of air using filter cassette, (3) by chemical reaction or absorption through the bubbling of air in a liquid medium in an impinger, and (4) by adsorption of chemicals in the air on the adsorbing materials. The *SUMMA canister* (6 or 15 liters) uses the SUMMA electro-polishing process to form chromium and nickel oxides on the internal surface so that adsorption of VOCs is minimized. Air samples are collected by allowing air to bleed into or be pumped into the canister. SUMMA canisters are used to measure low to ultra low concentration of gaseous compounds, typically in the ppt–ppb range. The *Tedlar bag*, albeit with the same principle, are simple and effective means for contaminants of high concentrations (>10 ppm$_v$). An *impinger* is used to collect compounds in condensable concentrations or those that can be readily retained by absorption or reactions with the liquid in the impinger. *Sorbent tubes* are used primarily in industrial hygiene. There are two types of sorbing materials – the thermally desorbed media including Tenax and carbonized polymer and the solvent-extracted media such as XAD-2 polymer, activated carbon, and Tenax. Air sampler using PUF is a combination of filter (glass or quartz) and a polyurethane foam to

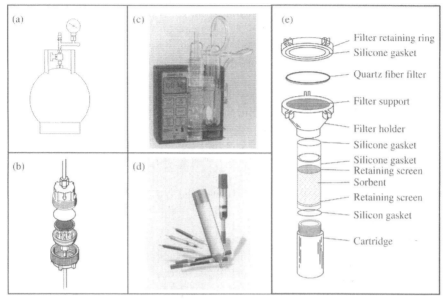

Figure 4.9 Common air sampling collection media/device: (a) Rigid whole air collection canister (SUMMA canister), (b) Filter cassette for particulate sampling, (c) Impinge, (d) Sorbent tube/cartridge, and (e) Polyurethane foam (PUF). Courtesy of SKC, Inc. except for E (U.S. Army Corps of Engineers, 2001)

collect trace organics in both the vapor and aerosol forms. PUF is used in conjunction with PS-1 sampler introduced above.

4.2 TECHNIQUES FOR SAMPLING VARIOUS MEDIA: PRACTICAL APPROACHES AND TIPS

No single sampling guideline or procedure applies universally to every case due to the wide variations of sampling situations. This is also why a site- and project-specific sampling plan should be developed and implemented. To succeed in sampling, both experience and common sense approach are essentially important. In this section, we discuss some important practical approaches and tips that might be helpful for beginners to perform a successful environmental sampling of various matrices.

Practical tips

- A common piece of advice from experienced field crews is to have good office preparation prior to field sampling. Review the SOP carefully and arrange a prior trip to the sampling locations as needed. Communicate with lab personnel for any doubt regarding sample amount, preservation, containers, and sampling tools.

- Preparations include, but not limited to, cleaning, labeling and organizing sample containers, preservation chemicals, decontamination of sampling equipment and device, and calibrating field monitoring equipment. A great deal of time is spent on in situ measurement such as for water/groundwater quality (pH, oxidation-reduction potential, screening methods), hydrogeological and/or meteorological parameters.

- Having the right tools and other essential items will make your sampling work less frustrating and more productive. While the needed items may vary from project to project, below is a list of an "Ideal Tool Kit" adapted from Bodger (2003):

  - The container: A sturdy plastic container (e.g., Rubbermaid$^{\circledR}$) or a metal toolbox kept in your vehicle can be used to organize your tools and supplies.
  - The tool contents: The box should contain a few important tools such as a wrench, screwdriver, plier, hammer, duct tape, wire stripper, tape measure, Teflon tape, electrical tape, flashlight, mirror, hand cleaner, gloves, clipboard, paper towel, and permanent marker.
  - The field book and the Chain-of-Custody form: These are the crucial documents that can be used in a court of law. The EPA guidelines recommend not only that field books are pre-numbered, but also that there are lines for your signature and the date on each page.
  - Other supplies related to: (a) decontamination (brushes, plastic sheeting, bucket, solvent spray, aluminum foil, soap, distilled water), (b) waste disposal (trash bag, liquid waste containers), and (c) health and safety equipment (safety glasses, hard hat, steel-toe boots, first-aid kit, ear plugs).
  - Always carry a map, a global positioning system (GPS), and a camera with you. Have important telephone numbers and emergency contacts handy.

4.2.1   Surface Water and Wastewater Sampling

Fresh surface waters are commonly separated into three groups: flowing waters (rivers, streams, and creeks), static waters (lakes, lagoons, ponds, and artificial impoundments), and estuaries. Waters in general also include tap water, rain, fog, snow, ice, dew, vapor, and steam. Wastewaters include mine drainage, landfill leachate, and industrial effluents, and so forth. These waters differ in their characteristics, therefore sample collection must be adapted to each. Because of this great variability and each may represent its own specific practices and procedures, we focus on several important types of surface waters in the following discussions.

The sizes of the stream or river and the amount of turbulence have a major influence on the representativeness of water samples to be collected. Identifying sampling locations that are well-mixed vertically or ones that are horizontally stratified are useful prior to sampling. In *small streams* (<20 ft wide), it is possible to select a location where a grab sample can represent the entire cross-section. This location can be at the mid-depth in moving water at the main flow line. This main flow line is not necessarily the center of the stream, but can be identified by observing the flow patterns across the channel. For *larger streams and rivers*, multiple samples across the channel width are required at a specific sampling location. As a minimum, one vertical composite (consisting of grab locations from just below the surface, mid-depth, and just above the bottom) at the main flow line would be necessary.

For *fast flowing rivers and streams*, it may be difficult to collect a mid-channel sample; health and safety concerns must dictate where to collect the sample. When sampling nearby a point pollution source, two samples from channel mid-depth are typically drawn: one upstream and one adjacent to, or slightly downstream of the point of discharge.

In *small ponds and impoundments*, a single vertical composite at the deepest point is adequate to characterize the water body. This deepest point of a naturally formed pond is generally near the center although this may need to be determined. For *lakes and larger impoundments*, stratification from temperature difference is often present in these bodies and is more prevalent than in rivers or streams. Therefore, several vertical aliquots should be collected. In determining the degree of stratification, measure DO, pH and temperature in an incremental depth of the water body.

Estuaries are areas where inland fresh water (both surface water and groundwater) mixes with oceanic saline water of a higher density. Estuaries are generally categorized as mixed, salt wedge, or oceanic, depending upon inflow and mixing properties (EPA, 1995). These must be considered in determining site-specific sampling locations. In addition, estuarine sampling should be conducted in two phases, that is, during both wet and dry periods. When water depth is less than 10 ft, samples are collected at the mid-depth unless the salinity profile indicates the presence of salinity stratification. In estuaries where water depth is greater than 10 ft, water samples may be collected at the one-foot depth, mid-depth, and one foot from the bottom. In addition, a horizontal sampler such as van Dorn sampler is often used rather than the vertical Kemmerer sampler (Fig. 4.3).

Practical tips

- When sampling rivers, stream, or creeks, locate the area that exhibits the greatest degree of cross-sectional homogeneity. Select a site immediately downstream of a mixing zone. In the absence of mixing zones, select a site without any immediate point sources such as tributaries, industrial, and municipal effluents.

- In many cases, compositing is needed to reduce the number of sample, but not to composite samples for VOC, SVOC, TPH, and oil and grease analysis. For the same reason, no headspace and vacuum filtration should be allowed for VOCs.

4.2.2 Groundwater Sampling

Groundwater sampling is probably the most complex of all and it is inherently difficult. The first and the most obvious is the installation of a sampling well in a manner that does not change the integrity of the surrounding water. Fortunately, well installation is often the task of specialized persons and is beyond the contents of this text. The most challenging tasks in routine groundwater sampling, among many others, are the following work:

- To characterize groundwater flow
- To purge and stabilize groundwater prior to sampling
- To minimize cross-contamination due to well materials and sampling devices

In Chapter 2, we introduced the basics of groundwater flow, well characteristics, and water level measurement. Here we further discuss the flow characterization, well purging prior to sampling, and special precautions on the cross contamination issues unique to groundwater sampling.

Groundwater Flow Direction

The determination of a groundwater flow is illustrated in the example provided below. To help understand this example, we need to know the concept of *hydraulic gradient*, which is defined as the slope of water table measured from a higher point of the water table to the lower point of water table across a site. This concept is important because the groundwater flow is in the direction of the gradient and the flow rate is proportional to the gradient. A related term is the *hydraulic head*, which is the vertical distance (in feet or meter) from some reference datum plane (usually taken to be sea level) to the water table. If we imagine two wells directly in line with the groundwater flow, the gradient would be the difference in head divided by the horizontal distance between wells.

$$\text{Hydraulic Gradient} = \frac{\text{Difference in Hydraulic Head}}{\text{Distance between Two Wells}} \qquad (4.1)$$

EXAMPLE 4.1. Two wells are drilled 300 m apart along an east-west direction. The east well and west well have hydraulic heads of 20.4 m and 20.0 m, respectively. The third well is located 200 m due south of the east well and has a head of 20.2 m. Schematically determine groundwater flow direction and estimate hydraulic gradient (Masters, 1998).

General Step (refer to Fig. 4.10):

1. Draw a line (AB) between two wells with the highest (A) and lowest (B) heads (i.e., head: A > C > B)

2. Locate the point (D) on line AB that corresponds to the well with intermediate head (C). Draw the line CD. This line is an *equipotential line*, meaning the head anywhere along the line is constant.

3. Draw a line EB perpendicular to line CD. Line EB is the flow line, meaning the groundwater flows in a direction parallel to this line. The water flows

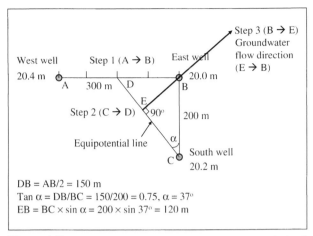

Figure 4.10 Determine the groundwater flow direction and hydraulic gradient

from the higher to the lower hydraulic head along line EB, meaning that the water flows from E to B rather than from B to E.

4. Determine the hydraulic gradient:

$$\text{Hydraulic gradient} = \frac{\text{Head of well C} - \text{Head of well B}}{\text{Distance from B to the equipotential line (CD)}}$$
$$= \frac{20.2 - 20.0}{120} = 0.00167 \ (Dimensionless)$$

Well Purging

Purging is used to remove stagnant water in the well borehole and adjacent sandpack so that the water to be sampled will be stabilized prior to sampling and representative groundwater can be obtained. Various methods for determining the necessary extent of well purging have been recommended.

The USGS recommends pumping the well until water quality parameters are stabilized. A "stabilized" condition can be determined by in situ measurement of the following parameters: DO: $\pm$ 0.3 mg/L; turbidity: $\pm 10\%$ (for samples greater than 10 NTUs); specific conductivity: $\pm 3\%$; oxidation-reduction potential (ORP): ± 10 mV; pH: ± 0.1 unit; temperature: $\pm 0.1^\circ$C. This approach is also specified in the U.S. EPA's Groundwater Guidelines for Superfund and RCRA Project Managers (EPA, 2002). The most commonly used parameters are pH, temperature and conductivity (Popek, 2003). The U.S. EPA also recommends the removal of three well-casing volumes prior to sampling. A handy formula in English measurement unit system is as follows (Bodger, 2003).

$$V = 7.48 \ \pi r^{2*} h \tag{4.2}$$

where V the volume (gallons) of well, r the radius of monitoring well in ft, and h the difference between depth of well and depth to water, or the height of the water column in ft; 7.48 is the conversion factor (the gallons of water per ft^3 of water).

Cross Contamination in Groundwater Sampling

In Section 4.1.4, we briefly discussed the potential contamination problem from the sampling container in general. Groundwater sampling has many other potential contamination sources, including contamination during well construction (well casings), well purging (pump and tubing), and sampling-related equipments (sampling tools, water level indicator, monitoring probes). Table 4.2 is a list of potential contaminants contributed to water samples by materials used in sampling devices or well casings (Cowgill, 1988). Cowgill (1988) recommended that all sampling devices or well casings should be thoroughly cleaned using steam prior to sampling.

Table 4.2   Contaminants contributed to water samples by materials used in sampling devices, well casings, and other items

Material	Contaminants contributed
PVC-threaded joints	Chloroform
Stainless steel pump and casing	Cr, Fe, Ni, and Mo
Polypropylene or polyethylene tubing	Phthalates and other plasticizers
Polytetrafluoroethylene (Teflon) tubing	None detected
Soldered pipes	Sn and Pb
PVC-cemented joints	Methyl ethyl ketone, toluene, acetone, methylene chloride, benzene, ethyl acetate, tetrahydrofuran, cyclohexanone, three organic Sn compounds, and vinyl chloride
Glass container	B and Si
Marker used in labeling	Volatile chemicals
Duct tape for all purposes	Solvent

Adapted from Cowgill (1988).

Practical tips

- Groundwater sampling requires more equipment and containers in numerous wells. A good office preparation will help reduce field time by an increased efficiency. Such preparations include containers, labels, coolers, freeze-packs, preservatives, calibration of instruments used for well purging and stabilization (pH, conductivity), sampling tools (pump, water level indicator), calibration solution, spare batteries, ropes, baggies, borehole logs, monitoring well development forms, sample collection forms, and many others.

- Unlike surface water sampling, you cannot just get to a well and collect the groundwater sample. You need to first purge the stagnant water by monitoring conditions of water (pH, temp, level, etc.) until the water is stabilized.

- Always keep in mind the potential contamination for the trace analysis of contaminants. Misrepresentative data can be a result of a combination of several sources due to well materials, pump, tubing, and containers.

- Try to minimize the aeration during sampling. Bailers are often used for both purging and sampling, but they cause mixing, collection of unwanted particulates from well bottom or casing, and aerating or degassing VOCs from samples. Bladder pumps are generally the best under many circumstances.

- Sampling sequences follow the order of VOCs, SVOCs, metals, PCBs/pesticides/herbicides, and water quality parameters.

4.2.3 Soil and Sediment Sampling

Soil sampling at a shallow depth is relatively an easy task. Sediment sampling is analogous to soil sampling at some extent, for example, with regard to sample compositing and post-sampling pretreatment (homogenizing, splitting, drying, and sieving). Both soil and sediment can be collected with a grab sampler (scoop and trowel) for surficial samples or a core sampler if a vertical integrity needs to be preserved. For sediment sampling, the corer is to further retain fine particles. It is critical to remove extraneous materials such as rocks, gravels, and debris such as plant leaves, roots, shells, and insects. But note that not all external materials are extraneous such as fly ash or waste that are part of the sample characteristics. In such a case, collect samples of any material thought to be a potential source of contamination. Composite sampling during field sampling is common for both soil and sediment because of the suspected heterogeneity and the need for savings of analytical costs. The only exception is for VOCs analysis since mixing, homogenizing, and composting of soil or sediment samples containing VOCs are not permitted. Another common error can be introduced in sample sieving and mixing procedures is to inappropriately discard nonsoil/sediment or nonsieved materials (EPA, 1989).

Additional concerns for sediment sampling include factors that affect sample representativeness, such as waterway width, flow, and bottom characteristics. Sediments are typically heterogeneous in composition and are subject to variation in texture, bulk composition, water content, and pollutant content. Therefore, large numbers of samples may be required to characterize a small area. Generally, samples are collected at quarter points along the cross-section of the water body and then are combined into a single composite sample. For small streams, one single sediment sample collected at the main flow line of the water body may be sufficient. Whenever possible, a sediment sample is collected at the same location as a surface water sample.

Sediment samples from lakes, ponds and reservoirs should be collected in approximately the deepest point of the water body. This is especially applicable to reservoir formed by impoundments of rivers or streams. Coarser grain sediments are found near the headwaters of the reservoir, while bed sediments are composed of fine-grained materials that have elevated contaminant concentrations. Contaminants tend to concentrate in the fine-grained sediment in depositional zones.

Practical tips

- Before digging soils, provide a detailed site map and check with local municipalities if the possibility of underground lines exists.
- Use sampling tools made of stainless steel or Teflon tools only. They are expensive, but it is less costly than re-sampling. The U.S. EPA states "equipment constructed of plastic or PVC should not be used to collect samples for trace organic compounds analyses" (Bodger, 2003).
- Do not use equipment made of galvanized metal as it contains Zn. Garden tools and cheap kitchen items are not recommended for sampling purpose as they are often chrome-plated.
- Make sure to scrape off part of the soils/sediment in direct contact with the sampler. For metal analysis, use stainless steel tools and plastic bucket to homogenize.
- For sampling deep soils (60–100 ft), a direct push Geoprobe ® method developed for EPA is desired. This direct push device has reduced costs in comparison with traditional drilling methods (HMTRI, 2002).
- When decant water after the collection of sediment samples, make sure that you do not lose the fines within the sediment.
- Sampling devices must be carefully cleaned prior to and between each sample to avoid cross contamination.
- Plan ahead, air-dry, and sieving could take a long time. Not to forget subsample for moisture measurement of soil and sediment.

4.2.4 Hazardous Waste Sampling

Wastes come from a variety of sources, including drums, storage tanks, lab packs, impoundments, waste piles, surfaces, and debris. Sampling approaches, therefore, vary considerably for each type of these wastes. Of critical importance is to first research all available documents, packing slips, and labels for the contents and associated health and safety precautions. Use proper PPE when open containers and sample wastes from *drums*, *lab packs*, or *waste tanks* with known chemical identities. For unknown wastes in drum and lab packs, open them remotely. Do not move drums with unknown chemicals since some chemicals may be potentially shock-sensitive, explosive, or reactive after a period of storage time. If the contents have multiple phases, sample each phase separately.

For *waste impoundments*, contaminant concentrations mostly will be stratified by depth and vary horizontally from the point of waste entry to the further end of the impoundment. For large impoundments, the logistics, health and safety concern of sampling will usually dictate whether to use boats, manlifts, or safety lines.

There is not established sampling protocol for *debris*, since the irregular and/or large pieces of materials such as metal, plastic, and wood present challenge both in sampling and subsequent sample preparation. Sampling from *waste piles* such as impoundment dredge or slag piles would be relatively easy. Depending on the sampling objective, either grab sampling using a scoop/trowel or interior sampling using an auger is appropriate. For large waste piles, EPA recommended a 3-ft depth for representative samples.

Surface sampling is unique because of the technique and the way to express the results. Surface sampling includes wipe, chip, and dust sampling depending on the types of surface and materials to be collected. *Wipe sampling* is usually conducted in a 100-cm^2 area on relatively smooth, nonporous surfaces such as wall, floors, ventilation ducts and fans, empty transformers, process equipment, and vehicles containing nonvolatile compounds. It is also typically used to determine if decontamination has been effective or not (e.g., disposal facility, sampling equipment). A piece of sterile medical gauze soaked with proper solvents (hexane, methanol, water, nitric acid) is used. To collect, you wipe back and front, and then side to side, to make sure you get the full area. The results are expressed in μg/100 cm^2. The second method, *chip sampling*, is used to collect nonvolatile species from porous surfaces including cement, brick or wood in areas such as floors near process areas, storage tanks, and loading dock areas. Chip sampling is usually performed with a hammer and chisel or with an electric hammer for a 1/8-inch sampling depth. Subsequent laboratory analysis may require special grinding or extraction procedures. Lastly, the *dust sampling* is used to collect metal and SVOCs in residues or dust found on porous or nonporous surface in areas such as bagging, processing, or grinding facitities where powdery contaminants and dust may have been accumulated. Unlike the other two surface sampling methods, the results from dust sampling are reported in weight/weight (mg/kg).

Practical tips

- In almost all waste sampling cases, you need to get 40-h HAZWOPER training before you can handle any hazardous materials. Safety and health precautions are the most important! It is not worth it to do any work if the work is not safe. Select the appropriate level of PPE and know how to use them.

- Special tools are needed to open drums. Use spark-free drum openers and other suitable tools.

- Sample with extreme cautions when dealing with containers of unknown chemical identities. In some cases, the labels may not reflect the true contents as drums may be recycled or chemicals have been degraded.

- Cross contamination due to sampling tools and containers is of less concern since analytes in hazardous materials are often present at high concentrations.

4.2.5   Biological Sampling

Biological sampling rests on the responsibility of trained ecologist or biologist. In Section 4.1.5, we discussed very briefly on the sampling tools for biota samples used in chemical residual analysis. In addition to residual analysis, biota sampling may be aimed at toxicological testing on contaminated sediment and surface water, or even community response assessment based on sampling results from terrestrial vegetation, benthic organisms, or fish, to just name a few. Readers in this specialized area should consult the detailed guidelines by the U.S. EPA, APHA, and the USGS provided at the end of this chapter (APHA, 1998; USGS 1993a–e).

Practical tips

- Biological samples are not easy to collect in many cases. Insufficient sample size due to the lack of specimen availability could result in an invalid statistical inference.
- If an improper biota sampling protocol was implemented, data may later find to be flawed. Causes could be the neglect of vast size differences between species, variations with a study population, tissue differentiations, growth stage, and habitat.
- Biological samples are noted for being much more susceptible to the decomposition of organic analytes than are other types.

4.2.6   Air and Stack Emission Sampling

Common types of air samples include ambient (outdoor) air, indoor or workplace air, and air from stacks or other types of emission exhausts such as automotive exhausts. *Soil air* is sometimes collected for special purpose such as from landfills or Superfund sites. *Ambient air* sampling represents two critical challenges. First, the concentrations for most atmospheric pollutants in ambient are very low. The analysis of low level organic compounds in the air requires large sample volumes. Second, the large variations in analyte concentrations can occur within in a short period of time due to the meteorological effects. Ambient air samples should be collected at a location that most closely approximates the actual human exposure.

Indoor air sampling requires some special considerations (Keith, 1991). The ventilation systems due to heating or cooling equipment can alter air flow and add pollutants to the air. The location of a sampler within a room will influence the results obtained. Household chemicals (cleaning agents, pesticides) could add compounds to the air that must be documented and considered.

Sampling air in a duct or a stack is another matter entirely. The location should be approximately three-quarters of the way down a long, straight run of ductwork (Bodger, 2003). The flow will be relatively calm in such a location and, as a result, the sample is more representative. For stationary sources such as *stack emission*, the sampling site and the number of traverse points used for the collection affect the quality of the data. Depending on the stack geometry, a cross-section of the stack perpendicular to the gas flow is divided into specified number of equal area. The required number and location of traverse points can then be determined for gas velocity measurement and sample extraction.

Practical tips

- Note the difference in sampling regulations and methodologies of various agencies (mainly EPA and OSHA) for various projects dealing with ambient air, workplace air, or emission from a smoke stack (ACGIH, 1989).

- Like testing of other media, "informal" testing differs from "compliance" testing. Use a common sense approach with sample equipment for the former. For the regulatory compliance testing, which are mostly conducted by certified professional testing labs, it requires expensive equipment and complex methodologies.

- The choice of sample locations is extremely important for the accuracy of air sampling. The key is to find a place that is either representative of the local air quality, or depending on your objectives, represent a worst case. When ambient air is collected, it is most useful to pick a location that most closely approximates actual human exposure.

- It is relevant to emphasize the importance of a good air-flow measurement. When mass emission rate of a compound (lb/hr, for example) is reported, the final calculation involves multiplying a pollutant concentration by the air-flow. The best concentration measurement in the work can be horribly skewed by an incorrect air flow determination (Bodger, 2003).

- Document sampling conditions at the collection time and site, including meteorological (wind direction, wind speed, temperature, atmospheric stability, atmospheric pressure, humidity, and precipitation), and topographical (building, mountains, hills, valleys, lakes, and sea) data. These factors will help later data interpretations.

REFERENCES

American Conference of Governmental Industrial Hygienist (ACGIH) (2001), *Air Sampling Instruments*, 9th Edition, Cincinnati, OH.

American Public Health Association, American Water Works Association, Water Environment Federation (1998), *Standard Methods for the Examination of Water and Wastewater*, 20th edition. Edited by: Clesceri, L.S., Greenberg, A.E., Eaton, A.D., Published By: American Public Health Association.

*Bodger K (2003), *Fundamentals of Environmental Sampling*, Government Institutes, Rockville, MD.

Christian GD (2004), *Analytical Chemistry*, 6th Edition, John Wiley & Sons Hoboken, NJ.

Cowgillum (1988), Sampling Waters: The Impact of sample variability on planning and confidence Levels, *In* Keith LH (1988), principles of Environmental sampling, American chemical society, Washing, DC.

Forestry Suppliers, Inc., http://www.forestry-suppliers.com; Ecotech Pty Ltd, http://www.ecotech.com.au/ (For pictures of common sampling tools).

Hazardous Materials Training and Research Institute (HMTRI) (2002), *Site Characterization Sampling and Analysis*, John Wiley & Sons Hoboken, NJ.

*Suggested Readings

Kᴇɪᴛʜ LH (1988), *Principles of Environmental Sampling*, American Chemical Society Washington, DC.

*Kᴇɪᴛʜ LH (1991), *Environmental Sampling and Analysis: A Practical Guide*, Lewis publishers, Boca Raton, FL.

Kʟᴇᴍᴍ DJ, Lᴇᴡɪs PA, Fᴜʟᴋ F, Lᴀᴢᴏʀᴄʜᴀᴋ JM (1990), Macroinvertebrate Field and Laboratory Methods for Evaluating the Biological Integrity of Surface Waters, USEPA. EPA/600/4-90/030.

Mᴀsᴛᴇʀs GM (1998), *Introduction to Environmental Engineering and Science*, 2nd Edition, Prentice Hall, Upper Saddle, NJ.

McDᴇʀᴍᴏᴛᴛ HJ (2004), *Air Monitoring for Toxic Exposures*, 2nd Edition, Wiley-Interscience Hoboken, NJ.

Pᴏᴘᴇᴋ EP (2003), *Sampling and Analysis of Environmental Chemical Pollutants: A Complete Guide*, Academic Press San Diego, CA.

Sᴍɪᴛʜ R-K (1997), *Handbook of Environmental Analysis*, 3rd Edition, Genium Publishing Corporation Amstordam, NY.

Texas Natural Resource Conservation Commission (1999), *Surface Water Quality Monitoring Procedures Manual* Austin, TX.

US Army Corps of Engineers (2001), Requirements for the Preparation of Sampling and Analysis Plans: Engineer Manual, EM 200-1-3 (http://www.usace.army.mil/).

US EPA (1989), *Soil Sampling Quality Assurance User's Guide*, 2nd Edition, EPA/600/8-89/046 (http://clu-in.org/download/char/soilsamp.pdf)

US EPA (1995), *Superfund Program Representative Sampling Guidance: Volume 1*, Soil, Interim final, EPA 540-R-95-141, OSWER Directive 9360.4-10, PB96-963207, December 1995.

US EPA (1995), *Superfund Program Representative Sampling Guidance: Volume 2*, Air (Short-Term Monitoring), Interim final, EPA 540-R-95-140, OSWER Directive 9360.4-09, PB96-963206, December 1995.

US EPA (1997), *Superfund Program Representative Sampling Guidance: Volume 3*, Biological, Interim final, EPA 540-R-97-028, OSWER Directive 9285.7-25A, PB97-963239, December 1995.

US EPA (1995), *Superfund Program Representative Sampling Guidance: Volume 4*, Waste, Interim final, EPA 540-R-95-141, OSWER Directive 9360.4-14, PB96-963207, December 1995.

US EPA (1995), *Superfund Program Representative Sampling Guidance: Volume 5*, Surface Water and Sediment, Interim final, OSWER Directive 9360.4-16, December 1995.

US EPA (1997), *Guidance for Data Quality Assessment, Practical Method for Data Analysis*, EPA QA/G-9.

US EPA (2001), *Methods for Collection, Storage and Manipulation of Sediments for Chemical and Toxicological Analyses: Technical Manual.* EPA-823-B-01-002, October 2001 (http://www.epa.gov/waterscience/cs/collectionmanual.pdf).

US EPA (2002), *Groundwater Guidelines for Superfund and RCRA Project Managers*, Office of Solid Waste and Emergency Response, EPA 542-S-02-001 (http://www.epa.gov/reg3wcmd/ca/pdf/sampling.pdf).

US Geological Survey (1993a), *Guidelines for the Processing and Quality Assurance of Benthic Invertebrate Samples Collected as Part of the National Water-Quality Assessment Program.* Open-File Report 93-407.

US Geological Survey (1993b), *Methods for Sampling Fish Communities as Part of the National Water-Quality Assessment Program.* Open-File Report 93-104.

US Geological Survey (1993c), *Methods for Collecting Benthic Invertebrates as Part of the National Water-Quality Assessment Program.* Open-file Report 93-406.

US Geological Survey (1993d), *Methods for Collecting Algal Samples as Part of the National Water-Quality Assessment Program.* Open-file Report 93-409.

US Geological Survey (1993e), *Methods for Characterizing Stream Habitat as Part of the National Water-Quality Assessment Program.* Open-file Report 93-408.

Wɪɢʜᴛ GD (1994), *Fundamentals of Air Sampling*, Lewis Publishers, Boca Raton, FL.

QUESTIONS AND PROBLEMS

1. Give a list of water quality parameters that must be measured immediately in the field while sample is being taken.

2. Which one of the following should be monitored in the field: (a) acidity, (b) DO, (c) pesticide, or (d) hardness?

3. Describe all possible physical, chemical and biological changes that can occur for the following compounds during the sample holding period: (a) Oil, (b) S^2 , (c) CN^-, (d) phenols.

4. Describe all possible physical, chemical and biological changes that can occur for the following compounds during the sample holding period: (a) chlorinated solvents, (b) PAHs, (c) metals.

5. Explain why: (a) HNO_3 rather than other acids is used for metal preservation; (b) amber bottle is preferred for PAHs; (c) zero-head space container for groundwater samples collected for tetrachloroethylene analysis.

6. What special precautions need to be aware of regarding sampling for VOCs analysis for each of the following: (a) Sampling sequence relative to other analytes; (b) Sample container; (c) sample compositing.

7. Explain potential contaminants that may leach from sample containers or sampling tools made of: (a) glass, (b) PVC, (c) plastics, (d) stainless steel.

8. Explain why glass containers are generally used for organic compounds whereas PVC-type containers are used for inorganic compounds?

9. Describe the difference between the Kemmerer sampler and the von Dorn sampler. Which one is particularly suitable for estuarine water sampling?

10. Which one of the following is used in the sampling of hazardous waste: (a) Ekman dredge, (b) Kemmerer, (c) Ponar, or (d) Coliwasa?

11. The maximum holding time for metal analysis (excluding Cr^{6+} and dissolved Hg) after acidification to pH < 2 is most likely: (a) 6 h, (b) 6 days, (c) 6 weeks, (d) 6 months.

12. Which of the following is used to preserve samples for the analysis of total metals (excluding Cr and Hg): (a) Cool, $4°C$, (b) NaOH to pH > 12, (c) H_2SO_4 to pH < 2, (d) HNO_3 to pH < 2?

13. A "thief" is a two-concentric tube tool used to collect: (a) granular, particles, and powdered waste, (b) liquid waste, (c) sticky and moisture waste, or (d) hard soils?

14. Which one of the following is used to collect air samples for the analysis of low concentrations of organic contaminants: (a) TSP sampler, (b) PM-10 sampler, (c) PS-1 sampler, (d) impinger.

15. Which of the following sampling device/media is based on absorption and/or chemical reaction: (a) impinger, (b) XRD sorbent, (c) Tedlar bag, (d) canister?

16. A bailer is commonly used to collect: (a) surface water sample, (b) groundwater well sample, (c) liquid hazardous waste sample, (d) none of above.

17. Which of the following is the best for VOCs sampling from groundwater well: (a) bailer, (b) air-lift pump, (c) peristaltic pump, (d) bladder pump? Explain.

18. Suppose you are recently hired as an entry level Environmental Field Specialist in a new firm dealing with groundwater remediation where a sampling and analysis plan (SAP) has not been developed. You are assigned as an assistant to a Project Manager for groundwater sampling, and are asked to do the office preparation for this sampling event. Make a list of items you may need in the field. Exclude containers and sampling tools in this list.

19. Three wells are located in x and y plane at the following coordinates: Well 1 (0,0), Well 2 (100 m, 0), and Well 3 (100 m, 100 m). The ground surface is level and the distance from the surface

to the water table in Well 1, 2, and 3 are in the order of 10.0 m, 10.2 m, and 10.1 m, respectively. Draw the well diagram and determine the flow direction and the hydraulic gradient.

20. A well has a water column depth of 10 ft and diameter of 6 in. (a) What is the well volume in gallons? (b) How much water need to be purged before samples can be collected?

21. Describe the special considerations in taking water samples from: (a) flowing waters (rivers and streams), (b) static waters (lakes and ponds), and (c) estuaries.

22. Describe why purging is important prior to groundwater sampling? What are the common approaches in groundwater purging?

23. Explain why cross contamination is especially important in groundwater sampling as compared with, for example, surface water sampling.

24. Explain which one of the following has the least potential for cross contamination: (a) PVC, (b) FRE, (c) Teflon, and (d) stainless steel?

25. Which surface sampling approach is appropriate in a tiled floor suspected of PCB contamination: (a) wiping sampling, (b) chip sampling, or (c) dust sampling?

Chapter 5

Methodology and Quality Assurance/Quality Control of Environmental Analysis

The U.S. EPA and many other agencies and organizations have promulgated a variety of standard methods for sampling and analysis. To select an appropriate method and understand the requirements for a particular project is an important first step, yet it is not an easy task. To many beginning environmental professionals, this could be a frustrating and discouraging experience. As Popek (2003) noted, "the analytical laboratory with its dozens of seemingly similar methods and arcane QA/QC procedures is a mysterious world that most professionals who work in consulting industry rarely visit or fully comprehend. The perplexing terminologies, the subtitles of analytical QA/QC protocols, and the minutia of laboratory methods usually intimidate those who come in contact with the analytical laboratory only occasionally." The purpose of this chapter is to serve as a roadmap in helping the reader to navigate the "maze" of sampling and analytical methodologies, as well as QA/QC procedures. We will discuss these methodologies and procedures promulgated by the U.S. EPA, as well as other agencies. While reading this chapter, the reader should not be distracted by a specific method number and its specific details, but rather should navigate the recommended Web sites, learn how to identify

Fundamentals of Environmental Sampling and Analysis, by Chunlong Zhang
Copyright © 2007 John Wiley & Sons, Inc.

the sources of these methods, and to familiarize oneself with the structures of these various resources.

5.1   OVERVIEW ON STANDARD METHODOLOGIES

A *method* is a body of procedures and techniques for performing an activity (e.g., sampling, chemical analysis, quantification), systematically presented, in the order in which they are to be executed. Before the introduction to various standard methods, the reader must be aware of the clear difference between regulatory methods and consensus methods. The *regulatory methods* are approved by the U.S. EPA and are mandatory under a certain program or a law. The *consensus methods* are published by professional organizations such as the American Society for Testing Materials (ASTM), the U.S. Geological Survey (USGS), and the Association of Official Analytical Chemists (AOAC). In some cases, the differences are not entirely clear because many of the consensus methods have been adopted by the U.S. EPA. A standard U.S. EPA method typically has the following components:

- Scope and application: overview and potential problems/variations
- Summary of method: overview of analysis
- Interferences: contamination of sample during handling
- Safety: protection of analyst
- Apparatus and materials: details of laboratory hardware
- Reagents: details of chemicals, preparation of standards
- Calibration: responses, calibration curves, and quantitation
- Quality control: proof that laboratory can meet specifications of method
- Sample collection, preservation, handling: details of sampling
- Extraction procedure: details of method to extract organics from matrix
- Instrumentation: type of instrumentation and operating procedures
- Qualitative identification
- Calculations: quantitative analysis
- Method performance: MDLs

5.1.1   The U.S. EPA Methods for Air, Water, Wastewater, and Hazardous Waste

Several offices within the U.S. EPA are responsible for generating or requiring the use of specific sampling and analytical methods. These include the Office of Air Quality Planning and Standards (OAQPS), the Office of Water (OW), and the Office of Solid Waste (OSW). These offices published many test methods primarily in print documents during the early years of the EPA. Starting in the mid-1990s, electronic versions of EPA air methods began to appear, and some water methods were distributed on floppy disks. In the late 1990s, EPA issued two compendium methods on CD ROM for waste and for water. The Internet access commenced with air methods and in 1998 the entire waste method (SW-846) was placed on the web. To help quickly search for a particular method and get an access to the method, two very useful indices are worth noting here:

- Index to EPA Test Method (Nelson, 2003; http://www.epa.gov/epahome/index)
- National Environmental Methods Index (NEMI) (http://www.nemi.gov).

Both method databases have a search option by analyte name or method number. The NEMI also includes methods from many other sources (e.g., USGS, ASTM) and a general online search can be performed by media (air, water, soil, and so forth), subcategory (organic, radiochemicals, and so forth), source (EPA, USGS, and so forth), and/or instrumentation. The NEMI can also be searched by the type of regulation, including National Pollutant Discharge Elimination System and National Primary/Secondary Drinking Water Regulations. An additional feature of this online database is direct Internet access to the actual methods for free (e.g., EPA and USGS) or for a fee (e.g., ASTM and AOAC).

Described below are major EPA test methods by the type of media, that is, air, water, and waste. It should be noted that while many methods were developed for a particular medium, there are other methods that are generally applicable to various media. Additionally, the EPA test methods are ever changing, so the reader should constantly check the most recent revision for any updates, additions, or deletions.

Air Test Methods

The U.S. EPA methods for atmospheric pollutants are under the authority of the Clean Air Act (40 CFR 50-99). Sources for air test methods are located in three different categories depending on the type of "air":

- Methods for ambient air: http://www.epa.gov/ttn/amtic/
- Methods for emission from stationary sources: http://www.epa.gov/ttn/emc
- Methods for workplace/indoor air: mostly associated with OSHA/NIOSH: http://www.cdc.gov/niosh/nmam/and http://www.osha.gov/dts/sltc/methods/toc.html.

The first two categories are the U.S. EPA air test methods, both of which are available from the Technology Transfer Network (TTN) site at http://www.epa.gov/ttn. To obtain *methods for ambient air*, select AMTIC (Ambient Monitoring Technology Information Center) from this Web site. There are three types of methods for monitoring ambient air:

- Method for *criteria pollutants*: These are the six air pollutants common throughout the United States, including particulate matter (TSP, PM_{10}, and $PM_{2.5}$), SO_2, CO, NO_2, O_3, and lead.
- Air Toxic (TO) Methods: These are for the measurement of 189 toxic organic chemicals called *hazardous air pollutants* (HAPs). These chemicals are known or suspected as carcinogens with high usage and emission in

many industries. There are 17 peer reviewed, standardized methods for the determination of volatile, semi-volatile, and selected toxic organic pollutants in the air. The methods are named from TO-1 to TO-17.

- Inorganic (IO) Compendium Methods: These are a set of 17 methods (in five categories) with a variety of applicable sampling methods and various analytical techniques for specific classes of inorganic pollutants. The methods are named as IO-1.x to IO-5.x.

To obtain *methods for emission from stationary sources*, select EMC (Emission Measurement Center) from the same TTN Web site. The EMC test methods can be divided into four categories based on the legal status and EPA's confidence in its application for the intended use in measuring air pollutants emitted from the entire spectrum of industrial processes causing air pollution:

- Category A: Methods proposed or promulgated in the Federal Register
- Category B: Source category approved alternative methods
- Category C: other methods
- Category D: Historic conditional methods

Category A contains mainly four types of important methods that have been proposed or promulgated in the Federal Register.

- Method 1 to 29: These are the methods associated with "New Source Performance Standard (NSPS)." They can be found in Appendix A to 40 CFR Part 60.
- Method 101 to 115: These are the methods associated with "National Emission Standards for Hazardous Air Pollutants (NESHAP)." They can be found in Appendix B to 40 CFR Part 61.
- Method 201 to 206: These are the methods associated with "State Implementation Plan (SIP)." They can be found in Appendix M to 40 CFR Part 51. Method 206 was renamed to CTM-027.
- Method 301 to 310: These are the methods associated with "Maximum Achievable Control Technology (MACT)" standards. They can be found in Appendix A to 40 CFR Part 63.

You can obtain parts of 40 CFR (Title 40—Protection of Environment) for free from the electronic Code of Federal Regulation (e-CFR) Web site: http://www.gpoaccess.gov/ecfr. A hardcopy may be obtained from the Government Printing Office at http://www.gpoaccess.gov/cfr/index.html.

Water Test Methods

The U.S. EPA water test methods are under the water program (The Federal Water Pollution Control Act, 40 CFR 100-140) and drinking water program (The Safe Drinking Water Act (SDWA), 40 CFR 141-149). Important methods and documents are as follows:

- Method 100 to 400 Series: Published as "Methods for Chemical Analysis of Water and Wastes" (EPA 600/4-79-020, Revised 1983). Also abbreviated as the MCAWW Methods, including Method 100 (physical properties), Methods 200 (metals), Methods 300 (inorganic and nonmetallics), and Method 400 (organics).
- Method 500 Series: Published as "Methods for the Determination of Organic Compounds in Drinking Water" (EPA-600/4-88/039, December 1988, revised July 1991). The Method 500 series was developed for the analysis of drinking water. It also covers drinking water, groundwater, or rather pristine (particulate free) natural water. It is a subset of the 600 series methods.
- Method 600 Series: Published as "Test Methods for Organic Chemical Analysis of Municipal and Industrial Wastewater" (EPA-600/4-82-057, July 1982). It was developed for the analysis of water and wastewater, but covers natural waters and drinking water as well.
- Method 900 Series: Published as "Prescribed Procedures for Measurement of Radioactivity in Drinking Water" (EPA-600/4-80-032, 1980).
- Method 1000 Series: A set of microbiological and toxicity procedures used for compliance monitoring under CWA, including the 1600 series methods (Whole Effluent Toxicity Methods).

The U.S. EPA has recently proposed to withdraw most of the 100–400 series methods in the MCAWW manual published in 1983, as many of these methods are already outdated. An example is the Freon (CFC-113) based extraction method for Oil and Grease measurement. Where MCAWW methods are withdrawn, alternative test procedures published by the EPA or other organizations (e.g., ASTM, Standard Methods; see Section 5.1.2) will be provided.

In locating the above-mentioned EPA water test methods, the readers can first search the online NEMI database (http://www.nemi.gov) for printable files in Adobe Acrobat format. The EPA's Office of Water has also taken the initiative to provide its methods and guidance documents on CD-ROM. This volume contains more than 330 drinking water and wastewater methods and guidance from over 50 EPA documents, including Methods for Chemical Analysis of Water and Wastes (MCAWW), revised in March 1983; Metals, Inorganic and Organic Substances in Environmental Samples; 40 CFR Part 136 Appendix A, B, C, and D; Series 500, 600, and 1600 methods. A powerful search engine allows users to search by method number, analyte name, or keywords in the text. All text, tables, diagrams, flowcharts, and figures are included. The user can electronically jump from the Table of Contents to documents of interest and look up chemicals on the analyte list and jump via hypertext link.

Waste Test Methods

The U.S. EPA waste test methods are commonly referred to as SW-846 (Test Methods For Evaluating Solid Waste, Physical/Chemical Methods). These methods are under U.S. EPA's solid waste program of the Solid Waste Disposal Act of 1965 as amended by the Resource Recovery Act of 1970 (40 CFR 240-299). The complete sets of methods are available online at http://www.epa.gov/epaoswer/hazwaste/test/main.htm. The manual presents the state-of-the-art routine analytical tests adapted for the RCRA program. It contains procedures for field and laboratory quality control, sampling, determining hazardous constituents in wastes, determining *hazardous characteristics of waste (toxicity, ignitability, reactivity, and corrosivity)*, and determining physical properties of wastes. The manual also contains guidance on how to select appropriate methods. SW-846 contains over 200 documents, including the Table of Contents, Disclaimer, Preface, Chapters 1 through 13, and specific methods from 0010 to 9000 series (Table 5.1). To locate a particular method such as Method 1311 in the 1000 series method, just click on the phrase "1000 Series," which will take you to a listing of the SW-846 methods in the 1000 series.

In addition to know the method structure as shown in Table 5.1, it is also helpful to note the following general guidance regarding SW-846 methods:

- Unlike EPA 500 and 600 series methods for chemical analysis, the SW-846 separates analytical methods into two basic subsets: (a) extraction, concentration, digestion; (b) instrumental methods. A sample is usually first treated by a method from (a), and the extract aliquot/digestate is analyzed by a technique in subset (b).

- For metal analysis, typical SW-846 methods include series 3000 (digestion) followed by either series 6000 or 7000. The multielement ICP method (6000 series) is becoming more common in place of the single element method in series 7000 using atomic absorption spectroscopy.

Table 5.1   Illustration of EPA SW-846 methods by method series numbers

Method series	Description of methods
0010–0100	Methods for gaseous and volatile compounds from stationary sources.
1000	Methods for hazardous characteristics of wastes, including toxicity, ignitability, reactivity, and corrosivity.
3000	Methods for sample preparation for metals and organic analysis, including digestion (3000) for metals, extraction (3500), and clean-up (3600) for SVOCs.
4000	Immunoassay screening methods for various compounds.
5000	Methods for sample preparation for VOCs (headspace, purge-and-trap, azeotropic distillation, vacuum distillation, volatile organic sampling train).
6000	Instrumental methods for metal analysis using inductively coupled plasma atomic emission spectroscopy (ICP).
7000	Instrumental methods of metal analysis using atomic absorption with direct aspiration and furnace techniques.
8000	Instrumental methods of organic analysis using GC, HPLC, and IR.
9000	Other methods for inorganic anions, total organic, radioisotopes, and so forth.

- For semivolatile organic compound (SVOCs) analysis, methods typically include various extraction procedures using series 3500 (liquid–liquid extraction, Soxhlet extraction, ultrasonic extraction) followed by clean-up and then analysis by series 8000 methods using GC, HPLC, or IR.

- For volatile organic compounds (VOCs), typical procedures include series 5000 method using headspace, purge-and-trap, azeotropic distillation, vacuum distillation, and volatile organic sampling train (VOST), followed by instrumental analysis with various GC methods.

- The *toxicity characteristic leaching procedure* (TCLP) (Method 1311) is perhaps the most important of all 1000 series methods. The EPA has maximum concentrations for a list of contaminants. If the TCLP test determines that the contaminant concentration exceeds the regulatory level, this waste is deemed to be toxic and hence "hazardous." The hazardous waste is then designated with an EPA hazardous waste number. For instance, 2,4-DNT has a regulatory level of 0.13 mg/L. If the TCLP leachate of a waste contains 5 mg/L 2,4-DNT, this is hazardous waste and EPA HW No. D016 is assigned to this waste.

- *Immunoassays* (series 4000 methods) are commercially available test protocols that are rapid, simple, and portable. Methods have been developed for various compounds such as pentachlorophenol (PCP), 2,4-dichlorophenoxyacteic acid (2,4-D), PCBs, petroleum hydrocarbons, PAHs, toxaphene, chlordane, DDT, TNT, and hexahydro-1,3,5-trinitro-1,3,5-triazine (RDX).

- Methods 0010–0100 are not as commonly used as other SW-846 methods. Series 2000 is reserved for future addition.

5.1.2 Other Applicable Methods: APHA/ASTM/ OSHA/NIOSH/USGS/AOAC

APHA Methods

Entitled "Standard Methods for the Examination of Water and Wastewater," this single volume method manual is available in print or CD-ROM. These methods are probably the most widely used for water and wastewater analysis. The APHA method was first published in 1905 (long before the creation of the U.S. EPA), and it is currently the 21st edition jointly published by APHA (American Public Health Association), AWWA (American Water Works Association), and WEF (Water Environment Federation). The method is known as the "mother" of all methods and is commonly referred to as the *standard method* (SM). Its method systems are structured into the following:

- Part 1000 Introduction
- Part 2000 Physical and aggregate properties
- Part 3000 Metals

- Part 4000 Inorganic nonmetallic constituents
- Part 5000 Aggregate organic constituents
- Part 6000 Individual organic compounds
- Part 7000 Radioactivity
- Part 8000 Toxicity
- Part 9000 Microbiological examination
- Part 10,000 Biological examination.

Aggregate properties include regularly monitored water quality parameters such as color, turbidity, odor, taste, acidity, alkalinity, hardness, conductivity, salinity, solids, and temperature. *Aggregate organic constituents* include important water and wastewater parameters such as BOD, COD, TOC, oil and grease, phenols, and surfactants. Many of the APHA methods have been adopted by the U.S. EPA, particularly trace elements and individual and nonindividual (aggregate) organic compounds. Since the EPA does not have any counterpart microbiological testing methods, in many of the environmental projects, the APHA method manual for microbiological testing is commonly used.

ASTM Methods

Entitled "Annual Book of ASTM Standards," the ASTM methods manual has a total of 77 volumes with more than 12,000 standard methods covering methods used in various industries as well as in the environmental field. It is updated yearly by the American Society for Testing and Materials (ASTM). Its once wide use has recently decreased because the expensive method manual has to be purchased from the ASTM. The EPA approved many of the ASTM methods for compliance monitoring under the CWA. Environmental professionals often use some of the methods for remediation-related methods, such as soil properties including soil organic content, porosity, permeability, soil grain size, or the properties of petroleum products contained in the subsurface including viscosity, density, and specific gravity.

The ASTM methods are named as Dxxxx. For instance, D4700-91 is standard guide for soil sampling from the vadose zone. Important methods related to environmental analysis are located in Section 11 of the Water and Environmental Technology, including the following:

- Volumes 11.01–11.02 contain standard procedures for assessing water
- Volume 11.03 contains sampling and analysis of atmospheres (ambient, indoor air, and workplace atmospheres), and occupational health and safety aspects of working in various environments
- Volume 11.04 contains environmental assessment (Phase I and Phase II), hazardous substances and oil spill response, and waste management (e.g., sampling and monitoring, physical and chemical characterization).

OSHA/NIOSH Methods

Both OSHA (Occupational Safety and Health Administration) and NIOSH (National Institute for Occupational Safety and Health) were created under the Occupational Safety and Health Act of 1970 for a common goal of protecting workers' safety and health. Although OSHA and NIOSH were created by the same Act of Congress, they are two distinct agencies with separate responsibilities. OSHA is administrated by the U.S. Department of Labor and is responsible for creating and enforcing workplace and health regulations. NIOSH is a research agency and is a part of the Centers for Disease Control and Prevention (CDC) in the Department of Health and Human Services. The OSHA and NIOSH methods mostly used in air sampling and analysis are probably the most important methods for industrial hygiene professionals.

Entitled "NIOSH Manual of Analytical Methods (NMAM) (4th edition, vol. 1–7, Department of Health and Human Service, U.S. Government Printing Office, Washington, D.C., 1995), the NMAM is a collection of methods for the sampling and analysis of contaminants in workplace air, bodily fluids (blood and urine of workers), biological, bioaerosols, and bulk samples (NIOSH, 2003). NIOSH methods numbers are assigned by sampling techniques as described below, and the details of their methods can be downloaded free of charge from http://www.cdc.gov/niosh/nmam/either by method number or by chemical name:

- 0001–0899: General air sampling
- 0900–0999: Bioaerosols
- 1000–1999: Organic gases on charcoal
- 2000–3499: Organic gases on other solid supports
- 3500–3999: Organic gases on other samplers
- 4000–4999: Organic gases on diffusive samplers (e.g., liquids, direct-reading)
- 5000–5999: Organic aerosols
- 6000–6999: Inorganic gases
- 7000–7999: Inorganic aerosols
- 8000–8999: Biological samples
- 9000–9999: Bulk samples.

There are two sets of OSHA methods that can be accessed on the Internet at http://www.osha.gov/dts/sltc/methods/. The first is OSHA's Analytical Methods Manual (2nd edition, Parts I and II, U.S. Department of Labor, Occupational Safety and health Administration, Salt Lake City, Utah, January 1990). OSHA has also a list of partially validated methods in the IMIS (integrated management information system) series, which is also available in printed form or CD-ROM (OSHA Chemical Sampling Information Database, 1997, OSHA Analytical Laboratory, Salt Lake City, Utah). Examples of OSHA methods and numbers are 7048 for cadmium sampling and analysis, and 2539 regarding procedures for aldehydes sampling and analysis.

USGS Methods

The U.S. Geological Survey (USGS) has published a large number of reports and methods. Many of them are now being served online for free access of files in PDF format. Sampling and analytical methods are related to the Techniques of Water Resource Investigations (TWRI), which can be accessed at http://pubs.usgs.gov/twri. The TWRI reports include 9 books:

- Book 1: Collection of Water Data by Direct Measurement
- Book 2: Collection of Environmental Data
- Book 3: Applications of Hydraulics
- Book 4: Hydraulic Analysis and Interpretation
- Book 5: Laboratory Analysis
- Book 6: Modeling Techniques
- Book 7: Automated Data Processing and Computations
- Book 8: Instrumentation
- Book 9: Handbooks for Water-Resources Investigations.

Book 5 of TWRI addresses laboratory analysis for: inorganic substances, minor elements, organic substances, radioactive substances in water and fluvial sediments, and the collection and analysis for aquatic biological and microbiological samples. Many of these methods are approved for use in analysis of wastewater compliance monitoring samples.

The USGS methods are identified by *a letter followed by a four-digit number and then a year designator to indicate the revision date*. The letters are: P (physical characteristics), I (inorganic substance), O (organic substance), B (biological method), R (radioactive determination), S (sediment characteristic), and E (emission spectrographic method). *The first digit of the four-digit code for the organic and inorganic methods indicates the type of determination*:

- 0 Sample preparation
- 1 Manual method for dissolved parameters
- 2 Automated method for dissolved parameters
- 3 Manual method for analyzing water-suspended sediment mixtures
- 4 Automated method for analyzing water-suspended sediment mixtures
- 5 Manual method for analyzing bottom material
- 6 Automated method for analyzing bottom material
- 7 Method for determining suspended parameters
- 9 Method for fish and other materials

The last three digits of the four-digit code identify the parameters. For instance, a method identified as 1-7084-85 is a flameless atomic absorption spectrometry (FLAA) method for barium analysis performed on suspended material, last updated in 1985.

AOAC Methods

AOAC was founded in 1884 as the Association of Official Agricultural Chemists, under the auspices of the U.S. Department of Agriculture (USDA), to adopt uniform methods of analysis for fertilizers. AOAC changed its name to the Association of Official Analytical Chemists, and then to the Association of Analytical Communities to reflect its wider membership including microbiologists, food science personnel, as well as chemists. Entitled "Official Methods of Analysis of AOAC International" (18th edition, 2005), this AOAC method is referred to as the "Bible" by its users in food science, beverage purity control, agriculture, pharmaceuticals, microbiology, and the environmental field. This two-volume set contains over 2700 collaboratively tested, internationally recognized methods. Most of the AOAC methods related to agricultural and food testing are the most authoritative sources of analytical methods used worldwide.

Soil Methods

These are the method books jointly published by the American Society of Agronomy (ASA) and the Soil Science Society of American (SSSA). It is the most authoritative and systematic methods used by soil scientists and engineers. Environmental professionals use these methods for soil-related physical, chemical, biological, and mineralogical analysis. The methods include four volumes:

- Methods of Soil Analysis, Part 1: Physical and Mineralogical Methods, 3rd edition, 1986.
- Methods of Soil Analysis, Part 2: Microbiological and Biochemical Methods, 3rd edition, 1994.
- Methods of Soil Analysis, Part 3: Chemical Methods, 3rd edition, 1996.
- Methods of Soil Analysis, Part 4: Physical Methods, 2002.

Other Methods

These include methods established by many other professional organizations and agencies at various states and federal levels as well as countries worldwide. The Department of Energy (DOE) and the National Oceanic and Atmospheric Administration (NOAA) are examples of organizations that have not been included in the above discussions. Many states have their own methods approved for use under a variety of monitoring programs. An example is the petroleum Leaking Underground Storage Tank (LUST) program—the total petroleum hydrocarbons (TPH) are tested by GC in California's method, while the Wisconsin method tests for gasoline range organics (GRO) and diesel range organics (DRO). Other countries have their own standard governmental and nongovernmental methodologies, but for the most part, they share a great deal of resemblance in the technical aspects to the U.S. methods.

To summarize, Table 5.2 illustrates the complex nature of various analytical methodologies for selected compounds. Detailed cross-references of various compounds under different programs (CWA, SDWA, RCRA, CAA) can be found in Lee (2000) and Smith (1997).

Table 5.2   Cross-references of selected analytes

Parameter	EPA	Std method	ASTM	USGS	AOAC
BOD_5	405.1	5210 B	—	I-1578-78	973.44
Lead, total (AA Direct aspiration)	239.1	3111B or C	D3559-90 (A or B)	I-3399-85	974.27
Mercury, total (Manual cold vapor)	245.1	3112 B	D3233-91	I-3462-85	977.22
Cr^{6+}, dissolved (Colorimetric)	—	3500-Cr D	D1687-92(A)	I-1230-85	—
Pyrene	610 (GC), 625, 1625 (GC/MS), 610 (HPLC)	6410B, 6440 B	D4657-87	—	—
Tetrachloroethene	601 (GC), 624, 1624 (GC/MS)	6230B, 6410B	—	—	—

5.2 SELECTION OF STANDARD METHODS

Given a wide variety of sampling and analytical methods (Section 5.1), it appears to be a daunting task for anyone to select the proper methods in meeting a project's particular need. Perhaps this is only partially true. At first, professionals involved in sampling and analysis should have their own preferred method of choice depending upon their disciplines and/or jurisdiction. Such boundaries are generally clear, for example, EPA methods are for environmental compliance or regulatory monitoring, USGS methods are for water resource related survey, OSHA/NIOSH methods are for industrial hygiene testing, and AOAC methods are for food and agricultural products. In other cases, the differences are not much clear and several methods may apply. For example, one can choose existing EPA, APHA, or USGS methods for stream water quality monitoring. Even within the EPA methods, several protocols are likely applicable. The investigator should then consider other factors, such as the project objective, the ease of operation, the availability of instruments, MDLs, and analytical cost.

Discussed below is a general guide for beginners in choosing standard methods based on the major types of analytical parameters. There is no set rule as to which method is preferred. In the meantime, the reader should note that a current trend within the EPA is to break down the barriers using new monitoring techniques with a program called *performance based measurement system* (PBMS). The PBMS conveys "what" needs to be accomplished, but not prescriptively "how" to do it. With this new approach, the EPA promotes flexibility in method selection.

5.2.1 Methods for Sample Preparation

Sample preparation is always the required first step prior to any trace analysis. With the exception of analyzing very clean matrices (such as groundwater or pristine natural water), environmental samples should always be "pretreated" first and then directly introduced into the instrument for quantitative analysis. We will briefly introduce three major types of methodologies for metals, SVOCs, and VOCs in liquid and solid waste samples. The technical details of sample preparation will be the subject of Chapter 7.

EPA 3000 Series: Digestion Methods for Metal Analysis

Samples for metal analysis must have an *acid digestion* procedure using heated HCl, HNO_3, H_2SO_4, $HClO_4$, or the combined use of these acids to release metals from sample matrix, so that the instrument will respond. Several EPA 3000 series methods in SW-846 will be the methods of choice. Detailed digestion methods can also be found in APHA 3030. Although SW-846 is proposed for solid waste, the methods are also applicable for other matrices, such as the examples given below:

- Method 3005A: Digestion of water for analysis by FLAA or ICP
- Method 3010A: Digestion of water for analysis by FLAA or ICP
- Method 3015: Microwave digestion for aqueous samples
- Method 3020A: Digestion of water for analysis by GFAA
- Method 3031: Digestion of soils for analysis by AA or ICP
- Methods 3050B: Digestion of sediments, sludge, and soils.

Note, slight variations exist depending on the types of metals to be analyzed or the instrument to be used. For example, Method 3005A is for total recoverable or dissolved metals, while Method 3010A is used for preparation of aqueous samples and wastes that contain suspended solids. *Total recoverable metals* refer to the concentration of metals in an unfiltered sample treatment with hot dilute mineral acid. Samples are acidified with HNO_3 at the time of collection. *Total dissolved metals* refer to the concentration of metals determined in a sample after filtration through a 0.45-μm filter. As HCl can cause interference during GFAA (graphite furnace atomic absorption) analysis, the digestate can be analyzed by only FLAA or ICP.

EPA 3500 Series: Extraction of Liquid and Solid Samples for Nonvolatile/Semivolatile Compounds

Several common extraction methods for nonvolatile or semivolatile compounds are listed in Table 5.3. The classical liquid–liquid (L–L) extraction is usually carried out with a 1-L aqueous sample using methylene chloride (CH_2Cl_2) as the extraction solvent. The volume of extract is reduced to ~1 mL by evaporation, so that a factor of about 1000 in analyte concentration is achieved. For solid samples, the benchmark method is the Soxhlet extraction, which is tedious and consumes a great deal of solvent. As shown in Table 5.3, new methods have evolved from the classical L–L extraction (aqueous samples) and Soxhlet extraction (solid sample), including continuous L–L, automated Soxhlet, sonication, solid phase extraction (SPE), and supercritical fluid extraction (SFE). These new methods are commonly used because the amount of solvent needed and the extraction time are both reduced.

Table 5.3   Sample extraction methods for semivolatiles and nonvolatiles

Method Series	Matrix	Extraction type	Analytes
3510	Aqueous	Separatory funnel liquid–liquid extraction	Semivolatile and nonvolatile organics
3520	Aqueous	Continuous liquid–liquid extraction	Semivolatile and nonvolatile organics
3535	Aqueous	Solid phase extraction (SPE)	Semivolatile and nonvolatile organics
3540	Solids	Soxhlet extraction	Semivolatile and nonvolatile organics
3541	Solids	Automated Soxhlet extraction	Semivolatile and nonvolatile organics
3542	Air sampling train	Separatory funnel and Soxhlet extraction	Semivolatile
3545	Solids	Pressurized fluid extraction (ASE)	Semivolatile and nonvolatile organics
3550	Solids	Ultrasonic extraction	Semivolatile and nonvolatile organics
3560/3561	Solids	Supercritical fluid extraction (SFE)	Semivolatile petroleum hydrocarbons and PAHs

EPA 5000 Series: Headspace or Purge-and-Trap for Volatile Compounds

Extraction methods shown in Table 5.3 cannot be used for volatile organics because open containers are involved in the extraction procedure. Sample preparation methods for volatile organic compounds are described in SW-846's series 5000 methods (Table 5.4). Headspace analysis and purge-and-trap (P & T) are probably the most commonly used methods. Separation in P & T is carried out by purging the sample (sometimes while heating) with an inert gas (N_2) and then trapping the volatile materials. The trap is designed for rapid heating so that it can be desorbed directly into GC.

Table 5.4 Sample preparation methods for volatile organic compounds

Method series	Matrix	Extraction type	Analytes
5021	Solids	Automated headspace	Volatile organics
5030	Aqueous	Purge-and-trap	Volatile organics
5031	Aqueous	Azeotropic distillation	Polar volatile organics
5032	Aqueous and solids	Vacuum distillation	Nonpolar and polar volatile organics
5035	Solids, organic solvents, oily waste	Closed system purge-and-trap	Volatile organics
5041	Air sampled by VOST	Purge-and-trap from VOST	Volatile POHCs

*Refer Chapter 7 for details. VOST = Volatile organic sampling train; POHCs = Principal organic hazardous constituents. They are specific hazardous waste compounds identified by the EPA, which are selected for monitoring during trial burn of a hazardous waste incinerator.

5.2.2 Methods for Physical, Biological, and General Chemical Parameters

EPA 100 and 300 Series: Physical Properties, Inorganic, and Nonmetallics

Common physical properties include conductance, pH, solids (SS, TDS, TS, VS), temperature, and turbidity. Inorganic and nonmetallic constituents refer to anions of various elements (Br, Cl, CN, F, I, N, P, S), total N and P, acidity, alkalinity, and dissolved oxygen (DO). Measurement methods include potentiometric (pH), gravimetric (solids), thermometric (temperature), titrimetric (alkalinity, acidity), membrane electrode (dissolved oxygen), or iodometric (Winkler method for dissolved oxygen). Analytical methods of these parameters are well established and can be found mostly in the EPA 100 and 300 series. Additional methods, in the order of importance, can also be located in APHA method 2000 and 4000 series, USGS, ASTM, and AOAC.

EPA 200, 6000, and 7000 Series: Metals

With the exception of few metals (e.g., Hg), three instrumental techniques are commonly used for the analysis of most metals: (a) flame atomic absorption; (b) graphite furnace atomic absorption (flameless); and (c) inductively coupled plasma–atomic emission spectroscopy (ICP). EPA 200 methods developed by the Office of Research and Development and EPA's SW-846 series 6000 (ICP) and 7000 methods (Table 5.1) are the methods of choice. APHA 3000 series and USGS I-xxxx-xx series also have the complete set of methods for the analysis of metals.

EPA 400 Series: Aggregate Organic Analytes

These are the gross properties of waters and wastewaters in representing certain organic compounds rather than individual organics. Important monitoring parameters for aggregate organics include chemical oxygen demand (COD), biochemical oxygen demand (BOD), total organic carbon (TOC), total recoverable petroleum hydrocarbons (TRPH), anionic surfactants (as MBAS), and oil and grease. These measurements are located in EPA 400 series, and non-EPA methods include primarily APHA 5000 series and parts of the USGS O-xxxx-xx methods.

APHA 8000–10000 Series: Biological Parameters and Testing

The U.S. EPA has few guidelines and procedures on biological testing. The AHPA method (commonly regarded as the standard method) is by far the most complete and therefore is the true "standard method" with regard to biological parameters and testing. The examples are as follows:

- APHA 8000 toxicity testing: Using sediment pore water, algae, *Daphnia*, fish, fathead minnow, and mutagenesis
- APHA 9000 microbiological examination: Heterotrophic plate count, microbial count, *E. Coli*, and total coliform
- APHA 10000 biological examination: Periphyton, macrophyton, benthic macroinvertebrates, nematological examination, and fish.

Detailed measurement principles of selected parameters will be introduced in Chapter 6 (general physical properties, inorganic, nonmetallics, and aggregate organics) or Chapter 9 (metal analysis).

5.2.3 Methods for Volatile Organic Compounds (VOCs)

The EPA methods applicable for the analysis of individual VOCs are parts of the EPA 500, EPA 600, and EPA 8000 series. In addition to EPA's methods, methods for VOCs analysis can also be found in APHA 6000 and USGS O-xxxx-xx. Of EPA 500 and 600 series methods, the most noted is Method 624 for the analysis of a number of VOCs using GC–MS. These compounds are purgeable with an inert gas bubbling through a 5-mL sample so that the volatile organics are transferred from the water phase to the vapor phase. Due to their volatile nature, VOCs cannot be analyzed by high performance liquid chromatography (HPLC). Operational principles of GC with various detectors will be discussed in Chapter 10.

5.2.4 Methods for Semivolatile Organic Compounds (SVOCs)

Like VOCs, methods for SVOCs can also be found in EPA 500, 600, and 8000 as well as APHA 6000 or USGS O-xxxx-xx methods. Unlike VOCs, however, many SVOCs (e.g., PCB, PAHs, pesticides) can be analyzed by either GC or HPLC. The counterpart of Method 624 is Method 625 for nonpurgeable but extractable semivolatile organic compounds (SVOCs). To begin SVOC analysis using Method 625, the pH of a 1-L sample is adjusted to a pH greater than 11. The sample is then extracted with methylene chloride to obtain a *base/neutral fraction*. The pH of the sample is then adjusted to less than 2 and is again extracted with methylene chloride to form the *acid fraction*. Each fraction is chemically dried to remove the water and concentrated by distillation to 1 mL. The fractions are then analyzed by GC/MS. Note that the acid fraction compounds are mostly phenolic compounds such as phenol, halogenated phenols, and nitrophenols. The base/neutral compounds include PAHs, ethers, di- and tri- halobenzene, and phthalates.

5.2.5 Methods for Other Pollutants and Compounds of Emerging Environmental Concerns

What we have described so far are the standard analytical methods for contaminants regulated by the U.S. EPA. Many contaminants have not been regulated by the U.S. EPA. As concern for these contaminants emerge, testing methods will have to be developed. The EPA has not established analytical methods for these contaminants so the responsibility rests within the research community.

Under the Safe Drinking Water Act, the EPA has initiated a process to periodically propose a *contaminant candidate list* (CCL) to include those chemicals that may require regulation in the future. The EPA uses this list of unregulated contaminants to prioritize research and data collection efforts to help to determine whether it should regulate a specific contaminant. As of February 2005, EPA has published the second CCL of 51 contaminants (http://www.epa.gov/ogwdw/ccl/cclfs.html). For example, algal toxins, organotins, and methyl *tert*-butyl ether (MTBE) are among these chemicals, any of which may enter into drinking water through various sources. Table 5.5 is a list of selected chemicals from the CCL along with several other emerging contaminants to illustrate the analytical complexity of these compounds.

It is apparent from Table 5.5 that testing for these compounds requires state-of-the-art instrumentation that many labs cannot afford to purchase. We will

Table 5.5   Selected emerging contaminants and their analytical methods

Emerging contaminant	Analytical method
Disinfection by-products	GC/MS, GC–ECD, LC/ESI–MS, IC, IC/APCI–MS, IC/ICPMS
Pharmaceuticals and endocrine disrupting compounds	LC/MS, LC/MS/MS, SPME-GC/MS, GC/MS, SPE–GC/MS/MS
Chiral contaminants	Chiral GC/MS
Methyl *tert*-butyl ether	GC/MS (EPA 8240B/60B), GC–FID (ASTM D4815)
Algal toxins	LC/ESI–MS, LC/ESI–MS/MS, MALDI/TOF–MS
Organotins	SPME–LC/ESI–MS, SPME–GC/MS, SPME–GC/ICPMS, TOF–MS, GC/ICPMS
Perchlorate	IC (EPA 314.0), ESI–MS, ESI–MS/MS
Natural organic matter	ESI–MS, APCI–MS/MS, FT–ICR–MS, ESI–FI–ICR–MS

Acronyms: APCI = atmospheric pressure chemical ionization; API = atmospheric chemical ionization; ECD = electron capture detection; EI = electron ionization; ESI = electrospray ionization; FT = Fourier transform; GC = gas chromatography; IC = ion chromatography; ICP = inductively coupled plasma; ICR = ion cydotron resonance; IR = infrared spectroscopy; LC = liquid chromatography; MALDI = matrix-assisted laser desorption/ionization; MS = mass spectrometry; SPE = solid-phase extraction; SPME = solid-phase microextraction; TOF = time-of-flight.

discuss the principles of some of these instruments in Chapter 12. Readers interested in the analysis of these emerging chemicals are referred to several excellent review articles by Clement and Yang (2001) and Richardson (2001, 2002, 2003).

EXAMPLE 5.1.   Selection of standard methods of the U.S. EPA and other agencies. Briefly answer the following questions:

(a) Which EPA method you would use for the analysis of volatile solvents (e.g., trichloroethene, chloroform, toluene) in groundwater samples at µg/L concentrations?

(b) Which EPA method you would use for the analysis of industrial wastewater sludge for cadmium, mercury, and chromium to evaluate whether or not the sludges can be disposed in a nonhazardous waste landfill?

(c) Which EPA method you would use for the analysis of semivolatile and nonvolatile pesticides in wastewater from a manufacturing facility to evaluate whether the wastewater is in compliance with the facility's discharge permit?

(d) Which method you would use for blood and urine samples of workers exposed occupationally to lead?

SOLUTION: (a) EPA 624 method would be the method of choice as the analytes are volatile organic compounds in groundwater. (b) SW-846 method would be preferred as the sample matrix is sludge. In particular, because the purpose is to evaluate whether the sludge would leach and pose hazardous, a TCLP (i.e., Method 1311) would be used to obtain the extract, followed by the analysis using 6010B (ICP) or 6020 (ICP/MS) depending on the availability of instrument. (c) EPA 625 would be the first choice as the chemicals are SVOCs and the matrix is wastewater. (d) For the bodily fluids samples such as blood and urine, an OSHA method would be the first option, which can be searched from OSHA's online Sampling and Analytical Methods database for the particular method.

EXAMPLE 5.2.   The use of National Environmental Method Index (NEMI) (http:// www. nemi.gov/). (a) Find a list of COD in water methods by the U.S. EPA and other agencies; (b) What is the EPA method 508? Locate this method for a printable hardcopy.

SOLUTION: Use quick/advanced search on the NEMI Web site, type "COD" as the analyte name and "water" as the media name. Eight methods are found—EPA 410.1, 410.2, 410.3, 410.4, standard method (APHA Method) 5220D, 5220C, USGS I-3561, and ASTM D1252B. (b) Use general search, type "508" as the method number. This method is entitled "chlorinated pesticide in water using GC–ECD." A printable method file in PDF format can be located by clicking the method number on the search result. Only the governmental methods by the EPA and USGS are downloadable free of charge.

5.3 FIELD QUALITY ASSURANCE/QUALITY CONTROL (QA/QC)

In Chapter 1, we defined the concept of QA/QC. Recall that QA is an integrated *management system* to ensure that QC system is in place and working as intended. This management system may include viewing QC data, evaluating parameters, taking corrective actions, planning for process, and personnel involved (data generator, data reviewer, QA manager, and data user). QC is a set of routine *technical activities* performed whose purpose is essentially error control with the use of QC samples such as blanks and spikes. Together, QA and QC help us to produce data of a known quality (e.g., precision, accuracy) and enhance the credibility of reported results.

A formal QA/QC program is written in a document called *Quality Assurance Project Plan* (QAPP). QAPP should be prepared and approved before sample collection begins. It should address the following essential components: project description, project organization and responsibility, QA objectives for measurement data in terms of precision and accuracy, sampling procedures, sampling custody, calibration procedures and frequency, analytical procedures, data reduction/validation and reporting, internal quality control checks and frequency, performance and system audit and frequency, specific routine procedures used to assess data precision/accuracy and completeness, corrective actions, and QA report and management.

In the following discussions on QA/QC, we first describe field QA/QC, particularly the types and number of field QC samples that are required in various instances. The laboratory counterpart of QA/QC, further divided into stages of sample preparation and sample analysis, will be introduced in the next section. Note that QA/QC samples (QC sample in short) are the extra samples collected during the sampling and analytical processes. These QC samples are fundamentally different from the true *environmental samples* used to represent the nature of the sampling matrix. The QC samples rarely provide useful information on the project decision-making process (e.g., to tell the extent of contamination if this is the project goal), but it is used to identify the error, diagnose what caused the error, and, if an error occurred, how to correct it. If the error is acceptable within the specified data quality objective (Section 2.2.3), the results from QA/QC samples will help to direct the reporting process so that the data user will know the level of confidence in the data.

Practical tips

- By law, any EPA-funded monitoring project must have an EPA-approved QAPP before it can begin collecting samples. However, even programs that do not receive EPA money should consider developing a QAPP, especially if data might be used by state, federal, or local resource managers.
- QA/QC is required for all aspects of a monitoring program, including field sampling, laboratory analysis, and data review and reporting.
- The QA/QC samples required by EPA in some cases can account for 30% of the total project cost. Efforts should be made in order to minimize the number of QA/QC samples while maintaining data quality.
- Personnel training is of utmost importance as the efficacy of QA/QC depends on how it is executed by the field and lab personnel.
- Always keep in mind that it is better to have a few reliable data points than many unreliable ones!

5.3.1 Types of Field QA/QC Samples

Errors occurring in the field (sample collection, preservation, transportation, and storage) may have originated from a single source or a combination of several sources. These include: analyte carryover from sampling equipment, incomplete decontamination of sample equipment between samples, cross-contamination between samples, and absorption of volatile chemicals from air during transportation and storage. To test the absence or presence of these errors, the following field QC samples should be collected:

- *Equipment (rinsate) blanks:* Equipment blanks are used to detect any contamination from sampling equipment. These blanks are prepared in the field before sampling begins, by using the precleaned equipment and filling the appropriate container with analyte-free water. If equipment is cleaned on site (e.g., nondisposable bailers, nondedicated pumps and tubing, augers), then additional equipment blanks should be collected after equipment decontamination for each equipment group. Contaminants frequently found in equipment blanks include common laboratory contaminants such as phthalates, methylene chloride, and acetone. Occasionally, oily materials, PCBs, and VOCs retained by groundwater pumps are found in equipment blanks after samples with high contaminant concentration have been taken (Popek, 2003).

- *Field blanks:* A field blank sample undergoes the full handling and shipping process of an actual sample. It is designed to detect sample contamination that can occur during field operation or during shipment. Field blanks are prepared in the field using certified clean water (HPLC-grade carbon-free water for organic analysis or deionized water for inorganic analysis), preserved in the same manner as other collected samples, and then submitted to the laboratory for analysis. For soil samples, field blank samples can be prepared with certified clean sand or soil rather than clean water. For high-concentration samples such as wastes, the usefulness of field blanks is limited. Parts per billion (ppb) or low parts per million (ppm) error has little significance when identifying high concentration wastes in the hundreds of ppm range. In air sampling, the preparation of field blanks varies depending on the sampling device/adsorbent media used during sampling (see Fig. 4.9, Section 4.1.5). For example, when using adsorbent media that is sealed by the manufacturer, the field blank is opened with the other sample media, resealed, and carried through the sample handling process.

- *Trip (travel) blanks:* Unlike field blanks, the trip blanks are used for VOCs analysis only. In addition, trip blanks are prepared *prior to* going into the field by filling containers (VOC vials) with clean water (HPCL-grade) or sand. The sample containers are kept closed, handled, transported to the field, and then returned to the laboratory in the same manner as the other samples. Trip blanks are used to evaluate error associated with shipping and handling, and analytical procedures. They are used in conjunction with field blanks to isolate sources of sample contamination already noted in previous field blanks. Like field blanks, trip blanks are not typically used during waste sampling because of the high concentrations in most waste samples. Special precautions must be exercised for trip blanks used in air sampling such as absorbent tubes, canister, and impinger-based sampling methods (EPA, 1995).

Several other field QC samples may be required depending on the specific data quality objectives (DQOs) for a particular project. *Field replicate samples*, also referred to as *field duplicates* and *split samples*, are field samples obtained from one sampling point, homogenized, divided into separate containers, and treated as separate samples throughout the remaining sample handling and analytical processes. These field replicate samples are used to assess error associated with sample heterogeneity, sample methodology, and analytical procedures. There are no "true" field replicates for biological samples. However, biological samples collected from the same station are typically referred to as replicates. Samples can also be collected next to each other (e.g., within a three-foot radius of the selected sampling site); these are called *collocated samples* because they are "co-located." Unlike field replicates, collocated samples are not composited and used as discrete samples in order to assess site variation in the immediate vicinity of the sampling area.

If the project purpose is to quantify true changes in contaminant concentrations due to a contaminated source or site, *background samples* are often collected as the QC sample. Background samples are collected upstream of area(s) of contamination (either on or off site), where there is little or no chance of migration of the contaminants of concern. This provides the basis for comparison between contaminated and uncontaminated areas. Background sampling is appropriate when sampling soil, surface water, groundwater, and air. For example, background water samples must often be collected to verify groundwater plume direction, ambient conditions, and attribution of sources. However, background sampling has less application to waste sampling. In biological sampling, *reference samples* may be taken from a reference area outside the influence of the site. The reference area should be close to the site and have habitats, size, and terrain similar to the site under investigations. These samples should be of the same species, sex, and developmental stage as the field site samples.

5.3.2   Numbers of Field QA/QC Samples

- *Equipment blanks*: Collect one equipment blank per type of sampling device per day. At least one equipment blank should be collected for every 20 samples per parameter group and per matrix. The number of equipment blanks can be reduced or eliminated if dedicated equipment (e.g., pump and tubing) or disposable sampling tools are used.
- *Field blanks*: Submit one field blank per day.
- *Trip blanks*: A trip blank should be included with each shipment. At least one trip blank should be prepared for each VOC method, and analyzed for each cooler used for storing and transporting VOC samples.
- *Field duplicates*: During each independent sampling event, at least one sample per matrix type or 10% of the samples, whichever is greater, must be collected for duplicate. In Superfund clean-up projects, when the goal is to determine total error for critical samples with contamination concentrations near the action level, a minimum of eight replicate samples is required to have a valid statistical analysis.
- *Collocated samples*: Determine the applicability on a site-by-site basis.
- *Background samples*: Collect at least one per matrix (surface water, sediment).

5.4   ANALYTICAL QUALITY ASSURANCE/QUALITY CONTROL

5.4.1   Quality Control Procedures for Sample Preparation

Significant errors can occur during sample preparation, that is, digestion for metals and extraction for semivolatile or nonvolatile organics. These errors are likely due to cross-contamination from glassware or chemicals used, contaminant loss owing to sorption or volatilization, matrix effects or interference, and incomplete digestion or extraction of the analyte as a result of an improper procedure. Two types of "spiking" QC samples are mainly used for the QA/QC purpose to assess sample preparation procedures.

- *Spiked sample/matrix spike:* A small quantity of a known concentration of analyte stock solution is added (spiked or fortified) to the sample (or aliquot of the sample) before the sample preparation. This sample then undergoes digestion or extraction and subject to analysis. The concentration of stock solution should be high enough so that only a small quantity of the stock solution is needed and the volume change of the sample becomes negligible. A general guideline should be followed regarding how much spiking chemical should be added. In general, the spike addition should produce a minimum level of 10 times and a maximum level of 100 times of the instrument detection limits (IDL). The concentration of each analyte in the spiking solution should be approximately 3–5 times the level expected in the sample (Csuros, 1994). A practical approach is to spike the concentration to reach near the midrange calibration point. Results of spiked samples are evaluated based on the percentage recovery for the analytical accuracy. Refer to Eq. 2.5 in Chapter 2 and the example below for the calculation of percentage recovery. The volume of spike stock solution that should be added to the sample can be calculated according to:

$$C_1 V_1 = C_2 V_2 \tag{5.1}$$

  where C_1 and C_2 are the concentration of the spike stock solution and the desired spike concentration, respectively. $V_2 =$ volume of the sample to be spiked, and V_1 is the volume of stock solution to be calculated.

- *Surrogate spikes:* Surrogates are organic compounds (nontargeted compound) that are similar to analytes of interest in chemical composition, extraction, and chromatography, but that are not normally found in the environment. Surrogate spikes are used to trace organic determination methods such as GC, GC/MS, and HPLC. These compounds are spiked into all blanks, standards, samples, and spiked samples prior to purging or extraction according to the appropriate methods. The analyst should make sure that surrogate standard should not interfere with target analytes. Surrogate samples are used to assess retention times, percentage recovery (accuracy), and method performance.

5.4.2   Quality Control Procedures During Analysis

Types of Laboratory QA/QC Samples

In principle, there are mainly three types of QC samples used for the QA/QC purpose during laboratory analysis. They are: (a) blanks used to assess any potential contamination; (b) spikes used to obtain the percentage recovery and therefore the accuracy; and (c) replicates (duplicates) used to determine the analytical precision. Details are described as follows for several commonly used laboratory QC samples.

- *Reagent blanks* are analyte-free water analyzed with the samples. Such blanks will detect any introduction of contaminants and artifacts into samples. *Preparation blanks/method blanks* are analyte-free matrix (reagent water) that are carried throughout the entire preparation and analysis and contain all of the reagents in the same volumes as used in the processing of the samples. They are used to detect any contamination (false positive results) during the sample preparation and analytical process. *Instrument blanks* are the analyte-free reagents (water or solvent) inserted in the analytical run, normally injected between high concentration and low concentration samples. Instrument blanks are used to flush the system in trace analysis and used to determine whether memory effects (carryover) are present in the analytical system.

- *Reagent water spikes* are analyte-free water spiked with the analyte of interest and are prepared in the same way as the samples. Such spikes monitor the effectiveness of the method. The level of the spike addition is recommended to be five times the method quantification limit.

- *Performance evaluation (PE)/laboratory control samples* (LCS) contain known identities and concentrations of target analytes in certified clean water. PE samples are usually prepared by a third party and submitted "blind" to analytical laboratories, while the LCS samples are prepared by the laboratory so that the laboratory knows the contents of the sample. Neither PE nor LC samples are affected by matrix interference, and thus they can provide a clear measure of laboratory error. The EPA has a guideline regarding the minimal number of analytes to be used in spikes (Popek, 2003).

- *Matrix spikes and matrix spike duplicates* (MS/MSDs) are field samples spiked in the laboratory using the same spiking chemicals described for LCS. Same as PE and LCS, predetermined quantities of stock solutions of certain analytes are added to a sample matrix prior to sample extraction/digestion and analysis. Samples are split into duplicates, each of the analytes detected. The percentage recovery is calculated to assess the accuracy, and the relative percent difference (RPD, see Eq. 2.9) is calculated and used to assess analytical precision. Unlike PE and LCS sample, MS/MSDs also indicates the degree to which matrix interference will affect the analysis.

- *Laboratory duplicates* are aliquot of the same sample that are prepared and analyzed at the same time, but submitted and analyzed as separate samples (see also field duplicates). The analyst does not know that they are duplicates. Discrepancies in duplicate samples indicate poor analytical reproducibility (the analytical precision) and/or the poor homogenization in the field.

- *Reference materials* (blind QC check samples) are obtained from an independent source with known analytical level(s). The analyst is not told the true concentration of the sample. These QC samples measure and validate the analytical system and the performance of the analyst.

- *Calibration standards* are the standard solutions used to obtain calibration curves, including a *calibration blank* and a series of several concentrations. For example, one calibration blank and four concentration levels are normally required for inorganic standard calibration curves (Section 2.1.5). If an internal calibration procedure is used, *internal standards* are added to all standards, QC sample, and all other samples. Internal standards should be the chemicals not normally present in the sample and interfere with the compounds of interest.

- *QC check standards* (or calibration verification standard, CVS) are standard solutions with known concentrations. They should be certified (purchased from the EPA or from other sources), or independently prepared standard from a source other than the calibration standard. They are used to verify that the standards and the calibrations are accurate and also to confirm the calibration curve. The value is accepted within $\pm 10\%$ deviation from the 100% recovery.

Number of Analytical (Laboratory) QA/QC Samples

The required number of laboratory QC samples is better discussed with a concept called batch, which is the basic unit of analytical QC. It can be defined as "a group of samples that behave similarly with respect to the sampling or testing procedure being employed and that are processed as a unit." A batch, by default, is considered to be a group of 20 samples that can be logically grouped together (same lots of reagents used in sample preparation, analyzed together with the same instrument within a sequence, or the same time period or in continuous sequential time periods); thus various types of batches exist, such as *extraction or digestion batches, matrix spike batches, clean-up batches*, or *analysis batches*. Keep in mind that the *one per twenty* (5%) *frequency* is a default that has been used in many EPA programs for many years. This means that for each batch or at least one per twenty samples there should have a QC sample including blanks, spikes, and duplicates of various types. Adding all these QC samples together, the total number of required QC samples should be around 25% of the samples sent out for laboratory analysis. Keep also in mind that internal and surrogate standard should be added to every blank, field, and QC sample.

Practical tips

- Having a proper number of QC samples during sample preparation and analysis will help to diagnose problems if errors are suspected. For most part, the analyst may not have to do the analysis over again from the beginning if proper QA/QC protocols are followed, thus saving time at the end.

- When loading a batch of samples in an auto-sampler tray, always keep calibration standard, external standard, blank in certain sequence in one analytical batch. Whenever possible, always do a calibration curve within that analytical batch on a daily basis.

- A perfect 100% recovery is probably not achievable in the real world for most analytes in complex matrices. Whether a certain percentage recovery is satisfactory depends on the analyte, the sample matrix, and the method. For example, the acceptable surrogate standard limit for VOCs in water and soil samples is approximately 85–115% and 80–120%, respectively. The same limits for base/neutral SVOCs are approximately 30–120% and 20–125% for water and soil samples, respectively (Popek, 2003).

- For qualitative purposes (chemical identification), it is suggested that comparison to an authentic standard is the only acceptable criteria for positive identification. Otherwise, "tentative" identification is the term used when an authentic standard is not used or it is unavailable, and the analysis is based solely on retention times for chromatography and spectral libraries.

EXAMPLE 5.3.   Calculation of recovery from a matrix spike QC sample. Pyrene in soil samples was extracted by sonic extraction followed by analysis using HPLC. A stock solution of pyrene (50 µL, 100 mg/L) in acetonitrile was spiked in a clean soil (0.5 g) and was extracted with 10 mL hexane. A portion of the extract (10 µL) was injected to HPLC and the concentration of the extract was determined to be 0.45 mg/L. Calculate the percentage recovery of the extraction.

SOLUTION: First, we calculate the amount of pyrene added (true value) as follows,

$$\text{True value} = \frac{100\,\text{mg}}{\text{L}} \times 50\,\mu\text{L} \times \frac{1\,\text{L}}{10^6\,\mu\text{L}} = 0.005\,\text{mg}$$

We then calculate the amount of pyrene measured (analytical value):

$$\text{Analytical value} = \frac{0.45\,\text{mg}}{\text{L}} \times \frac{10\,\text{mL}}{10^3\,\text{mL}} = 0.0045\,\text{mg}, \text{ therefore we use Eq. 2.5:}$$

$$\text{Percentage recovery} = \frac{\text{Analytical value}}{\text{True value}} \times 100 = \frac{0.0045\,\text{mg}}{0.005\,\text{mg}} = 90\%$$

REFERENCES

American Society of Testing and Materials (2005), Annual Book of ASTM Standards (a total of 77 volumes) West Conshohocken, PA.

CLEMENT R, YANG PW (2001), Environmental Analysis. *Anal. Chem.*, 73:2761–2790.

CSUROS M (1994) *Environmental Sampling and Analysis for Technicians*, Lewis Publishers, Boca Raton, FL.

*LEE CC (2000), *Sampling, Analysis, and Monitoring Methods*, 2nd edition, Government Institute, Rockville, MD.

National Water Quality Monitoring Council (2005), *National Environmental Methods Index* (NEMI) (http://www.nemi.gov).

NELSON P (2003), *Index to EPA Test Methods*, April 2003 revised edition. US EPA New England Region 1 Library, Boston, MA (http://www.epa.gov/region01/info/testmethods/pdfs/testmeth.pdf).

NIOSH (2003), *NIOSH Manual of Analytical Methods* (NMAM), 4th edition, 3rd supplement (http://www.cdc.gov/niosh/nmam).

OSHA (2006), *Sampling and Analytical Methods* (http://www.osha.gov/dts/sltc/methods).

POPEK EP (2003), *Sampling and Analysis of Environmental Chemical Pollutants: A Complete Guide*, Academic Press, San Diego, CA.

RICHARDSON SD (2001), Water Analysis. *Anal. Chem.*, 73:2719–2734.

RICHARDSON SD (2002) Environmental mass spectrometry: Emerging contaminants and current issues. *Anal. Chem.*, 74:2719–2742.

RICHARDSON SD (2003), Water analysis: Emerging contaminants and current issues. *Anal. Chem.*, 75: 2831–2857.

*SMITH R-K (1997), *Handbook of Environmental Analysis*, 3rd edition, Genium Publishing Corporation Amsterdam, NY.

*US EPA. *Test Methods for Evaluating Solid Wastes Physical/Chemical Methods*. SW-846 Method On-Line: http://www.epa.gov/epaoswer/hazwaste/test/main.htm.

US EPA. *Superfund Analytical Services/Contract Laboratory Program* (CLP) http://www.epa.gov/oerrpage/superfund/programs/clp/analytic.htm.

US EPA. *Analytical Methods* (Water). http://www.epa.gov/ost/methods/.

US EPA. *Test Methods: Performance-Based Measurement System* http://www.epa.gov/epaoswer/hazwaste/test/pbms.htm.

US EPA. *Methods for Chemical Analysis of Water and Wastes*, EPA-600/4-79-020.

US EPA. *Methods for Organic Chemical Analysis* (600 Series and Methods 1624/1625) http://www.epa.gov/waterscience/methods/guide/methods.html.

US EPA. *Approved Methods for Organic Chemicals* (EPA 500 Methods) http://www.epa.gov/ogwdw/methods/orch_tbl.html.

US EPA. *Approved Methods for Inorganic Chemicals and Other Contaminants* (EPA 200 and 300 Methods) http://www.epa.gov/ogwdw/methods/inch_tbl.html.

US EPA (1994), *Ambient Measurement Methods and Properties of the 189 Clean Air Act Hazardous Air Pollutants*, EPA 600/R-94/098.

US EPA (1999), *Compendium of Methods for the Determination of Toxic Organic Compounds in Ambient Air*, 2nd edition. EPA/625/R-96/010b (Methods TO-1 to 17).

US EPA (1995), *Superfund Program Representative Sampling Guidance: Volume 1, Soil*, Interim final, EPA 540-R-95-141, OSWER Directive 9360.4-10, PB96-963207, December 1995.

US EPA (1995), *Superfund Program Representative Sampling Guidance: Volume 2, Air* (Short-Term Monitoring), Interim final, EPA 540-R-95-140, OSWER Directive 9360.4-09, PB96-963206, December 1995.

US EPA (1997), *Superfund Program Representative Sampling Guidance: Volume 3, Biological*, Interim final, EPA 540-R-97-028, OSWER Directive 9285.7-25A, PB97-963239, December 1995.

US EPA (1995), *Superfund Program Representative Sampling Guidance: Volume 4, Waste*, Interim final, EPA 540-R-95-141, OSWER Directive 9360.4-14, PB96-963207, December 1995.

US EPA (1995), *Superfund Program Representative Sampling Guidance: Volume 5, Surface Water and Sediment*, Interim final, OSWER Directive 9360.4-16, December 1995.

QUESTIONS AND PROBLEMS *(Hint: Use appropriate Web sites whenever applicable)*

1. What are the hazardous characteristics described in SW-846? Which method in SW-846 is used to test toxicity characteristic of wastes?

2. Which of the following is used for the acid digestion of samples containing metals: (a) EPA 3000 series; (b) EPA 3510 series; (c) EPA 3540 series; (d) EPA 3560 series.

*Suggested Reading

3. Which of the following is related to the sample preparation and analysis of metals in SW 846 method: (a) Series 3000; (b) Series 6000; (c) Series 7000; (d) All of above.

4. Cold vapor atomic absorption spectroscopy is commonly used for the analysis of: (a) Cr^{6+}; (b) Cr^{3+}; (c) Hg; (d) As.

5. Which of the following is used for sample preparation of volatile compounds: (a) EPA 5030; (b) EPA 3560; (c) EPA 500 series; (d) EPA 600 series.

6. Which of the following are *not* directly used for the sample preparation and analysis of organic contaminants in SW 846 method: (a) Series 3500; (b) Series 4000; (c) Series 3600; (d) Series 8000.

7. Which of the following is *not* related to the analysis of organic chemicals: (a) LC; (b) ICP; (c) GC; (d) ECD.

8. Describe the difference between EPA method 624 and 625.

9. Which EPA method you would use for the analysis of volatile solvents (e.g., trichloroethene, chloroform, toluene) in groundwater samples at µg/L concentrations.

10. Which EPA method you would use for the analysis of industrial wastewater sludge for cadmium, mercury, and chromium to evaluate whether the sludges can be disposed in a nonhazardous waste landfill.

11. Which EPA method you would use for the analysis of semivolatile and nonvolatile pesticides in wastewater from a manufacturing facility to evaluate whether the wastewater is in compliance with the facility's discharge permit.

12. Describe the difference and similarity between OSHA methods and NIOSH methods.

13. Which of the following is likely the ASTM method: (a) 239.2; (b) D3086-90; (c) TO-2; (d) 4500-Cl D.

14. From the online database from National Environmental Method Index (NEMI) (http://www.nemi.gov/). (a) Find a list of TOC in water methods by all agencies; (b) What is EPA method 425.1? Download a hardcopy of method 425.1.

15. From the online database from National Environmental Method Index (NEMI) (http://www.nemi.gov/). (a) Find a list of immunoassay methods for organics in soil/sediment; (b) Find a list of available method for dioxin under the National Pollutant Discharge Elimination System (NPDES).

16. What is the primary difference between quality assurance (QA) and quality control (QC)?

17. Describe the difference between field blanks and trip blanks.

18. Describe the difference between PE/LCS and MS/MSD.

19. Explain what type of the field QC samples should be used in the following cases:

    (a) To know if a bailer is contaminated in the groundwater sampling.

    (b) To know if airborne contaminants will be responsible for sample contamination during shipment from and to the lab.

    (c) To know if an error (trace contamination) could occur during the entire sampling operation, shipment, and handling.

    (d) To know if any sample variation exists in the field.

(e) To compare if the soil pollution source is due to the atmospheric deposition of an upwind smelter.

20. Explain what type of the laboratory QC samples should be used in the following cases:

    (a) To determine the effects of matrix interferences on analytical accuracy of a sample.

    (b) To determine the analytical precision of a sample batch.

    (c) To establish that laboratory contamination does not cause false positive results.

    (d) To determine whether memory effects (carryover) are present in an analytical run with an instrument.

    (e) To know whether an extraction procedure is appropriate or not.

21. The table below has a list of QC parameters and the primary means of general QC procedure. Match the right column of each corresponding QC procedure to the left column. Explain.

QC parameter	Place a letter here	QC procedure
1. Accuracy		a. Analysis of blanks
2. Precision		b. Analysis of matrix spikes
3. Cross-contamination		c. Analysis of replicate samples
4. Extraction efficiency		d. Analysis of reference materials or samples of known concentrations

22. A 100-μL stock solution containing 1000 mg/L Pb was added to a 100-mL groundwater sample as the matrix spike and acid digested. The Pb concentration measured by ICP–MS was 0.095 mg/L after a tenfold dilution. Calculate the percentage recovery.

23. Chromium (Cr) concentration can be determined by colorimetric spectrometry. The calibration curve was determined to be: $y = 0.045 \, x$ (μg/mL) $+ 0.0005$, where y is the absorbance reading. A 10-mL surface water sample was spiked with a Cr stock solution to reach a concentration of 6 mg/L. The matrix spike has an absorbance of 0.250 against a DI water as the blank. This original surface water sample has an absorbance reading of 0.012. Calculate the percentage recovery.

24. Discuss how the chemical used in a surrogate spike is different from a spike used for recovery test?

25. Twenty soil samples have been collected and sent to the laboratory for the analysis of lead contents. (a) What would be the appropriate analytical method. (b) What QC samples are required for this analysis?

26. Based on the information in the text and literature search, recommend the test methods by filling in the blanks with appropriate method numbers.

27. There is a need to determine if a manufacturing facility can landfill its waste solids from the manufacturing of plastic resins. Compounds suspected to be in the waste solids include semivolatile organic compounds such as 2,4,6-trichlorophenol, pentachlorophenol, pyridine, and cresols. The required EPA procedure is TCLP of SW 846 method 1311. You are responsible for the development of this procedure in your new lab. Explain briefly the TCLP procedure and give a list of the equipment and chemical reagents needed to develop such a method.

Analytes	EPA method 100–600 series	SW-846 method	APHA method 20th edition	USGS method
General physical properties of water				
Metals				
VOCs				
SVOCs				

28. You are designing a sampling program and you are in charge of developing the field sampling QA/QC plan. Your crew will be on site at an abandoned landfill for two days to collect the following samples: (a) Fourteen water samples from groundwater monitoring wells. The samples will be collected by pumping each well with a portable peristaltic pump and 25–50 feet of Teflon tubing into clean sampling containers. (b) Twenty-nine soil samples that will be taken using a hand auger from depths of 1–3 m. The hand auger is essentially a rotating shovel, and the samples will be removed from the auger by scraping them into clean sampling containers.

    How many and what type of QA/QC samples are necessary? Fill in the following table by: (1) labeling the type of QA/QC samples across the top; and (2) listing the correct number of QA/QC samples necessary for each matrix type.

	Type and number of QA/QC samples					
Matrix						
Ground Water						
Soil						

Chapter 6

Common Operations and Wet Chemical Methods in Environmental Laboratories

This chapter signifies a shift of course content from *field* sampling to the *laboratory* analysis. It also introduces some of the basic laboratory techniques required to perform advanced instrumentation to be introduced in Chapters 8 to 12. The purposes of this chapter are (i) to introduce some of the essential operations in environmental laboratories performing wet chemical analysis along with safety and waste disposal activities, and (ii) to help understand the chemical principles involved in the measurement of common environmental parameters using wet chemical methods.

The importance of wet chemical methods is self-explanatory. The gravimetric or volumetric based wet chemical method (i.e., the classical method as defined in Chapter 1) does not require expensive equipment and therefore can be routinely employed in many environmental laboratories. The wet chemical methods are not inferior but are equally important to instrumental analysis. For some monitoring parameters, they are absolutely necessary for sample preparation prior to instrumental analysis. Even with sophisticated instrumental analysis, the data

Fundamentals of Environmental Sampling and Analysis, by Chunlong Zhang
Copyright © 2007 John Wiley & Sons, Inc.

accuracy sometimes depends on the careful calibration using classical wet chemical methods or the common techniques described in this chapter.

A unique feature of this chapter is the description of underlying chemical principles of common wet chemical methods. The chemical reactions and stoichiometry are often excluded in standard cookbooks by the U.S. EPA and APHA. The comprehension of these fundamental principles is not only important for the acquisition of quality data, but also important to enhance the work efficiency. In many cases, this knowledge helps solving the problem when complications arise while following the recipes of standard methods, or when modifications are needed because of sample-specific interferences.

Csuros (1997) made a similar argument regarding the importance of understanding the chemistry rather than just following the step-by-step cookbook. As she stated, "A true chemist, or analytical chemist, has several characteristics. He or she has knowledge of the methods and instruments used for analysis, and understands the principles of the analysis. Laboratory analysts should have a chemistry background adequate to understand and correctly apply all of the laboratory rules and to evaluate and interpret the results of their analysis. They have to know how to plan and organize laboratory work so the time is used efficiently. Thus, a laboratory technician should be a skilled, well-trained chemist—in sharp contrast with the so-called 'determinators' who simply twist the dials of an instrument or follow 'cookbook' analytical procedures".

6.1 BASIC OPERATIONS IN ENVIRONMENTAL LABORATORIES

6.1.1 Labware Cleaning Protocols for Trace Analysis

Proper cleaning techniques are critical to obtain quality data. This is particularly essential for trace environmental analysis. A thorough cleaning will reduce the chance of obtaining erroneous results and simplify troubleshooting procedures. Even though there has been an increased use of disposable labwares, not all labwares will be disposable and laboratory cleaning can account for a significant portion (up to one-third) of an analyst's time in trace analysis. In a multi-users' lab, such as the case in academic settings, it is imperative not to share other people's glassware and plasticware. In many cases, it appears clean to the naked eye, but in fact, it contains enough residue to completely overwhelm the amount of a trace chemical in an environmental sample.

The cleaning technique varies depending on the analyte of interest—the optimal cleaning method for analyzing trace organics will not be the same as the method for trace metal analysis. A general rule of thumb is that an *acid wash* is required if trace metals are the analytes of interest, and an *acetone wash* is minimally required if trace organic contaminants are the target analytes. The following are basic guidelines for routine cleaning, but they may be not relevant for every type of analysis:

- Immediately after each use, soak dirty glassware in a 2% hot detergent solution. Do not use household detergent, but use commercially available laboratory detergents such as Alconox or Liquinox for a *detergent wash*. Brush thoroughly to clean-up obvious residuals but avoid scratch. A detergent wash should not be used if a surfactant is the analyte of interest.

- For metal anlaysis, prepare an *acid bath* using 10–15% HNO_3 (except for N analysis), or 10–15% HCl for labwares. An acid rinse should not be applied to metallic items such as caps, spatulas, and brushes with metal parts.

- For trace organic analysis, use wash bottles filled with pesticide-grade isopropanol, acetone, or methanol to thoroughly rinse the glassware. Acetone is the most commonly used.

- In between detergent wash and acid or acetone wash, use running hot tap water to allow the water to run into and over the labware for a short time, and then fill each item with water, thoroughly shaking and emptying at least three times. After acid or acetone wash, use reagent water (Section 6.1.2) to rinse at least three times.

- Once labware is properly cleaned, dried, and stored in a location that prevents contamination from airborne dust or other particles, air-dry test tubes, culture tubes, and flasks are hanged in a basket with their mouths facing downward over folded paper towels. If oven drying is used, do not exceed over 140°C for glassware or 60°C for plasticware. Never oven dry burets, pipets, or volumetric flasks as their measuring accuracies will be affected.

- In biological work, glassware (serology tubes, Petri dishes) contaminated with blood clots or bacteria must be sterilized with 2% disinfectant solution or boiling water prior to cleaning. If virus or spore-bearing bacteria are present, *autoclaving* is absolutely necessary.

- For special types of precipitates, *aqua regia* (concentrated HCl: HNO_3 at 3:1 ratio) is the preferred approach. This is a very corrosive substance and should be used only when required. For extremely dirty glassware and those with coagulated organic residuals, a *chromic acid wash* is very effective, but it has been deprecated in many countries. The chromic acid cleaning solution can be prepared by adding 20 g of technical grade powder sodium dichromate into 300-mL concentrated H_2SO_4. This solution is now replaced with other non-chromium-based solutions, such as Nochromix^{R}.

6.1.2 Chemical Reagent Purity, Standard, and Reference Materials

1. Reagent water

Water is the most widely used analytical solvent for the preparation of analytical blanks, samples and standard solutions. Poor water quality is one of the easiest problems to fix, but its quality may be also one of the least understood factors in many analytical laboratories. There are many ways to prepare pure water to meet quality requirements for the analysis of organic, inorganic, or biological parameters. Reagent water has been defined at three or four distinct purity levels or types by various professional organisations including National Committee for Clinical Laboratory Standards (NCCLS), the College of American Pathologists (CAP), and the American Society for Testing and Materials (ASTM). Table 6.1 is a set of water purity specifications established by ASTM. For example *Type I water* has the highest quality and can be used for test methods requiring minimum interference and bias. It is prepared by distillation, deionization, or reverse osmosis followed by polishing with a mixed bed deionizer and passing through a 0.2-μm pore size membrane filter. The polishing steps remove trace organics, particulate matter, and bacteria. *Type II water* is prepared by distillation or deionization and can be used if bacterial presence can be tolerated. *Type III water* can be used for glassware washing, preliminary rinsing of glassware and a feed water for the production of higher purity waters.

Table 6.1   ASTM reagent-grade water specification

	Type I	Type II	Type III	Type IV
Total Matter, max. mg/L	0.1	0.1	1.0	2.0
Electrical conductivity, max. μmho/cm at 25°C	0.06	1.0	1.0	5.0
Electrical Resistivity, min. M/cm at 25°C	16.67	1.0	1.0	0.2
pH at 25°C	NS	NS	6.2–7.5	5.0–8.0
Min. color retention time of $KMNO_4$, minutes	60	60	10	10
Max. soluble silica	ND	ND	10 μg/L	No limit

* NS = not specified. The measurements of pH in Type I and Type II reagent waters are meaningless, since electrodes used in this test contaminate the water. ND = not detected.

Distilled water is excellent with regard to particulates and bacteria, good in dissolved organics, but may be poor in dissolved ionized gases. *DI water* is excellent in terms of ionized species, but poor in dissolved organics, particulate and bacteria. The highest purity reagent-grade water is equivalent to the Type I water. The *reagent water* is the water with no detectable concentration of compound or element to be analyzed at the detection limit of the analytical method (APHA, 1998). However, this definition of reagent water has limited use.

In practice, a commonly used indicator for high quality water is the high resistivity (>10 megaohm-cm at 25 °C) or low conductivity (<0.1 μmho/cm at 25 °C). These parameters, however, do not tell the amount of organics or non-ionized contaminants, nor will they provide an accurate assessment of ionic contaminants at the ppb level. Fortunately, with most commercially available water purification systems, obtaining a desired quality of water is not as problematic as before if proper maintenance is followed. This, however, may not be true for certain trace instrumental analysis. As Mabic et al. (2005) pointed out, between 70 and 80% of HPLC performance problems are attributable directly to the quality of water used in preparing HPLC eluents, standards, and samples.

2. Common Acid and Alkaline Solutions

Table 6.2 lists the properties of commercial acids and bases commonly used in environmental laboratories. To prepare a dilute solution, cautiously add the required amount of concentrated acid/alkaline as received, and mix to a designated volume of the proper type of distilled water. Dilute to 1 L and mix thoroughly.

Table 6.2 Preparation of Acid and Alkaline Solutions

Acid/Base	HCl	H_2SO_4	HNO_3	H_3PO_4	NH_4OH
Specific gravity (room temperature)	1.18	1.84	1.42	1.70	0.90
% active ingredient in conc. reagent by mass	36	96	71	85	28
Molarity (normality) of conc reagent	11.6 (11.6)	18.0 (36.0)	16.0 (16.0)	14.7 (44.1)	14.8 (14.8)
Volume (mL) needed to prepare 1 L of 1 N solution	83.3	27.8	64.1	22.7	66.7

3. Reagent Purity

When making a solution, one must first decide what degree of chemical purity is needed on the basis of its analytical purpose and on the cost of the chemical. Generally reagents with the highest grade available are purchased for environmental trace analysis, particularly for the preparation of standard solutions. Low-grade chemicals can only be used for special purposes, such as acetone for washing glassware. Three common chemical grades, in order of decreasing purity, are ACS, reagent, and technical. The *ACS grade* meets the requirements of the American Chemical Society Committee on Analytical Reagents. The *reagent grade* is suitable for most analytical work and is more than adequate for general lab use. For these analytical (research) grade chemicals, the specified impurities are often at 10–1,000 ppm levels. The *technical grade* is suitable for general industrial uses and its appropriateness for lab use must be checked when substituting for higher grades.

Since there are no universal specifications on chemical grades, various other grades can be seen on the container labels depending on the types of chemicals, the end uses of the chemical, and the manufacture. This may become a confounding factor if one wants to compare the purity of chemicals from various sources.

Following are a few more examples of chemical grades: (a) HPLC grade solvents are normally recommended for use in HPLC and all related chromatographic applications; (b) SpectrAR is used for spectrophotometry; (c) ChromAR solvents are specially purified for use in chromatography; (d) ScitillAR is used in liquid scintillation for radioactive analysis; and (e) USP denotes pharmaceutical chemicals tested according to United States Pharmacopeia.

4. Standard and Reference Materials

Standards are materials containing a known concentration of an analyte. They provide a reference to determine unknown concentrations or to calibrate analytical instruments. There are two types of standard solutions used in titration-type analysis. The *primary standard* is prepared from a reagent that is extremely pure, stable, and has no waters of hydration. Without these properties, an accurate mass determination of the standard is difficult to obtain. Chemicals that absorb water or CO_2 readily from air make poor candidates for primary standards. Reactions between the standard and other chemicals must be rapid, complete and stoichiometric. Important ones used in environmental analysis are sodium carbonate (Na_2CO_3, MW = 105.99), potassium hydrogen phthalate ($KHC_8H_4O_4$, MW = 204.23), potassium dichromate ($K_2Cr_2O_7$, MW = 294.19), and potassium permanganate ($KMnO_4$, MW = 158.03). *A secondary standard* is a reagent whose concentration is usually standardized (e.g., via titration) against a primary standard. Examples are H_2SO_4, HCl, NaOH, and sodium thiosulfate ($Na_2S_2O_3$). These chemicals cannot be prepared accurately from their original forms due to various reasons (indefinite concentration for concentrated H_2SO_4 and HCl, moisture absorption for NaOH, and instability for $Na_2S_2O_3$). Because they are not as stable as primary standards, most secondary standards have a short shelf life.

  Standard reference materials (SRM) are particular standards in that their concentrations are known in either liquid or sample matrix (e.g., sediment). They are typically purchased from a reputable vendor or certifying organization. SRMs are used to establish the accuracy of analytical instrumentation. The EPA requires that standards used for calibration and other purposes be of known purity and traceable to a reliable reference source. While the EPA and the National Institute for Standards and Technology (NIST) have provided administered repositories for standards in the past, common practice at the present generally relies on commercial vendors for SRMs. These commercially available materials can generally be traced to reference standards maintained by either EPA or NIST, and virtually all vendors provide certification documents for the purity and traceability of the materials that they supply.

6.1.3 Volumetric Glassware and Calibration

Volumetric glasswares are subject to error if used improperly. Keep in mind that beakers, Erlenmeyer flasks and graduated cylinders are never used for accurate volume measurement in preparing solutions. Only volumetric flasks, burets, and pipets are accurate enough in measuring volume. *Volumetric flasks* are calibrated to a specified volume (5, 10, 25, 50, 100, 250, 500, 1000 mL) when filled to the line etched on the neck. They are commonly used in preparing standard solutions for calibration purpose. *Burets* are specialized graduated cylinders where a stopcock at the bottom is used as a valve to control the release of solution into a receiving flask. They are particularly used for titration. Pipets are used to transfer known volumes of liquid from one container to another, which come in two main size ranges and the prices differ considerably: (a) micropipets with volumes smaller than 1 mL (1, 10, 50, 100, 500 μL), which include syringe pipets and automatic pipets that come with disposable pipet tips. (b) regular glass or plastic pipets with a larger volume (1, 2, 5, 10, 25, 50, 100 mL), which can be further divided into two types (Fig. 6.1):

- *Volumetric pipets* deliver fixed volumes (1, 2, 5, 10, 25, 50, 100 mL). Volumetric pipets labeled as Class A have a certain time in second imprinted near the top. This is the time allowed to elapse from the time the finger is released until the pipet is empty. When emptying a volumetric pipet, the liquid is allowed to drain out, but is not forced out. Hence, the volumetric pipets are not blow-out pipets. They are also more accurate than the measuring pipets.

- *Measuring pipets* have graduation lines that are calibrated in convenient units so that any volume up to the maximum capacity (1, 2, 5, 10, 25, 50, 100 mL) can be delivered. They are not as accurate as volumetric pipets due to the fact that any imperfection in their internal diameter will have a greater effect on the volume delivered. Measuring pipets are divided into *serological pipets* (the graduations marks continue to the tip) and *Mohr pipets* (the graduations end before the tip). Note that serological pipets can be blow-out pipets if they have a frosted band or two thin rings on the top. This means that after all the liquid has been allowed to drain out by gravity, the last drop remaining in the tip is also forced out. Most of the disposable pipets nowadays are the serological "blow-out" type. The last drop of the solution should be blown out with a pipet bulb.

Figure 6.1 Different types of pipets (from the top): 10-mL volumetric pipet, 10-mL serological pipet, and two Mohr pipets (Courtesy of Dr. Ruth Dusenbery, The University of Michigan-Dearborn)

Volumetric glassware is calibrated either to contain (TC) or to deliver (TD). Glassware designated TD will do so with accuracy only when inner surface is clean so that water wets it immediately and forms a uniform film upon emptying. This thin film will not be a part of the delivered volume. If the glassware is designated TC, it is calibrated that the film should be part of the contained volume. Pipets marked with TC are designed to transfer usually viscous solutions, and a thin film remaining inside should be flushed out with a suitable solvent.

Practical tips

- To calibrate a volumetric device, weigh a certain volume of water on analytical balance. For instance, a 1-mL water should be close to 1.0000 g.

- To minimize error with measuring pipets, it is a good practice not to use the tip portion of the pipet. For instance, use only a maximum of 4 mL for a 5-ml pipet. Alternatively, since almost all serological pipets have an additional mL marked above the 0 mL line, the accuracy can be improved by not using the 1 mL liquid near the tip.

- Do not use disposable pipet for certain hydrophobic solvents such as hexane and chloroform. These harsh solvents will dissolve disposable pipets, therefore, glass pipets are recommended.

- Do not submerge the marked portion of the pipets in the liquid as they may contaminate the solution.

6.1.4 Laboratory Health, Safety, and Emergency First Aid

A safe and healthful practice is no doubt the most important and should be an integral part of all activities in environmental sampling and analysis. Although the victims, as a result of unsafe and health hazardous operations are the people who are directly involved in the first place, the presence of any safety and health hazard should be prevented and is the responsibility of all parties including the organization, the project manager and the lab manager. In the United States, the minimum standard of practice for health and safety activities is detailed in government documents (OSHA 29 CFR 1910.1450). Prior training for all parties is required and a written laboratory hygiene plan (LHP) or chemical hygiene plan (CHP) should always be developed and implemented within an organization.

Prior to sampling and analytical work, one should always make efforts in identifying any potential safety and health hazards and should be knowledgeable about the methods of *prevention*. These safety and health hazards could be related to *chemical hazards* (strong acids/bases, perchloric acid and other highly reactive chemicals, organic solvents and reagents), *physical hazards* (electrical, mechanical, radiation- or compressed gases-related), or *biological hazards* (infectious biological samples). For chemical hazards, a good lab practice is to know your chemicals through the *material safety data sheet (MSDS)* for their potential toxicity and other hazards.

Many chemicals routinely used in laboratories do not appear to pose immediate (acute) health hazards, but they may accumulate or contribute to a long-term (chronic) health problem. Examples of such chemicals are benzene, carbon tetrachloride, chloroform, and perchloroethylene, which are all suspected carcinogens. Volatile chemicals should be of particular concern from a health and safety perspective. These include many common solvents used particularly in environmental laboratories during sample extraction. Volatile chlorinated hydrocarbons, if inhaled, cause narcosis and damage to the central nervous system and liver. The vapor of volatile hydrocarbons can also cause fire or explosion. Keep volatile solvents in small containers and away from any heat or spark source (e.g., hotplate or open flame). The toxic and flammable volatiles should be used under a fume hood with adequate ventilation.

Ignorance of even a small part of safety rules can sometimes lead to catastrophic accidents such as fire, explosion, or death. Electricity and water can be a very dangerous combination. For example, when hotplate or other heating devices are used with cooling water such is the case for reaction and sample preparation (reaction vessels, water bath, and oil bath), care should be exercised to secure water from leakage. In such cases, never leave experiment (reaction) unattended. Compressed gas cylinders are another example of potential danger. Transport cylinders by means of a suitable hand truck. Do not roll and always keep caps closed when in transport. Special safety procedures need to be followed for hydrogen and pure oxygen cylinders.

Besides all active preventive measures, one should be fully aware of an adequate passive *protection* devices and supplies including emergency first aid to reduce the consequences to a minimal level in the even of an incident. Know the use of adequate level of *personal protection equipment (PPE)* for the prevention of eyes, head, hearing, feet, legs, and so forth. These may include goggles, gloves, earplugs, protection hard hats, and safety footwear. Become thoroughly acquainted with the location and use of safety facilities such as artificial respiration, safety showers, eyewash fountains, fire extinguishers, and exits. Always know where the first aid kit is located in the lab and finally keep the telephone numbers and contact persons (safety officials, hospital, police department) handy.

Practical tips

- If you suspect safety and health hazards but do not know how to deal with it, do not hesitate to ask your supervisor and get proper training.

- If a sampling and analysis procedure was proved to be safe and hazard-free by other labs, other persons, or other times, do not assume it will be the same under your control. Take extra precaution.

- Before the beginning of a sampling and analysis procedure, ask yourself "what would happen if ..." questions.

- It is better not to work alone and certainly one should not work alone if the procedure to be conducted is hazardous.

- Remove gloves before leaving the work area and before handling such things as telephones, doorknobs, writing instruments, and notebooks.

- There are three types of fire extinguishers: (a) Water (air-pressurized water) extinguishers are used for Class A (wood, paper and cloth) fire only. Do not use water or foam extinguishers for electrical fires. (b) CO_2 extinguishers are designed for Class B and C (flammable liquid and electrical, respectively). (c) Dry chemical (powder) extinguishers come in a variety of types which are always clearly labeled as ABC (for Class A, B and C), BC (for Class B and C), and D (for Class D fires involving metals such as K, Na, Al, and Mg). Do not use dry chemical (powder) extinguishers for electronics (instrument) as they will make clean-up difficult and possibly destroy the electronics. Use CO_2 extinguishers only in these situations.

6.1.5   Waste Handling and Disposal

Source reduction and waste minimization is always the best strategy before developing any disposal measures. Source reduction can be achieved simply through the purchase and use of smaller quantities of chemicals, which can be justified by the high disposal cost for unused or expired reagents if larger quantities were purchased. Sometimes a hazardous reagent can be substituted with a non-hazardous chemical or even eliminated by the use of an alternative procedure.

Have separate containers for general trash, broken glass, and used needles. Laboratories typically have various containers for other liquid and solid chemical wastes (inorganic, organic, acid, alkaline, oxidants, etc.). While very dilute liquid waste can be discharged into laboratory sinks, there are exceptions for certain chemicals, including concentrated corrosive acids/bases, highly toxic chemicals, malodorous and lachrymatory chemicals, and other chemical reagents that may develop potential fire hazard or upset sewage treatment plants.

Special rules apply for certain highly toxic or reactive chemicals, biological wastes, and radioactive chemicals. Wastes containing PCBs, dioxins, and asbestos require consultation with federal and state officials before disposal. Peroxides present special problems in the laboratory because they can be violently reactive or explosive upon shock, decomposition, and/or reactions with air and water (ACS, 2003). Treat radiochemicals with extreme care and minimize the exposure if at all possible. Handling radiochemical wastes often requires special training within an organization. Dilution methods are acceptable for low activity radioactive wastes, such as ventilation for airborne and flushing into sink for liquid radiochemical wastes. Biohazard wastes carrying pathogenic microorganisms and virus should be collected separately in specially marked biohazard bags and autoclaved (121 °C, 103 kPa) for a minimum of 30 min prior to disposal. Used pipets should be disinfected in discard jars by using hypochlorite disinfectants at the highest possible concentration.

6.2 WET CHEMICAL METHODS AND COMMON TECHNIQUES IN ENVIRONMENTAL ANALYSIS

6.2.1 Gravimetric and Volumetric Wet Chemical Methods

The essential measurement involved in gravimetric methods is weighing with an analytical balance. Gravimetric methods are rarely used in environmental contaminant analysis because their concentrations are often low in the range of minor to trace level. Gravimetric methods are used only for a few parameters such as solids, moisture, sulfate, and oil and grease. Volumetric (i.e., titrimetric) analysis, on the contrary, still finds wide applications for the measurement of a variety of important parameters. Although titrations may be performed automatically with an autotitrator (mostly electrochemically based instrument), the simplicity of the buret-based manual titration points to many applications employed in environmental analysis.

Titration is based on the simple stoichiometric relationships of chemical reactions. While titration can be performed by many electrochemical titration devices (Chapter 11), the classical titrimetric methods rely on a color change using indicators. The chemistry underlying color change is based on one of following four major reactions: (a) acid-base reaction, (b) oxidation-reduction (redox) reaction, (c) complexation/chelation reaction, and (d) precipitation reaction. Each of these reactions has its unique applications in environmental measurements (Table 6.3). For instance, redox-based titrations are applicable to transition metals or elements that possess more than one oxidation states. Acid-base titrations can be employed for acidic/basic inorganic or organic compounds, and the indicator changes color at a specific pH range during titration. Table 6.3 is a list of common wet chemical

Table 6.3 Wet chemical methods commonly used in environmental analysis

Parameters	Method/Operation*	EPA	APHA (SM$_{20}$)
Moisture	G/Oven Drying	-	-
Solids	G/Filtration	160.1 to 160.4	2540A to G
Acidity	T (acid-base)	350.1	2310 B
Alkalinity	T (acid-base)	310.1 & 310.2	2320 B
Hardness	T (complexation)	130.1 & 130.2	2340 C
DO	T (redox)	360.2	4500 O B to F
BOD	T (redox)	405.1	5210 B
COD	T (redox)/Reflux	410.1 to 410.3	5220 C
Oil and Grease	G/Extraction/Distillation	413.1	5520 B
Residual Chlorine	T (redox)	330.2 to 330.4	4500 D
Chloride	T (precipitation)	325.3	4500-Cl$^-$ B
Ammonia	T (acid-base)/Distillation	350.2	4500-NH$_3$ C
Cyanide	T (complexation)/Distillation	335.1	4500-CN$^-$ D
Sulfide	T (redox)/Distillation	376.1 & 9030 A	4500-S^{2-} F

* G: Gravimetric; T: Titrimetric; SM$_{20}$: Standard Method, 20th Edition.

methods used in environmental analysis. The method numbers of both EPA and APHA are also given so the reader can refer to the references for details.

6.2.2   Common Laboratory Techniques

The basic laboratory techniques introduced here are not only related to the wet chemical methods shown in Table 6.3, but also the basics to all phases of environmental lab work. Several key points or even pitfalls for each of these techniques are briefly introduced. Readers can locate these materials in traditional analytical chemistry books. Several sample preparation techniques, such as extraction and digestion, will be further described in Chapter 7.

Filtration

Filtration is used to remove (collect) materials from a liquid or air matrix in which the materials are suspended. The filter media can be filter papers of various types or under gravity filtration or vacuum (suction) filtration using a water aspirator or a vacuum pump. Filtration is used for the analysis of suspended solids in water, or the analysis of dissolved phase contaminant after the removal of suspended solids. In air sampling, the filter medium is connected with a sampling pump to collect atmospheric particulates. For quantitative analysis of solids, use *ashless quantitative filter papers* such as Whatman and Schleicher & Schuell. To remove (collect) small particles with pore sizes of 0.025 to 14 μm, use only *membrane filter* such as Millipore and Gelman. The 0.22 μm is the most common in removing bacteria or collecting colloidal particles. The membrane filters have a thin polymeric structure with extraordinarily fine pores. When vacuum filtration is used, use *hardened-grade papers* and thick wall flasks. The hardened-grade papers have great wet strength to sustain vacuum.

Centrifugation

A *centrifuge* is used to separate solids or liquid particles of different densities or sizes by rotating them in a centrifuge tube. If colloidal particles or macromolecules (protein, nucleic acids) are to be separated, an *ultracentrifuge* is used at a higher speed. The low speed centrifuge can run up to 5, 000 revolutions per minute (rpm) and the high speed ultracentrifuge can run up to 100,000 rpm. Another form of common centrifuge is the *microcentrifuge* used with 0.5 or 1.5 mL disposable plastic tubes, which can generate 10,000 to 13,000 rpm. Following are the common precautions: (a) Always use balanced pairs of centrifuge tubes. Balancing is extremely important to avoid excessive heating and vibrations or even damaging the centrifuge. Turn off immediately if vibrations or noise occur. (b) Never fill centrifuge tubes above the maximum that is recommended by the manufacturer. Use only centrifuge tubes with protective rubber cushions. The regular test tubes will not sustain centrifugal force and will break. (c) Fully secure the rotor and rotor seal before operation. Rotors on high-speed centrifuge or ultracentrifuge units are subject to powerful mechanical stress that can result in rotor failure. In the extreme case, improper loading and balancing of rotors can cause the rotors to break loose while spinning, resulting in total damage of centrifuge, flying metal fragments, and shock wave.

Distillation

Distillation is the procedure needed for ammonium, cyanide, sulfide, and phenol analysis. The distillation device consists of a distilling flask, a cooling water condenser, and a receiving flask to collect distillate. When the distillation flask containing volatile chemical(s) is heated, the most volatile component vaporizes at the lower temperature, and the vapor passes through a cooled tube (condenser), where it condenses back into its liquid state in the receiving flask. Use vacuum distillation if the compound will decompose at atmospheric pressure before its boiling point is reached. Cautions include the following: (a) Make sure all connections in the apparatus are tight. A leakage during distillation can be the major error of analysis. (b) Use proper tubing for the cooling water and clamp the tubing so it will not be in contact with the heating device. Do not leave distillation unattended. (c) Adding a few boiling stones always ease the bumping.

Reflux

Reflux is used in COD analysis where organic compounds are oxidized by a strong oxidizing agent under heated and acidic conditions. A reflux apparatus allows a reaction mixture to be heated for an extended period of time without a loss of water or solvent. The water or solvent vapor is refluxed back to the reaction mixture by a condenser that is vertically attached to the heated flask. A reflux device is similar to distillation in a sense, hence operational and safety rules described above also apply to reflux.

Concentration

Concentration apparatus is common in all laboratories performing trace analysis. The apparatus can be as simple as a mild N_2 stream from a gas cylinder, or an all-glass apparatus such as *Kuderna-Danish (K-D) evaporative concentrator* or a rotary evaporator (rotavap) (Fig. 6.2). Their purpose is to remove excess solvent from the mixture of analyte and solvent so that the concentration of analytes will be sufficiently high to be detected.

In a *K-D concentrator*, boiling chips should be added and the assembly set over a vigorously boiling water bath. The water level should be maintained just below the lower joint and the apparatus mounted so that the lower rounded surface of the flask is bathed in steam. Volatile solvent are allowed to escape under the hood, and the analyte refluxes until the final concentrate is collected in the lower tube of approximately 1–2 mL.

In a *rotavap*, the rotating distilling flask is lowered into a water bath where the water temperature should not exceed the boiling point of the solvent. The solvent is

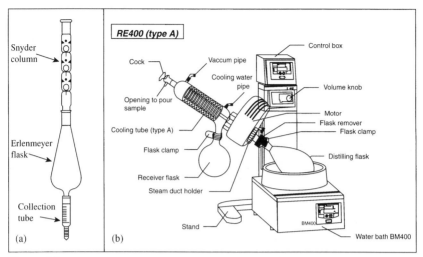

Figure 6.2   Apparatus used to concentrate: (a) Kuderna-Danish (K-D) evaporative concentrator (Courtesy of Kontes Glass Company), and (b) Rotary evaporator (rotavap) (Courtesy of Yamato Scientific Co., Ltd, Japan)

removed under vacuum (with a simple water aspirator or a vacuum pump), trapped by a condenser, and is collected in the receiver flask. If samples contain very volatile solvent such as diethyl ether or methylene chloride, some solvents in the receiver flask may be evaporated and lost by vacuum. To prevent this, a cooling bath on the receiver or dry ice condenser can be used.

Digestion

Digestion is commonly used for metals, total phosphorus, and total nitrogen under heated conditions. The digestion procedure breaks down an organically bound substance and converts the substance to the analyzed form by using liquid oxidizing agents such as H_2SO_4, HNO_3, $HClO_4$, HCl, or using oxidizing mixtures such as aqua regia. Sometimes, addition of bromine (Br_2) or H_2O_2 to mineral acids will increase their solvent action and hastens the oxidation of organic materials in the sample. If the decomposition of silicates is needed, HF is used, but never use glass containers with HF. Digestion is sometimes dangerous, particularly if the sample contains a high amount of organic compounds. Always wear goggles, lab coat, and use heavy-duty gloves in case of acid spills during acid digestions.

Extraction

Chemicals have different solubilities in different solvents, which can be used to selectively remove a solute from a mixture. This extraction process is often used as a sample preparation to concentrate trace organic compounds. Two classical extraction procedures are the liquid–liquid (L–L) extraction using a separatory funnel for liquid samples, and Soxhlet extraction for solid samples (soil, sludge).

Make sure that the correct phase is collected (discarded) in the L–L extraction. Diethylether, benzene, petroleum ether, and hexane are lighter than water, whereas chloroform, ethylene dichloride, methylene chloride, and tetrachloromethane are heavier than water. Always perform extraction under the hood and minimize any exposure to (inhalation of) solvents.

6.3 ANALYTICAL PRINCIPLES FOR COMMON WET CHEMICAL METHODS

Discussed in this section are some of the most common wet chemistry methods perhaps routinely employed in many environmental laboratories (Sanyer et al., 1994; Smith, 1997; Williams, 2001). We focus here on the underlying chemistry and operational principles for the measurement of these selected parameters summarized in Table 6.3. Details of these methods can be found in EPA 100–400 or APHA 2000, 4000, and 5000 series methods.

6.3.1 Moisture in Solid and Biological Samples

The moisture content of solid (soil, sludge, sediment) and biological samples must be measured so that contaminant concentration can be reported on a dry basis. This determination involves weighing a portion of the sample in a tarred crucible, drying to a constant weight at $104 \pm 1°C$ and then calculating the percent moisture by:

$$\% \text{ Moisture } = \left(\frac{\text{wet weight} - \text{dry weight}}{\text{wet weight}} \right) \times 100 \tag{6.1}$$

$$\text{Concentration (dry weight basis)} = \frac{\text{concentration on wet basis}}{100 - \% \text{ moisture}} \tag{6.2}$$

6.3.2 Solids in Water, Wastewater, and Sludge (TS, TSS, TDS, VS)

Solids are the matter suspended or dissolved in water or wastewater. Waters with high dissolved solids (minerals) are unsuitable for drinking as well as for industrial uses. Solids suspended in surface water and wastewater effluent are esthetically unsatisfactory and undesirable from a water quality standpoint. Solids (sludge) in wastewater are related to the biomass in the biological process and therefore are an important process control parameter. As described below, there are various types of solids that can be measured gravimetrically and volumetrically.

Total Solids (TS)

"Total solids" is the term applied to the material residue left in the vessel after evaporation of a sample and its subsequent drying in an oven at a defined temperature. Total solids (TS) includes total suspended solids (TSS) and total dissolved solids (TDS), i.e., TS = TSS + TDS.

Total Suspended Solids and Total Dissolved Solids

TSS is the portion of total solids retained by a filter (0.45 μm in pore size is the most common), while TDS is the portion that passes through the filter. The temperature at which the residue is dried has an important bearing on results, because weight losses due to volatilization of organic matter, mechanically occluded water, water of crystallization, gases from heat-induced chemical decomposition, as well as weight gains due to oxidation, depend on temperature and time of heating. Each sample requires close attention to desiccation after drying. Note the different temperatures used for TSS ($104 \pm 1°C$) and TDS ($180 \pm 2°C$) measurement. Because the temperature difference, in practice, the equation TS = TSS + TDS may not work out for some samples.

Volatile Solids (VS)

This is the difference between the dried solids and the same sample further ignited at $550 \pm 50°C$ in a muffle furnace. VS is useful in control of wastewater treatment plant operation because it approximates the amount of organic matter in the solid fraction. Thus, $TVS = TS - TS_{ash}$; $TVSS = TSS - TSS_{ash}$; $TVDS = TDS - TDS_{ash}$. "Ash" is the residue at $550 \pm 50°C$ in a muffle furnace.

Settleable Solids

Settleable solids can be measured either gravimetrically or volumetrically. The common volumetric method is defined as the volume (mL) of settleable solids per liter of wastewaters after 1 h settling under quiescent conditions (mL/L). It is normally performed in a *Imhoff cone*, which is simply a cone-shaped plastic container with a holding volume of 1 L and the side of the cone graduated in mL. Settleable solids can be used to estimate the *sludge volume index* (SVI), which represents the volume (mL) occupied by 1 g of activated sludge after 30-min settling, i.e, SVI (mL/g) = settleable solids (mL/L)/TSS (g/L). A high SVI (>100) indicate poor settling quality and possible bulking problem in secondary clarifier of the wastewater treatment plant.

6.3.3 Acidity, Alkalinity, and Hardness of Waters

1. Acidity

All waters having pH lower than approximately 8.3 should contain acidity. Acidity is the capacity of water to neutralize a base (OH^-). Sources of acidity in natural water include CO_2 (air, bacterial degradation), $H_2PO_4^-$, H_2S, protein, fatty acids, and salt of trivalent metals such as hydrated Fe^{3+} and Al^{3+} (e.g.,

$Al(H_2O)_6^{3+} \rightleftharpoons Al(H_2O)_5OH^{2+} + H^+$). In polluted water, acidity may be caused by free mineral acid (H_2SO_4, HCl) from metallurgical industry, acid mine drainage, acid rain, and organic acid waste.

To measure acidity, an alkaline solution (NaOH) and two indictors are used. The first indicator (methyl orange) changes color from red to yellow at approximately pH 3.7 (Eq. 6.3). This corresponds to *mineral acidity* or *methyl orange acidity*. If NaOH is continued to titrate, the second indicator phenolphthalein changes color from colorless to pink at pH 8.3 (Eq. 6.4). Titration to the phenolphthalein end point of pH 8.3 measures acidity from both mineral acids and weak acids. This *total acidity* is also termed *phenolphthalein acidity*.

$$\tag{6.3}$$

$$\tag{6.4}$$

To calculate mineral acidity or total acidity from the titration data, the amount of NaOH must be converted to the equivalent amount of $CaCO_3$ (in mg) per liter of water. Since the molecular weight (MW) of $CaCO_3$ is 100 g/mol or 1×10^5 mg/mol, its equivalent weight is ½ MW or 5×10^4. This is used as the conversion factor so that the acidity can be reported as mg $CaCO_3$/L:

$$\text{Acidity} = \frac{N \times V \times 5 \times 10^4}{V_S} \tag{6.5}$$

where N is the normality of base used in titration, and V_s the volume of water sample (mL), V the volume of base used in titration (mL). For mineral acidity, $V = $ mL NaOH needed to the first end point (pH 3.7) and for total acidity, $V = $ mL NaOH needed to the second end point (pH 8.3).

2. Alkalinity

Alkalinity is the capacity of water to neutralize an acid (H^+). In natural water, alkalinity is due to H_2CO_3 (CO_2 dissolved in water). Alkalinity from other materials is insignificant and may be ignored. In polluted waters, minor amounts of NH_3 and salts of weak acids such as borate, silicate, and phosphate (i.e., the conjugate bases of HBO_3, H_4SiO_4, H_3PO_4), and organic acids will contribute to alkalinity. The titration procedure can distinguish two types of alkalinity, that is, phenolphthalein alkalinity and total alkalinity (methyl orange alkalinity). The *phenolphthalein alkalinity* is the acid-neutralizing power of hydroxide and carbonate ions (OH^- and CO_3^{2-}) present in the water sample, while *total alkalinity* represents all the bases in it (OH^-, CO_3^{2-} and HCO_3^-).

The titration procedure for alkalinity is exactly the opposite of the procedure for acidity. Acid (H_2SO_4) rather than NaOH is used for titration and the two same indicators are used but in the reverse order. The procedure utilizes the first indicator, phenolphthalein, to signal the end points (pH 8.3) in titrating OH^- and CO_3^{2-}. Further titration with H_2SO_4 reaches the end point of the second indicator (methyl orange) at around pH 4.5, which measures the total alkalinity. Note that the pH for the color change is not at a fixed value, but rather a range, that is, pH 3.1–4.4 and pH 8.0–9.6 for methyl orange and phenolphthalein, respectively.

Like acidity, alkalinity is also commonly reported as mg $CaCO_3$/L, hence Eq. 6.5 can be used to calculate both phenolphthalein alkalinity and total alkalinity. For alkalinity, the volume (V) and concentration (N) refer to H_2SO_4 rather than NaOH. To interpret the acidity and alkalinity results, one should be aware of their physical meaning and their difference with commonly measured pH. Although water with a high acidity (mineral acidity) is corrosive and undesirable, water with a high alkalinity may not be totally unwanted. In some cases, an adequate level of alkalinity should be maintained to retain the buffering capacity of the water. Acidity or alkalinity are related to pH but they differ in that pH is an intensive parameter, which is independent of the volume of water.

3. Hardness

Water hardness is caused by multivalent metallic cations, primarily the divalent Ca^{2+} and Mg^{2+}. Anions and monovalent cations such as Na^+ and K^+ do not contribute to hardness. Hard water adversely affects the suitability of water for uses in domestic and industrial purposes. Hard waters produce scale in hot-water pipes, heaters and boilers ($Ca^{2+} + 2HCO_3^- \rightarrow CaCO_3(s) + CO_2(g) + H_2O$). They also require considerable amounts of soap to produce foam. If direct measuring instruments (i.e., AA or ICP) are available, water hardness can be calculated from the measured concentrations of Ca^{2+} and Mg^{2+}:

$$\text{Hardness (mg/L as } CaCO_3) = \sum M^{2+} \text{ (mg/L)} \times \frac{50}{\text{EW of } M^{2+}} \qquad (6.6)$$

where the equivalent weight (EW) of Ca^{2+} and Mg^{2+} are 20.0 and 12.2, respectively. In most laboratories, a simple titration procedure is used. In the presence of divalent cations (M^{2+}), Eriochrome Black T (an indicator) forms a weak complex with M^{2+}, turning from blue color to wine red.

$$M^{2+} + \text{Eriochrome Black T (Blue)} \rightarrow \text{M-Eriochrome Black T (Red)} \qquad (6.7)$$

When EDTA (ethlenediaminetetraacetic acid, or its sodium salt Na_2EDTA) is added to the water sample, EDTA forms a strong chelate with divalent cations. It disrupts the wine red complex and changes back to the blue color of the indicator (the end of titration):

$$M^{2+} + Na_2EDTA \rightarrow 2Na^+ + M\text{-}EDTA \text{ (Colorless)} \tag{6.8}$$

Hardness is also reported as mg $CaCO_3$/L, hence Eq. 6.5 can be used for the calculation of hardness, where the volume (V) and concentration (N) represent EDTA. Despite the simplicity of the standard procedure, a skilled analyst should further understand two potential problems associated with this procedure and the corresponding means to mitigate the problems. (a) The sharpness of the end point color change increases with increasing pH. However, Ca^{2+} and Mg^{2+} will likely precipitate out at high pH. A satisfied compromise of pH is in the range of 10.0 ± 0.1. (b) The presence of other metals will consume EDTA thereby overestimating the hardness. By adding other complexing agents such as CN^- and S^{2-}, such interference can be eliminated. This is an example of how important it is to understand the chemistry rather than just following the step-by-step cookbook procedure.

6.3.4 Oxygen Demand in Water and Wastewater (DO, BOD and COD)

1. Dissolved Oxygen

DO in natural water originates from the molecular oxygen (O_2) in the atmosphere. The atmosphere has 20.95% O_2 by volume of dry air, but DO in water is typically at the mg/L level due to the low solubility of O_2. Many aquatic organisms cannot survive if the DO is depleted below 4 mg/L. In aerobic biological process, DO is supplied by aeration whereas in anaerobic process DO is harmful to the bacteria and must be removed. The amount of DO in natural water depends on many physical, chemical, and biochemical factors–aeration, wind, water flow velocity, algae, temperature, atmospheric pressure, organic compounds, salt content, bacteria, animals, and so forth. Dissolved oxygen in polluted water is highly dependent on the amount and types of pollutants, and the presence of bacteria therefore, it is an important parameter for the assessment of water quality. Most often, oxygen-consuming pollutants are organic compounds (protein, sugar, and fatty acids) that are readily biodegraded by aerobic bacteria, although in some cases inorganic contaminants in reduced forms will also consume a significant amount of oxygen.

There are two methods for DO analysis: the classical *Winkler method* (iodometric method) and the polargraphic *membrane electrode method*. The membrane electrode procedure (refer to Chapter 11) is an electrochemical method based on the rate of diffusion of molecular oxygen across a membrane, which is the most common method for in situ measurement. The iodometric method is a titrimetric procedure based on the oxidizing property of DO. This iodometric test is the most precise and reliable method for DO analysis; in some case, it is used to calibrate the electrode method.

For the Winker method, dissolved oxygen is "fixed" by adding manganese (II) sulfate immediately after a sample is collected in a 300-mL BOD bottle in the field and the sample is transported to the lab for titration. The "fixation" reaction between DO and Mn^{2+} is as follows:

$$Mn^{2+} + 2OH^- + \tfrac{1}{2}O_2 \rightarrow MnO_2(s) + H_2O \tag{6.9}$$

DO rapidly oxidizes an equivalent amount of the dispersed divalent manganous hydroxide precipitate to oxides of higher valence state. The sample completely fills the 300 mL bottle to ensure no further oxygen is introduced. After being transported to the laboratory (within 6 h of sample collection), the sample is acidified with concentrated H_2SO_4. In the presence of iodide ions (I^-) in an acidic solution, the oxidized manganese reverts to the divalent state, with the liberation of iodine (I_2) equivalent to the original DO content:

$$MnO_2(s) + 2I^- \rightarrow Mn^{2+} + I_2 + 2H_2O \tag{6.10}$$

The released I_2 can then be titrated with standard solution of sodium thiosulfate ($Na_2S_2O_3$) using a starch indicator:

$$I_2 + 2S_2O_3{}^{2-} \rightarrow S_4O_6{}^{2-} + 2I^- \tag{6.11}$$

The titration end point can be easily detected visually from blue to colorless. By summing Eq. 6.9 to 6.11, the overall reaction of the Winkler method is:

$$2S_2O_3{}^{2-} + 2H^+ + \tfrac{1}{2}O_2 \rightarrow S_4O_6{}^{2-} + H_2O \tag{6.12}$$

The above reaction indicates that 1 mol of O_2 is equivalent to 4 mol of thiosulphate ($S_2O_3^{2-}$) in the final titration. The DO can be calculated by the following equation:

$$DO\,(mg/L) = \frac{N \times V \times 8000}{V_S} \tag{6.13}$$

where N and V are the normality and volume (mL) of $Na_2S_2O_3$, respectively, V_s is the sample volume (typically 200 mL is withdrawn from 300 mL sample).

A thorough knowledge of the above chemical reactions helps analysts to command the key step(s) for an accurate DO measurement. Understanding the reaction stoichiometry is also required to fully comprehend the formula (Eq. 6.13) used to calculate DO. For example, one may wonder how the conversion factor of 8,000 is included in Eq. 6.13. The example below is used to illustrate this point.

EXAMPLE 6.1. (a) From the reaction stoichiometry shown in Eq. 6.12, derive the conversion factor (8000) shown in Eq. 6.13. (b) The Standard Method (SM_{20}) does not provide any chemical reactions shown from Eq. 6.9 to 6.12, but calls for a equivalency of 1 mL 0.025 M $Na_2S_2O_3$ and 1 mg DO/L. Demonstrate this equivalency. (c) Explain whether a positive or negative interference would it cause if oxidizing agents or reducing agents are present in wastewater and effluent samples for DO measurement.

SOLUTION: (a) From $S_2O_3{}^{2-}$ to $S_4O_6{}^{2-}$, the oxidation number changes from $+2$ to $+2.5$. Since the net change is $+1$ per molecule, meaning one mole of electron per mole $S_2O_3^{2-}$ is lost to O_2. This also means that 1 mol of thiosulfate ($S_2O_3{}^{2-}$) is equal to 1 equivalent of the same species. $N \times V$ equivalents of $S_2O_3{}^{2-}$ will be equal to $N \times V$ mol of $S_2O_3{}^{2-}$ or $N \times V/4$ mol of O_2. Converting this number of moles into mass of O_2 in mg, this equals to $(N \times V/4) \times 32 \times 1000$ or $N \times V \times 8,000$, where 32 is the molecular weight of O_2.

(b) Note the stoichiometric mole ratio is 1:4 ($O_2 : S_2O_3{}^{2-}$) and the volume of water sample withdrawn from 300-mL BOD bottle in the standard method is 200 mL

$$1\,mL \times \frac{0.025\,mol\,S_2O_3{}^{2-}}{L} \times \frac{1\,mol\,O_2}{4\,mol\,S_2O_3{}^{2-}} \times \frac{32 \times 1000\,mg\,O_2}{1\,mol\,O_2} \times \frac{1}{200\,mL} = 1.0 \frac{mg\,O_2}{L}$$

(c) In general, oxidizing agents liberate iodine (I_2) from iodides (I^-), thereby consuming more $Na_2S_2O_3$ or overestimate DO (positive interference). Some reducing agents will reduce iodine to iodide, resulting in negative interference. This explains why the standard methods have established several *modified Winker methods*, including the use of NaN_3, $KMnO_4$, and alum flocculation for the removal of interfering NO_2^-, Fe^{2+}, and suspended solids, respectively.

2. Biochemical Oxygen Demand (BOD)

BOD refers to the amount of oxygen that would be consumed if all the biodegradable organics in one liter of water were oxidized by aerobic bacteria. The BOD test is based on the determination of DO. For the standard 5-day BOD_5 test, 300-mL BOD bottles (tapered water seal stopper) are used, saturated with DO to about 9 mg/L, and the initial DO is measured (DO_0). The parallel samples are incubated at 20°C in the dark for 5 days, and the DO is measured again (DO_5). The BOD_5 is calculated as:

$$BOD_5\ (mg/L) = DO_0 - DO_5 \tag{6.14}$$

The above direct method is for samples with BOD_5 of less than 7 mg/L. However, most wastewater samples exceed this value, and serial dilution is needed. The usable dilutions must have initial DO 7–9 mg/L and final values of > 1 mg/L, and a minimum depletion between initial and final DO of 2.0 mg/L is required to accurately calculate the BOD. It is also critical to assure the presence of essential nutrients (Ca, Mg, Fe salts, and phosphate buffer), absence of toxic materials, and inoculation with multi-micro organism seed.

3. Chemical Oxygen Demand (COD)

A BOD test provides the closest measure of the processes actually occurring in the natural water system. However, this test is very time-consuming (5 days) and involves many uncertain factors such as the origin, concentration, pollutants, the number, and viability of active microorganisms present to affect the oxidation of all pollutants. The COD test is used as a measure of the oxygen equivalent of the organic matter content of a sample that is susceptible to oxidation by a strong chemical oxidant as opposed to biological oxidation in the BOD test.

In a COD test, results can be obtained in 2 h or less. Also the method is simple and inexpensive. When wastewater contains only readily oxidizable organic matter and is free from toxins, the results of a COD test provide a good estimate of the BOD. One disadvantage of the method is that dichromate can oxidize materials that would not ordinarily be oxidized in nature. Therefore, the COD test is unable to differentiate between biologically oxidizable and biologically inert organic matter. The COD test can also generate a large volume of liquid hazardous waste (acid, chromium, silver, and mercury).

During COD measurement, the water sample is refluxed with excess of potassium dichromate ($K_2Cr_2O_7$) in concentrated sulfuric acid for ~2 h. The reaction involved in the usual case, where organic nitrogen is all in a reduced state (oxidation number of -3), may be represented in a general way as follows:

$$C_nH_aO_bN_c + dCr_2O_7^{2-} + (8d + c)H^+ \rightarrow nCO_2 + (a + 8d - 3c)/2\,H_2O \\ + cNH_4^+ + 2dCr^{3+} \tag{6.15}$$

where $d = 2n/3 + a/6 - b/3 - c/2$. The reaction requires strong acidic conditions and elevated temperature. As shown in Eq. 6.15, organic matter is converted to CO_2 and water, organic nitrogen in a reduced state will be converted to NH_4^+, whereas higher oxidation states will be converted to nitrates. After dichromate digestion is complete, the excess of dichromate is titrated with ferrous ammonium sulfate ($Fe(NH_4)_2(SO_4)_2$):

$$6Fe^{2+} + Cr_2O_7^{2-} + 14H^+ \rightarrow 6Fe^{3+} + 2Cr^{3+} + 7H_2O \tag{6.16}$$

The indicator used for the above equation is a chelating agent 1,10-phenanthroline (ferroin). When all $Cr_2O_7^{2-}$ is reduced, ferrous ions (Fe^{2+}) react with ferroin to form a red-colored complex:

$$Fe^{2+} + 3C_{12}H_8N_2 \rightarrow Fe\{C_{12}H_8N_2\}_3(red) \tag{6.17}$$

Note that $Cr_2O_7^{2-}$ has a yellow to orange brown color depending on the concentration, and Cr^{3+} has a blue to green color. So the color of the solution during the titration starts with an orange brown, and then a sharp change from blue-green to reddish brown, which corresponds to the color of $Cr_2O_7^{2-}$, Cr^{3+}, and $Fe\{C_{12}H_8N_2\}_3$, respectively. Calculation of COD is made using the following formula:

$$COD\ (mg/L) = \frac{N \times (V_0 - V) \times 8000}{V_s} \tag{6.18}$$

where N is the normality of $Fe(NH_4)_2(SO_4)_2$, V_s is the water sample volume (mL), and V_0 and V are the volumes (mL) of $Fe(NH_4)_2(SO_4)_2$ titrated for the blank and the sample, respectively. Note that the same conversion factor of 8,000 is shown above as previously described in Eq. 6.13 for DO. In Eq. 6.16, the net electron loss per mol Fe^{2+} is 1 mol.

6.3.5   Oil and Grease in Water and Wastewater

The term "oil" represents a wide variety of substances from low to high molecular weight hydrocarbons found in petroleum to light gasoline, heavy fuel, and lubricating oils. "Grease" represents higher molecular weight hydrocarbons and all glycerides of animal and vegetable origin. The term "oil and grease" is defined by the operational procedure rather than the representation of distinct chemicals or groups of compounds.

The partition-gravimetric method described herein includes acidification and extraction followed by a gravimetric measurement. Acidification is required before extraction if fatty acids are present in the sample. The fatty acids occur primarily in a precipitated form as Ca and Mg salts with soaps. As such, they are insoluble in the solvents. Acidification with HCl to PH$\sim$ 1.0 will release the free fatty acids for analysis. The reaction involved may be represented by the equation:

$$Ca(C_{17}H_{35}COO)_2 + 2H^+ \rightarrow 2C_{17}H_{35}COOH + Ca^{2+} \tag{6.19}$$

Hexane is then used to extract oil and grease in a separatory funnel. The extracted phase is drained to a tarred distilling flask through anhydrous Na_2SO_4 to remove residual water, and then hexane is separated by evaporation in a distillation device (boiling point is 69°C for hexane; maintain water bath temperature at 85°C). The tarred flask is weighed again and the gain in the weight of the tarred flask is due to oil and grease.

Modification of the above procedure is needed for solid samples such as sludge. The method for solid samples consists of weighing a definite amount of sample, acidifying it to release fatty acids, and then adding sufficient $MgSO_4 \cdot H_2O$ to combine with all free water by forming higher hydrated forms, $MgSO_4 \cdot 7H_2O$ to ease extraction with hexane. The extraction is then performed in a Soxhlet extractor in place of a separatory funnel for liquid samples.

6.3.6 Residual Chlorine and Chloride in Drinking Water

1. Residual Chlorine

Disinfection using gaseous chlorine (Cl_2) is the most widely used in domestic water supply and sewage treatment plants. When chlorine is added to water, it rapidly hydrolyzes to form hypochlorous acid (HOCl) and hydrochloric acid (HCl). HOCl is a weak acid that can be further dissociated into hypochlorite (OCl^-). The reactions are:

$$Cl_2 + H_2O \rightleftarrows HOCl + H^+ + Cl^- \tag{6.20}$$

$$HOCl \rightleftarrows H^+ + OCl^- \tag{6.21}$$

At equilibrium, aqueous phase Cl_2 normally is negligible above pH 3. Hence the two species formed by Cl_2 in water are primarily HOCl and OCl^-. These two species are termed as *free chlorine residuals* that have the disinfection capacity in killing bacteria and pathogens. Unlike HOCl and OCl^-, chloride (Cl^-) is not effective in disinfection at all.

In the presence of ammonia (sometimes it is intentionally added), the following reactions will occur to form mono-, di-, and tri-chloroamines:

$$NH_4^+ + HOCl \rightarrow NH_2Cl + H_2O + H^+ \tag{6.22}$$

$$NH_2Cl + HOCl \rightarrow NHCl_2 + H_2O \tag{6.23}$$

$$NHCl_2 + HOCl \rightarrow NCl_3 + H_2O \tag{6.24}$$

The chloroamines are called *combined chlorine residuals*. They are less effective in disinfection but can last longer, which is essential for continued disinfection in water distribution conduits.

The current EPA and APHA manuals list three titrimetric methods along with several other instrumental methods for the measurement of residual chlorines. These three titrimetric methods are (a) iodometric titration, (b) iodometric back titration, and (c) DPD-FAS titration. The iodometric method is applicable to the measurement of total chlorine residuals in natural and treated waters at concentrations greater than 0.1 mg/L. The iodometric back titration method is applicable to all types of waters but is primarily used for wastewater. Both back titration method and DPD-FAS method are suited to differentiating free chlorines and combined chlorines. Described below are the chemical principles of these three wet chemical methods commonly employed in laboratories because of their simplicities.

Iodometric Titration   Free chlorine (OCl^- and $HOCl$) and chloramines stoichiometrically liberate iodine (I_2) from potassium iodide (KI) at pH 4 or lower. The I_2 is titrated with a standard reducing agent sodium thiosulfate ($Na_2S_2O_3$) using a starch indicator. The end point is the disappearance of the blue color.

$$Cl_2 + 2KI \rightarrow I_2 + 2KCl \tag{6.25}$$

$$I_2 + starch \rightarrow blue\ color \tag{6.26}$$

$$I_2 + 2Na_2S_2O_3 \rightarrow 2NaI + Na_2S_4O_6 \tag{6.27}$$

Iodometric Back Titration   In this method, the liberated I_2 (Eq. 6.25) is immediately reacted with an excess amount of $Na_2S_2O_3$ (Eq. 6.26). The remaining $Na_2S_2O_3$ is back titrated with I_2 as the titrant (Eq. 6.28) instead of using $Na_2S_2O_3$ as the titrant.

$$I_2 + 2Na_2S_2O_3(excess) \rightarrow 2NaI + Na_2S_4O_6 \tag{6.28}$$

Compared with iodometric titration method, the iodometric back titration method causes immediate reaction of the I_2 generated so that any contact between the full concentration of liberated I_2 and the potential I_2-reducing substance in wastewater is minimized.

Note that molecular iodine (I_2) is not soluble in water to any appreciable degree. To obtain iodine in solution for titrations, the iodine is reacted with excess iodide (I^-) to form the triiodide ion (I_3^-), which is very soluble and reacts as if it were molecular iodine. A handy source of iodine is from potassium biiodate, $KH(IO_3)_2$, which can be prepared as a primary standard according to the following reaction in the presence of an acid:

$$KH(IO_3)_2 + 10\,KI + 11H^+ \rightarrow 6I_2 + 6H_2O + 11K^+ \tag{6.29}$$

It is possible to use the back titration method to measure separately free chlorine residuals and combined chlorine residuals by conducting a two-stage titration. Phenylarsine oxide (PAO) is the reducing agent normally used for this purpose. At pH above 7, PAO reacts with free chlorine residuals in a quantitative manner, whereas at pH below 4, only chloramines react with PAO.

DPD-FAS Titration   The N,N-diethyl-p-phenylenediamine (DPD) is added to a sample containing free chlorine residuals and an instantaneous reaction occurs, which produces a red-color (Eq. 6.30). At pH 6.2 to 6.5 the reaction is rapid and reversible. Then the solution is titrated with ferrous ammonium sulfate (FAS) $[Fe(NH_4)_2(SO_4)_2]$ until the disappearance of the red color as the end point. If a small amount of iodide is further added to the sample, monochloramine reacts to produce iodine, which in turn oxidizes more DPD to form additional red color. Titrate with FAS. If a large quantity of iodide is then added, dichloramines will react to form still more red color. Titrate with FAS again. By three successive titrations, free chlorine, monochloramine, and dichloramine residuals can be determined.

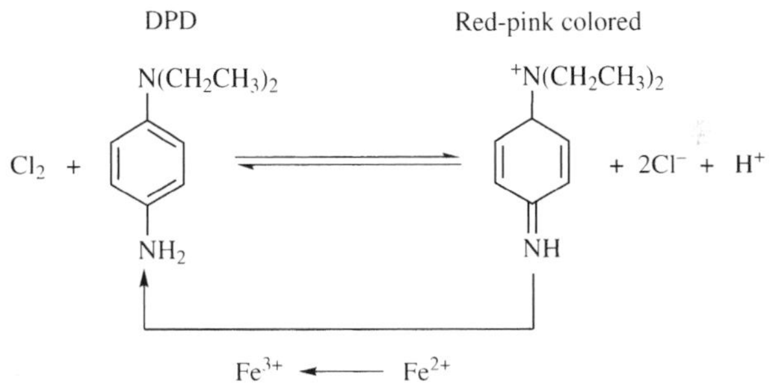

$$\tag{6.30}$$

2. Chloride

The measurement of chloride (Cl^-) is necessary to determine the suitability of water for domestic, industrial or agricultural uses. High chloride concentration may be a result of seawater intrusion due to excessive pumping of groundwater, evaporation of irrigated water leaving salts behind in the soil, or discharge of human excreta into municipal wastewater. In addition, chloride is a common tracer in chemical fate and transport studies, when a simple and reliable analytical method is required. Two wet chemical methods are available for the measurement of chloride. The selection is largely a matter of personal preference. The argentometric method

(Mohr method) is suitable for relatively clean water when 0.15 to 10 mg Cl^- is present in the portion titrated. The end point of the mercuric nitrate method is easier to detect.

Argentometric Method (Mohr Method)

In a neutral or slightly alkaline solution, potassium chromate (CrO_4^{2-}) can indicate the end point of the silver nitrate titration for a solution containing Cl^-. Silver chloride (AgCl, $K_{sp} = 3 \times 10^{-10}$) is precipitated quantitatively before orange silver chromate (Ag_2CrO_4, $K_{sp} = 5 \times 10^{-12}$) is formed.

$$AgNO_3 + Cl^- \rightleftarrows AgCl + NO_3^- \tag{6.31}$$

$$2Ag^+ + CrO_4^{2-} \rightleftarrows Ag_2CrO_4(\text{orange}) \tag{6.32}$$

Mercuric Nitrate Method

Cl^- is titrated with mercuric nitrate, $Hg(NO_3)_2$, because of the formation of soluble, slightly dissociated mercuric chloride. In the pH range from 2.3 to 2.8, diphenylcarbazone (DPC) indicates the titration end point by the formation of a purple complex with excess mercuric ions.

$$Hg^{2+} + 2Cl^- \rightleftarrows HgCl_2 \tag{6.33}$$

(6.34)

6.3.7 Ammonia in Wastewater

Ammonia is produced mostly by decomposition of organic nitrogen-containing compounds and by the hydrolysis of urea in wastewater. Ammonium concentration in surface and groundwater is usually low. The titrimetic method can be used directly in the case of drinking water or highly purified wastewaters. A preliminary distillation step is needed for other sample matrices, which transfers NH_3 into a receiving flask containing pH 9 borate acid (H_3BO_3) (Eq. 6.35). The ammonia in the distillate is then titrated with H_2SO_4 in the presence of a mixed indicator (methyl red and methylene blue). The sample is then titrated until it turns pale lavender (Eq. 6.36).

$$NH_3 + H_3BO_3 \rightarrow NH_4^+ + H_2BO_3^- \tag{6.35}$$

$$2NH_4^+ + 2H_2BO_3^- + H_2SO_4 \rightarrow 2H_3BO_3 + (NH_4)_2SO_4 \tag{6.36}$$

EXAMPLE 6.2.   Reaction stoichiometry. The Standard Method uses standard 0.02 N H_2SO_4 solution to titrate ammonia after distillation of water samples. The calculation formula was provided as:

$$\text{mg NH}_3 - \text{N/L} = \frac{(A - B) \times 280}{\text{mL Sample}} \tag{6.37}$$

where A is the volume of H_2SO_4 titrated for sample (mL), and B the volume of H_2SO_4 titrated for blank (mL). Verify this formula using stoichiometry involved in distillation and titration.

SOLUTION: Combine reactions Eqs 6.35 and 6.36 by eliminating the reaction intermediate, $NH_4^+ H_2BO_3$. This results in a balanced overall reaction as follows:

$$2NH_3 + H_2SO_4 \rightarrow (NH_4)_2SO_4 \tag{6.38}$$

The above reaction indicates that 2 mol NH_3 will stoichiometrically react with 1 mol H_2SO_4. Note that 0.02 N H_2SO_4 is equal to 0.01 M (mol/L). We then have:

$$0.01 \frac{\text{mol } H_2SO_4}{L} \times (A - B) \text{ mL } H_2SO_4 \times \frac{2 \text{ mol } NH_3}{1 \text{ mol } H_2SO_4} \times \frac{14,000 \text{ mg } NH_3 - N}{1 \text{ mol } NH_3}$$
$$\times \frac{1}{\text{ml Sample}} = \frac{(A - B) \times 280}{\text{mL Sample}} \frac{\text{mg } NH_3 - N}{L}$$

In the above calculation, NH_3-N denotes "ammonia as nitrogen," hence 1 mol NH_3 contains 14 g N or 14,000 mg N. This is a common practice in reporting as "N basis" or "P basis" for the concentration of anions such as NO_3^-, NO_2^- and PO_4^{3-}. In addition, the titrant volume (B) for the blank sample must be subtracted to minimize the error.

6.3.8   Cyanide in Water, Wastewater and Soil Extract

Cyanide refers to all of the cyanide groups in cyanide compounds that can be determined as the cyanide ion (CN^-), which is the conjugate base of the more toxic HCN gas. Cyanide is often present in the form of a metal complex, which is stable and less toxic. Organically bound cyanides, such as acrylonitriles (cyanoethene) are important industrial chemicals. Cyanides are used in plastics, electroplating, metallurgy, synthetic fibers, and chemicals. The titrimetric method is applicable if cyanide concentration is above 1 mg/L in the aqueous sample or the distillate from wastewater, soil or waste extract.

First, hydrogen cyanide (HCN) is liberated from an acidified sample by distillation and purging with air through a scrubbing solution containing NaOH. Cyanide concentration in the scrubbing solution is then titrated with standard silver nitrate ($AgNO_3$) to form the soluble cyanide complex, $Ag(CN)_2^-$. As soon as all CN^- has been complexed and a small excess of Ag^+ has been added, the excess Ag^+ combines with a silver-sensitive rhodanine indicator (p-dimethyl-aminobenzal-rhodanine, $C_{12}H_{12}N_2OS_2$), which immediately turns solution from yellow to brownish-pink. The titration is based on the following reaction:

$$2CN^- + Ag^+ \rightarrow [Ag(CN)_2]^- \tag{6.39}$$

The concentration of CN^- in $\mu g/L$ in the original sample can be calculated as follows. Again, as illustrated earlier in this chapter, the reader should be aware of the operational principles and the stoichiometry behind this formula.

$$CN^- \left(\frac{\mu g}{L} \right) = \frac{(A - B)}{C} \times D \times \frac{E}{F} \times \frac{2 \text{ mol } CN^-}{1 \text{ eq. } AgNO_3} \times \frac{26.02 \text{ g } CN^-}{1 \text{ mol } CN^-} \times \frac{1 \times 10^6 \mu g}{1 \text{ g}}$$

$$(6.40)$$

where A is the mL $AgNO_3$ for titration of sample, B the mL of $AgNO_3$ for titration of blank, C the mL of sample titrated, D the actual normality of $AgNO_3$, E the mL of sample after distillation, F the mL of original sample used for distillation.

6.3.9 Sulfide in Water and Waste

Sulfide is often present in groundwater due to the prevalent reducing conditions. In wastes and wastewaters, sulfide comes partly from the anaerobic decomposition of organic matter, and mostly from the bacterial reduction of sulfate (SO_4^{2-}). The term "sulfide" can include hydrogen sulfide (H_2S, dissolved and unionized), bisulfide (HS^-) and sulfide (S^{2-}) – all in dissolved forms. At an environmentally relevant pH, S^{2-} is often negligible. In addition, "sulfide" is also present as metallic sulfides – some of them are "acid-soluble" (ZnS) and others are "acid-insoluble" such as CuS and SnS_2. These metallic sulfides have very low water solubility, and therefore are the precipitates associated with suspended matter in water, soils or sediment.

Like several other parameters discussed previously, sulfide can also be determined by titration-based iodometric method. Recall also from Chapter 4 that sulfide samples are preserved by the addition of zinc acetate and NaOH to maintain pH > 9 (Table 4.1). The zinc sulfide precipitate is then oxidized to sulfur with the addition of excess iodine (I_2) in acidic solution. Then the remaining iodine is determined by titration with a standard solution of sodium thiosulfate until the blue iodine-starch complex disappears.

This titrimetric method is applicable to all aqueous, solid waste materials and effluents. The specific procedures may differ slightly depending on the matrices and the forms of sulfide species to be measured. Distillation is not needed for clean aqueous samples, but in most cases, it is required. Sulfide is separated from the sample by the addition of H_2SO_4, heated to 70°C and the resulting H_2S is carried by a stream of N_2 into zinc acetate gas scrubbing bottles, where it is precipitated as ZnS. This pretreatment will measure all the acid-soluble sulfide species. If this step is replaced by the addition of HCl, tin(II) chloride, and 100 °C, "acid-insoluble" sulfide can be released and determined semi-quantitatively. Tin (II) is added to prevent oxidation of sulfide to sulfur by the metal ions such as Cu^{2+}.

REFERENCES

American Chemical Society (2003), Safety in Academic Chemistry Laboratories Vol. 1, Accident Prevention for College and University students, American Chemical Society Committee on Chemical Safety, 7th Edition Washington, DC.

APHA (1998), Standard Methods for the Examination of Water and Wastewater, 20th Edition Washington, DC.

BERGER W, McCARTY H, SMITH R-K (1996), Environmental Laboratory Data Evaluation, Genium Publishing Corporation Amsterdam, NY.

*CSUROS M (1997), Environmental Sampling and Analysis: Lab Manual, Lewis Publishers, Boca Raton, FL.

HACH Company (1992), HACH Water Analysis Handbook, 2nd Edition, Loveland, CO (Procedures available: http://www.hach.com).

National Research Council (1995), Prudent Practices in the Laboratory: Handling and Disposal of Chemicals, National Academies Press, Washington, DC.

MABIC S, REGNAULT C, KROL J. 2005, The misunderstood laboratory solvent: Reagent water for HPLC. *LCGC North America*, 23(1):74–82.

Occupational Safety and Health Administration (1990), Laboratory Standard. Occupational Exposure to Hazardous Chemicals in Laboratories. 29 CFR 1910.1450.

SAWYER CN, McCARTY PL, PARKIN GF (1994), Chemistry for Environmental Engineering, McGraw-Hill, NY.

SMITH R-K (1997), Handbook of Environmental Analysis, 3rd Edition, Genium Publishing Corporation, Amsterdam, NY.

US EPA (1983), Methods for Chemical Analysis of Water and Wastes, EPA/600/4–79/020.

US EPA (1983), Test Methods for Evaluating Solid Wastes Physical/Chemical Methods. SW-846 Method On line: http://www.epa.gov/epaoswer/hazwaste/test/main.htm

WILLIAMS I. 2001. Environmental Chemistry, John Wiley & Sons, NY West Sussex, England.

QUESTIONS AND PROBLEMS

1. Define the following terms: aqua regia, acid bath, acetone wash, Type I water, reagent water, PPE, MSDS, primary standard, Mohr pipet, mineral acidity.

2. Illustrate how to prepare 2 N HCl and 2 N H_2SO_4 solutions from concentrated HCl and H_2SO_4.

3. Explain why $K_2Cr_2O_7$ has to be used to calibrate the concentration of $Na_2S_2O_3 \cdot 5H_2O$ which is commonly used in titration.

4. What does it mean if a pipet is labeled as "5 in 1/10 mL, TD 20°C"?

---

*Suggested Readings

5. Explain the operational difference for pipets labeled as "TD" or "TC".

6. List all active and passive measures for a healthful and safe lab practice.

7. What action would you take in the following situation?

   (a) Your labmate placed broken glass pipets (not the disposable ones) in the paper waste bin.

   (b) Your colleague went to a vending machine down stairs while he was performing a COD reflux experiment.

   (c) Your colleague is doing extraction using chloroform on her workbench.

   (d) You saw a facility personnel rolling a gas cylinder while it is not capped.

   (e) Your research assistant placed volumetric flasks in an oven to dry for the experiment next day.

8. What is the primary difference between reflux and distillation?

9. What is the major difference between liquid–liquid separatory funnel extraction and Soxhlet extraction in terms of the applicable type of sample matrices?

10. What oven temperature is used to dry samples for suspended solid measurement? Why this constant temperature is important? Why different temperatures of $103\,°C$, $180\,°C$ and $550\,°C$ are used in solid measurement?

11. A 100-mL water sample was used for suspended solid measurement. The mass of weighting bottle plus filter paper was $25.6257\,g$ before filtration and $25.6505\,g$ after titration. What is the TSS in mg/L?

12. A sludge sample was collected after sludge was dewatered from a belt-dewatering equipment, the moisture content was 12%. This sludge sample was further air-dried and measured for its copper content at $80\,mg/kg$ (dry basis). What is the copper concentration in mg/kg on a wet basis?

13. Explain why only a single titration step is needed for alkalinity measurement if the water pH is lower than approximately 8.3.

14. Explain why acidity is not present if a water sample has a pH of greater than approximately 8.5? What is the major difference between pH and acidity?

15. Discuss the major species responsible for the acidity in natural water and possible contributing species in polluted water?

16. Discuss the major species responsible for the alkalinity in natural water and possible contributing species in polluted water?

17. If $3.0\,mL$ of $0.02\,N\ H_2SO_4$ is required to titrate $200\,mL$ of sample to the phenolphthalein end point, what is the phenolthalein alkalinity as mg/L of $CaCO_3$? If additional $20.0\,mL$ of H_2SO_4 is needed to reach the methyl orange point, what is the total alkalinity as mg/L of $CaCO_3$?

18. Are Na^+ and K^+ included in the hardness calculation? Why or why not?

19. A water sample has Ca^{2+} and Mg^{2+} concentration of $25.0\,mg/L$ and $17.0\,mg/L$. Calculate the hardness in mg $CaCO_3$/L (atomic weight: $Ca = 40$, $Mg = 24$).

20. A 25-mL groundwater sample was added to 25-mL DI water for a two-fold dilution. The volume of $0.015\,mol/L$ EDTA to reach the titration end point was $6.28\,mL$. Calculate the water hardness in mmol/L and in mg $CaCO_3$/L.

21. To prepare EDTA solution for hardness measurement, 8.54 g EDTA disodium salt ($C_{10}H_{14}N_2O_8Na_2 \cdot 2H_2O$, MW = 372.2) was dissolved in 1000 mL DI water. After calibration, its concentration was measured to be 22.82 mmol/L. A 50-mL water sample consumed 6.06 mL EDTA solution. (a) What is the actual concentration of EDTA in mmol/L compared to the measured concentration of 22.82 mmol/L? (b) What is the hardness of this water sample in mg $CaCO_3$/L?

22. Write all chemical reactions involved in the measurement of DO using Winkler (iodometric) method.

23. Write the chemical reactions for the measurement of (a) COD, and (b) chloride using Mohr method, (c) residual chlorine using iodometric back titration, and (d) cyanide.

24. During the reflux process of COD measurement, if the solution turns into green color, what causes this and what is the solution to this problem?

25. Define the difference between chloride, free chlorines, and combined chlorines. How they differ in their disinfection capacity?

26. Explain what types of interference (positive or negative) will be caused in the presence of reducing agent or oxidizing agent for DO measurement?

27. List three substances that interfere with the Winkler method, and indicate which modification would be used to overcome each interference?

28. (a) Explain why the conversion factor in Eq. 6.13 is 8000. (b) If 200 mL sample was used for dissolved oxygen (DO) measurement, it requires 8.6 mL of 0.0275 N $Na_2S_2O_3$ solution. What is the DO in mg/L?

29. A standard procedure by APHA's method was followed to determine DO in water. The volume of 0.025 M $Na_2S_2O_3$ consumed was 6.8 mL in titrating 200-mL solution. What is the DO in mg/L?

30. Glucose is oxidized according to: $C_6H_{12}O_6 + 6O_2 \rightarrow 6CO_2 + 6H_2O$. It can be used as a standard of COD because its theoretical COD can be calculated from its known concentration. The oxidation reduction reaction involved in COD measurement is:

$$6Fe^{2+} + Cr_2O_7^{2-} + 14H^+ \rightarrow 6Fe^{3+} + 2Cr^{3+} + 7H_2O$$

(a) What is the theoretical COD value of a solution containing 0.50 g/L glucose ($C_6H_{12}O_6$; MW = 180) (*Hint:* 1 mole glucose consumes 6 moles of O_2)?

(b) Why the equivalent weight of $K_2Cr_2O_7$(MW = 294.2) is 294.2/6 = 49.03?

(c) To prepare 500 mL 1.0 N $K_2Cr_2O_7$, how many grams of $K_2Cr_2O_7$ are needed?

(d) List three major sources of error in COD measurement. How might they be eliminated?

31. Potassium hydrogen phthalate (abbreviated as PHK, $HOOCC_6H_4COOK$, MW = 204.23) is always used as the COD standard for its measurement. To prepare 1 liter of a PHK solution containing 500 mg/L COD, how many grams of PHK need to weigh. The oxidation of PHK is assumed: $2HOOCC_6H_4COOK + 15\,O_2 \rightarrow K_2O + 16CO_2 + 5H_2O$ (*Hints:* 2 mole PHK consumes 15 moles of O_2).

32. Describe the difference between BOD and COD in terms of the chemicals each method represents.

33. Explain in the BOD measurement why (a) diluted water, (b) bacterial seed, and (c) initial oxygen saturation are all needed.

34. Explain the difference between direct iodometric titration and iodomeric back titration. How is the color change different at the end point of titration?

35. A municipal wastewater sample contains 50 mg/L of ammonia (NH_3), what is the equivalent concentration of NH_3–N (i.e., NH_3 as nitrogen) in mg/L?

36. The NPDES permit limit for PO_4^{3-}–P (i.e., PO_4^{3-} as phosphoras) is 1 mg/L, what is the equivalent concentration for PO_4^{3-} in mg/L?

37. Describe the sample pre-treatment method for sulfide measurement using titration procedure.

38. Explain the following: (a) why acid is added for oil and grease measurement? (b) why cyanide is added for hardness measurement? (c) why zinc acetate is used for sulfide measurement? and (d) why NaOH is used during the distillation of cyanide?

Chapter 7

Fundamentals of Sample Preparation for Environmental Analysis

Sample preparation is the step after samples have been collected and preserved but before samples are introduced into instrument for further analysis. Very rarely can environmental samples be directly injected into instrument without any pretreatment. Sample preparation is a very important part of the sample measurement process and should not be underestimated. It is often the most labor-intensive and time-consuming fraction, a bottleneck, for the entire measurement process. The purpose of this chapter is to introduce the principles of sample preparation techniques, commonly used in trace analysis. The techniques introduced in this chapter cover main categories of environmental contaminants, including digestion for metals, extraction and post-extraction clean-up for SVOCs, derivatization for non-VOCs, and preparation for VOC and air samples. For brevity, those particular and recently evolved sample preparation techniques are omitted. For example, there are new sample preparation techniques developed recently that tend to be faster and

Fundamentals of Environmental Sampling and Analysis, by Chunlong Zhang
Copyright © 2007 John Wiley & Sons, Inc.

greener, such as nanoparticle, microchips, stir-bar sorption extraction, and robotics. The reader should consult further references (e.g., Mitra, 2003) for specific details.

7.1 OVERVIEW ON SAMPLE PREPARATION

7.1.1 Purpose of Sample Preparation

There are examples that sample preparations may not be needed at all. For instance, clean drinking water can be directly used for metal analysis without acid digestion. Several noninvasive techniques (e.g., X-ray fluorescence) need just a minimal sample preparation. Unfortunately, most samples need a more or less tedious preparation, which could become the rate-limiting step in trace analysis. Figure 7.1 is used to illustrate the percentage time spent in a typical chromatographic analysis. It was estimated that an average 61% of the total time spent in a typical chromatographic analysis is devoted to sample preparation.

The purpose(s) of environmental sample preparations can be one or a combination of the following. Note that not all of these techniques are described in this chapter.

- *To homogenize sample or remove moisture:* If the collected sample is heterogeneous or contains too much moisture, it needs air-drying, freeze-drying (if the chemical is unstable), homogenization, grounding (size reduction), and sieving. Doing so will remove water for convenient storage and handling and assure that a minor portion of subsample taken for analysis will be more likely representative.

- *To increase/decrease analyte concentration:* Increasing analyte concentration is often needed for almost all trace analysis of environmental chemicals. The applicable concentration apparatus such as Kuderna-Danish evaporative concentrator and rotary evaporator have been described in Chapter 6. Occasionally, dilution is used for the analysis of highly contaminated samples so the concentration falls within the calibration range.

- *To remove interfering chemicals:* Although interference is less an issue for instrumental analysis than wet chemical analysis, it could become the main

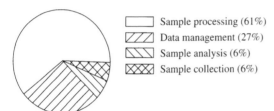

Figure 7.1   Survey results for the distribution of time that analytical chemists spend on sample analysis (Majors, 1991, Reprinted with permission)

issue for many trace organic compounds in complex matrices such as sludge and waste.

- *To change sample phase:* The type of sample phase needs to be changed to fit into a specific type of instrument. Few instruments can accept solid samples. Although both aqueous and solvent phase samples can be directly injected into a HPLC, GC normally accepts samples in either solvent phase or gaseous phase. Likewise, there are special sample requirements for IR and NMR.

- *To liberate analyte from sample matrix:* The analyte species need to be liberated from sample matrices so that the instrument detector will respond. The digestion of soil sample for metal analysis partially serves this purpose. Digestion will enable metals bound to soil minerals and organics to be dissolved in an acidic solution.

- *To modify chemical structure:* Chemical derivatization is one such example used to increase or decrease volatility and thermal stability for HPLC or GC analysis.

7.1.2 Types of Sample Preparation

In general, the types of needed sample preparation described in this chapter depend on the sample matrix, chemical properties, and the specific instrument being used subsequently. From Figure 7.2, it is clear that distinct preparation techniques exist between solid and liquid samples. For each matrix type, sample preparation differs fundamentally between organic and metal analysis. Sample preparation procedures

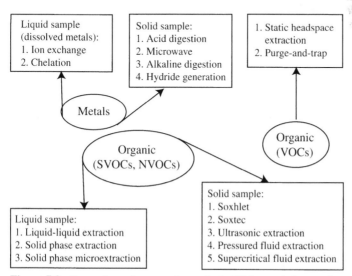

Figure 7.2 Types of sample preparation methods (VOCs = volatile organic compounds, SVOCs = semivolatile organic compounds, NVOCs = nonvolatile organic compounds)

for organic compounds, in turn, also vary with the volatility of the organic compounds (not shown in Fig. 7.2). The text following this section will introduce the important sampling preparation techniques for metals, SVOCs, and VOCs.

7.2 SAMPLE PREPARATION FOR METAL ANALYSIS

7.2.1 Various Forms of Metals and Preparation Methods

Metals in aqueous samples are present in various forms, which can be defined either chemically or operationally. *Dissolved metals*, defined chemically, include hydrated ions, inorganic/organic complexes, and colloidal dispersions, which are equivalent to operationally defined dissolved metals in an unacidified sample that pass through a 0.45-μm membrane filter. *Suspended metals* are chemically bonded to Fe-Mn oxides, Ca oxides, or sorbed to organic matter, which are operationally defined as those metals in an acidified sample that are retained by a 0.45-μm membrane filter. *Total metals* are operationally defined by the metals that are determined in an unfiltered sample after vigorous acid digestion (hence, total metal = dissolved metal + suspended metal). Given such definitions, one should bear in mind that acid preservation is required in the field for total metal analysis, whereas filtration is performed prior to acid preservation for dissolved and suspended metals. *Total recoverable metals* using hot dilute acid were defined in Section 5.2.1.

Similarly, metals in solid samples such as soils and sediments also have various forms (speciation). Although the EPA has not established such methods, species analysis is a subject of extensive investigations as a result of concern over the toxicity and bioavailability of certain species. The classical method for measuring metal speciations in soils was a *sequential extraction procedure* developed by Tessier et al. (1979). This early method is still in substantial use today, although several modifications of this initial method have been reported. According to this method, metals are fractioned into exchangeable and bound to carbonates, Fe-Mn oxides, organic matter, and residual.

Sample preparations for total metal analysis are always performed by one of the two common acid digestion procedures, that is, the classical hotplate digestion method and the microwave-assisted acid digestion method (Fig. 7.3). Both methods use mineral/oxidizing acid(s) and an external heat source to decompose the matrix and liberate metals in an analyzable form. The hotplate digestion is conducted under a ventilation hood, specially designed for minimal exposure to metals while allowing corrosive acids to be washed from the hood after the digestion is completed. The microwave-assisted acid digestion system uses a specially designed microwave that is acid-proof along with safety features for acid fume collection and programmable temperature and pressure control. This device can offer faster and more reproducible results than the conventional hotplate methods. If closed vessels are used, loss of volatile elements (Hg, Pb, As, Sb, Cr, Cu, Cd, Ca) can also be avoided.

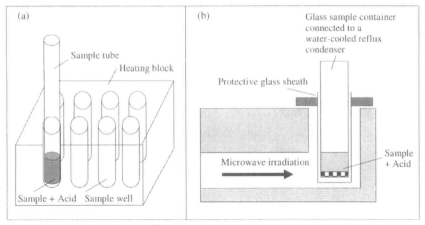

Figure 7.3 Schematic diagram of two common acid digestion apparatus: (a) heat block acid digestion system, (b) microwave-assisted acid digestion system. (Dean, JR, 2003, Methods for Enviromental Trace Analysis, © John Wiley & Sons Limited. Reproduced with Permission)

7.2.2 Principles of Acid Digestion and Selection of Acid

Acid digestion is often conducted on a hotplate under a ventilation hood especially designed for minimal exposure to metals, while allowing corrosive acids to be washed from the hood after the digestion is completed. The purpose of acid digestion is to dissolve metals from sample matrix so that metals can be in a measurable form. Various acids are used in acid digestion, which could be confusing in differentiating one protocol from the other in the standard methods (EPA 300 series methods for water and wastewaters and 3000 series method in SW-846 for liquid and solid wastes, Section 5.2.1). In selecting which single or combined acid(s) to use, one should first consider what instrument is available for metal analysis. In general, HCl is not preferred if graphite furnace atomic absorption spectroscopy (GFAA, see Chapter 9) is the method of analysis because of the interference from chloride. The second factor relates to the characteristics of the sample matrix. The general guidelines on acid selection are as follows: (a) For clean samples or easily oxidized materials, the use of HNO_3 alone is adequate, (b) For readily oxidizable organic matter, the HNO_3–HCl or HNO_3–H_2SO_4 digestion is adequate, (c) For difficult to-oxidize organic matter, HNO_3-$HClO_4$ is needed, and (d) If matrix contains silicates, then HNO_3–$HClO_4$–HF digestion is necessary. No acids other than HF will liberate the metal of interest from the silica matrix.

HNO_3 is an acid of preferred choice either alone or in combination with other acid(s). The only exception to use HNO_3 is when samples contain highly concentrated alcohols and aromatic rings that can form explosive compounds (e.g., nitro glycerine and TNT). There are several reasons why HNO_3 is preferred over other acids. First, HNO_3 is acting as both an acid and an oxidizing agent in the digestion process. As a strong acid, it dissolves inorganic oxides into solution (Eq. 7.1). As an oxidizing agent/acid combo, HNO_3 can oxidize zero valence inorganic metals and non-metals into ionic form (Eqs. 7.2 and 7.3):

$$CaO + 2H_3O^+ \rightarrow Ca^{2+} + 3H_2O \qquad (7.1)$$

$$Fe^0 + 3H_3O^+ + 3HNO_3(conc.) \rightarrow Fe^{3+} + 3NO_2(brown) + 6H_2O \qquad (7.2)$$

$$3Cu^0 + 6H_3O^+ + 2HNO_3(dilute) \rightarrow 2NO(clear) + 3Cu^{2+} + 10H_2O \qquad (7.3)$$

Nitric acid undergoes both one electron (Eq. 7.2) and three electron (Eq. 7.3) changes. The one electron change is observed with concentrated HNO_3, whereas the three electron change is observed when dilute HNO_3 is in reaction. The presence of brown fumes during digestion is indicative of reactions going by one electron.

$$H_3O^+ + HNO_3 + e^- \rightharpoonup NO_2\uparrow \text{ (brown)} + 2H_2O(\text{concentrated}) \qquad (7.4)$$

$$3H_3O^+ + HNO_3 + 3e^- \rightharpoonup NO\uparrow \text{ (clear)} + 5H_2O(\text{dilute}) \qquad (7.5)$$

The second advantage of using HNO_3 is that it does not form any insoluble compounds with metals and nonmetals. The same cannot be said for H_2SO_4, HCl, HF, H_3PO_4, or $HClO_4$. An additional advantage in using HNO_3 is that nitrate (NO_3^-) is the acceptable matrix for both flame and electrothermal atomic absorption and the preferred matrix for ICP–MS. When digesting highly aromatic samples, or samples containing –OH functionality, a pretreatment with concentrated sulfuric acid as a dehydrating agent is recommended as shown in the following equation:

$$R{-}CH_2{-}CH(OH){-}R' + H_2SO_4 \rightarrow R{-}CH{=}C(OH){-}R' + H_2O + H_2SO_3 \quad (7.6)$$

Practical tips

- Evaporate HNO_3 as much as possible, but not to dryness in the digestion container. Volatile metals such as Hg, Pb, Cd, Ca, As, Sb, Cr, and Cu are subject to loss if solution becomes dry. A complete digestion usually can be seen from the light color of the digestate or the digestate does not change in appearance with continued refluxing.

- Avoid metal contamination from any potential sources, including air borne dusts, acids, containers, and pipettes. Plastic pipet tips are often contaminated with Cu, Fe, Zn, and Cd. Avoid using colored plastics that can contain metals. Use certified metal-free plastic containers and pipet tips when possible. Avoid using glass if analyzing Al and Si. Use metal-free water and ultra-pure acids (mineral acids can be sources of many elements including Al, As, Mn, Cr, Ni, and Zn).

- Take extreme cautions for samples with high organic contents. Pretreat HNO_3 before adding $HClO_4$ to oxidize most of the organic matter. Do not use $HClO_4$ alone and never evaporate it to dryness. Use metal-free hood and perchloric acid hood (stainless steel or PVC liner) if $HClO_4$ is being used. Never heat $HClO_4$ in fume hood unless it has been designed for use with

$HClO_4$ with a functioning spray-nozzle wash-down system. Perchloric acid salts could be explosive!

- Wear goggles all the time during digestion and use rubber gloves when pouring or handling acids. Keep workbench and hood dry. Wipe any liquid that could be concentrated acid. Flush the hood system for at least 10 minutes at the end of acid digestion.

- Always include a preparation blank that has all the reagents except sample itself (see section 5.4.2). Any trace level of metals from the blank should be subtracted from sample results.

7.2.3 Alkaline Digestion and Other Extraction Methods

This section briefly introduces specific digestion/extraction procedures that are required for four metals and metalloids (Cr, Hg, As, Se). For these elements, speciations are more important than the total metal concentration because of the distinctly different environmental effects that exist among various species. In other words, a separate sample digestion/extraction is needed for these several elements.

1. Chromium (Cr)

The toxicity of Cr and its mobility in aquatic and terrestrial environments depend on its oxidation state. Two most stable oxidation sates of chromium in the environment are hexavalent, Cr(VI), and trivalent, Cr(III). In natural waters, *trivalent* chromium exists as Cr^{3+}, $Cr(OH)^{2+}$, $Cr(OH)_2^+$, and $Cr(OH)_4^-$. In the *hexavalent* form, chromium exists as CrO_4^{2-} and $Cr_2O_7^{2-}$. Chromium(III) is expected to form strong complexes with amines, and can be adsorbed by clay minerals. Chromium may exist in water supplies in both the hexavalent and the trivalent state although the trivalent form rarely occurs in potable water. Chromium is considered nonessential for plants, but an essential trace element for animals. Hexavalent compounds have been shown to be carcinogenic by inhalation and are corrosive to tissue.

The EPA method (3061A) calls for an alkaline digestion procedure for Cr(VI), which is distinctly different from the acid digestion described above for other metals. With this method, an alkaline solution containing 0.28 M Na_2CO_3/0.5 M NaOH is mixed with the sample and heated at 90–95°C for 60 min. This treatment extracts/dissolves the Cr(VI) from soluble (e.g., $K_2Cr_2O_7$), adsorbed, and precipitated forms (e.g., $PbCrO_4$) of Cr compounds in soils, sludges, sediments, and similar waste materials. The pH of the digestate must be carefully adjusted during the digestion to maintain the integrity of the Cr species (i.e., avoid reduction of Cr(VI) or the oxidation of Cr(III)).

If samples contain soluble Cr(III), its oxidation must be suppressed by the addition of Mg^{2+} in a phosphate buffer system. The concentration of soluble Cr(III) can also be measured by extracting the sample with DI water and analyzing the resultant leachate for both Cr(VI) and total Cr. The difference between the two values approximates Cr(III).

2. Mercury (Hg)

Mercury (Hg) is a nonessential element for both animals and humans and all forms of inorganic and organic Hg are considered to be toxic. The most notorious is, perhaps, the methyl-mercury because of its high toxicity and bioaccumulation in the food chain in aquatic systems. The digestion method discussed here is for the *total mercury* using H_2SO_4–HNO_3 in the presence of $KMnO_4$–$K_2S_2O_8$ (potassium permanganate–potassium persulfate). Both $KMnO_4$ and $K_2S_2O_8$ are oxidizing agents where $KMnO_4$ is used to maintain the oxidizing condition for Hg^{2+}. The excess $KMnO_4$ is removed with hydroxylamine hydrochloride (a weak reducing agent).

Mercury is a unique element from the analytical standpoint because this is the only metal in an elemental state that is a volatile liquid at room temperature, and can be purged directly from the sample for analysis by a specific atomic absorption called cold vapor atomic absorption spectroscopy (Chapter 9). Because of this, the acid digested sample containing Hg^{2+} is further reacted with stannous chloride ($SnCl_2$) to produce free Hg (Hg^0):

$$Hg^{2+} + Sn^{2+} \rightarrow Hg^0 + Sn^{4+} \tag{7.7}$$

3. Arsenic (As) and Selenium (Se)

Species of As and Se Both arsenic (As) and selenium (Se) are metalloids. They are nonessential for plants but As is essential for several animal species and Se is essential to most animals. Arsenic, in inorganic form, has two oxidation states, i.e., arsenate, designated as As (V), and the trivalent arsenite, designated as As (III). Arsenite is many times more toxic than arsenate. The predominant As (V) species in water is $H_2AsO_4^-$ at pH 3-7 and $HAsO_4^{2-}$ at pH 7-11. Under reducing conditions, the major As (III) species are $HAsO_2$ (aq) or H_3AsO_3. Organic As can also arise from industrial discharges, pesticides, and biological actions of inorganic As. Unpolluted fresh water normally does not contain organic arsenic compounds, but may contain inorganic arsenate and arsenite.

Inorganic selenium (Se) exists predominately as selenate ion (SeO_4^{2-}) designated as Se(VI), and selenite ion (SeO_3^{2-}) designated as Se (IV). Other common aqueous species include Se^{2-}, HSe^-, and Se^0. Se derived from microbial degradation of seleniferous organic matter includes selenite, selenate, and the volatile organic compounds dimethylselenide and dimethyldiselenide. Nonvolatile organic selenium compounds may be released into water by microbial processes.

Digestion for total As and Se Total arsenic or selenium can be measured by one of the following two digestion methods:

- *Total recoverable As/Se:* H_2SO_4–HNO_3–$HClO_4$ digestion is effective in destroying organics and most particulates in untreated wastewaters or solid samples, but does not convert all organic aresenics to As(V). Organic selenium compounds have rarely been observed in water.

- *Total As/Se:* H_2SO_4-$K_2S_2O_8$ digestion is effective in converting organic As to As(V) and organic Se to Se (VI) in potable and surface waters and in most wastewaters, the contents of As/Se thus measured are termed total As/Se.

The digestion for As is performed in a specific container called the Berzelius beaker, which allows for accepting acid digestion, adding liquid sodium borohydride reagents, purging with N_2 gas and stirring. The hydride (arsine, AsH_3) generated is transported to an atomic absorption atomizer for analysis (See Sec. 9.4). With EPA 270.3, Se in the sample is reduced from the oxidation state of +6 to +4 by the addition of $SnCl_2$. Zn is added to the acidified sample, producing H_2 and converting Se into the hydride (hydrogen selenide, SeH_2). This gaseous SeH_2 is swept into an argon–hydrogen flame of an atomic absorption spectrometer.

Measurement of As species   By incorporating proper sample preparation into an analytical scheme, it is possible to determine arsenite, arsenate, and total arsenic.

- *Arsenite:* Arsenite can be reduced selectively by an aqueous sodium borohydride ($NaBH_4$) solution to arsine (AsH_3) at pH 6.

$$3\,BH_4^- + 3\,H^+ + 4\,H_3AsO_3 \rightarrow 3\,H_3BO_4 + 4\,AsH_3 \uparrow + 3\,H_2O \qquad (7.8)$$

  Arsenate, methylarsonic acid, and dimethylarsenic acid are not reduced under these conditions. AsH_3 is swept by a stream of oxygen-free N_2 from the reduction vessel through a scrubber containing lead acetate $[(Pb(CH_3COO)_2]$ solution into an absorber tube containing silver diethyl-dithiocarbamate $[AgSCSN(C_2H_5)_2]$ and morpholine dissolved in chloroform, forming a red color compound that can be measured colorimetrically at 520 nm.

- *Arsenate:* The same sample described above is then acidified with hydrochloric acid (HCl) and another portion of sodium borohydride solution is added. The arsine formed corresponds to the amount of arsenate present in the sample.

- *Total inorganic arsenic:* A portion of subsample is reduced at a pH of about one followed by the same procedure as described above assuming the absence of methylarsenic compounds in the sample.

Measurement of Se species   The digestion described below has an increasing order of rigorousness to suit for the analysis of various Se species:

- *Persulfate ($S_2O_8^{2-}$) digestion:* A small amount of ammonium or potassium persulfate ($K_2S_2O_8$) is added to the mixture of sample and HCl to remove interferences from reducing agents and to oxidize relatively labile organic selenium compounds such as selenoamino acids and methanoseleninic acid. This method is adequate for most filtered groundwater, drinking water, and surface water.

- *Alkaline hydrogen peroxide (H_2O_2) digestion:* The H_2O_2 digestion is required if organic selenium compounds are present. This digestion is used to remove all reducing agents that might interfere, and to fully oxidize organic selenium to Se(VI). This method determines total selenium in unfiltered water samples where particulate Se is present.

- *Permanganate (MnO_4^-) digestion:* This digestion utilizes $KMnO_4$ to oxidize Se and remove interfering organic compounds. HCl digestion is included here because it is conveniently performed in the same reaction vial. The method has a good recovery even with heavily contaminated water samples that contain organic selenium compounds, dissolved organic matter, and visible suspended material.

7.3 EXTRACTION FOR SVOC AND NON-VOC FROM LIQUID OR SOLID SAMPLES

The primary sample preparation method for organic compounds is the extraction procedure (refer to Table 5.3 for a list of EPA methods). As most extraction procedures are performed in open containers, it can only be used for semi-volatile organic compounds (SVOCs) and nonvolatile compounds. Seven extraction methods are described below, including liquid–liquid (L–L) extraction using a separatory funnel or continuous device, solid phase extraction (SPE), solid phase microextraction (SPME), Soxhlet and automatic Soxhlet extraction, ultrasonic extraction, pressured fluid extraction (PFE; also known as accelerated solvent extraction, ASE), and supercritical fluid extraction (SFE). The focus here is to introduce the operational principles and their environmental applications rather than the detailed apparatus and how to use them.

7.3.1 Separatory Funnel and Continuous Liquid–Liquid Extraction (LLE)

Separatory funnel liquid–liquid extraction relies on the partitioning of the analytes between the water phase and an organic phase. Processing a large number of samples using a separatory funnel (Fig. 7.4a) is very labor intensive. The continuous liquid–liquid extraction automates the extraction process somewhat (Fig. 7.4b and c). The analyst is freed from manually shaking the phases so that simultaneous extraction of multiple samples is possible.

Despite the variations among different extraction devices, several general principles need to be appreciated for liquid–liquid extraction. These include the following: (a) Two phases must be immiscible and the density of two phases must be different; (b) The analyte must not be volatile; (c) The analyte must favorably distribute one phase (solvent) over the other (water).

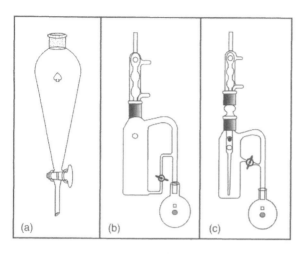

Figure 7.4   Liquid–liquid (L–L) extraction: (a) Conventional separatory funnel L–L extraction (Courtesy of Kimble Glass Inc.) (b) Continuous L–L, heavier than water extraction, and (c) Continuous L–L, lighter than water extraction (Courtesy of Kontes Glass Company)

Solvent Density and Immiscibility   The density difference is needed for a phase separation. The need for immiscibility of two phases is also apparent. For instance, water miscible solvents such as acetone, methanol, ethanol cannot be used for L–L extraction. Figure 7.5 is a solvent miscibility chart. One should be able to use this chart to explain which solvents can or cannot be used in L–L extraction if, for example, aqueous phase samples are to be extracted.

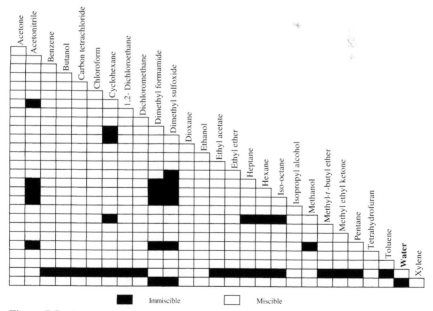

Figure 7.5   Solvent miscibility chart (Sadek, PC, 2002, Reproduced with Permission, John Wiley & Sons Limited)

Analyte Volatility   If an analyte is volatile, then clearly L–L extraction cannot be used. The separatory funnel and continuous extraction can only be used for semivolatile or nonvolatile compounds. In Chapter 2, we gave an approximate definition of what are VOCs and SVOCs. The volatility of chemical from a liquid can be evaluated from its *Henry's law constant (H)*. Mackay and Shiu (1981) provided the following useful guidelines for organic solutes in water.

- **Highly volatile:** volatilization is rapid if $H > 10^{-3}$ atm·m^3/mol
- **Volatile:** volatilization is significant if H is in the range of 10^{-3} to 10^{-5} atm·m^3/mol
- **Semivolatile:** volatilization is slow if H is in the range of 10^{-5}–3×10^{-7} atm·m^3/mol
- **Nonvolatile:** volatilization is not important if $H < 3 \times 10^{-7}$ atm·m^3/mol

Henry's constant used in defining volatility should not be confused with *vapor pressure*. Vapor pressure relates to the volatility from the pure substance into the atmosphere, whereas H refers to the volatility from liquid to the air. A compound with high vapor pressure (such as ethanol) could have very low volatility if ethanol is present in water, but it evaporates quickly when a drop of ethanol is spilled on to a table.

Analyte Partition Coefficient   The partition coefficient of an analyte is defined as: $D = C_s/C_w$, where C_s and C_w are the equilibrium concentration in solvent and water, respectively. The extraction efficiency (E) is independent of the initial analyte concentration, but is a function of the partition coefficient and water-to-solvent ratio (V_w/V_s):

$$E = \frac{C_s V_s}{C_s V_s + C_w V_w} = \frac{1}{1 + \dfrac{V_w}{V_s}\dfrac{1}{D}} \tag{7.9}$$

The residual concentration of analyte in the aqueous phase after a single extraction can be expressed as:

$$C_w^1 = C_w^0 \frac{V_w}{DV_s + V_w} = C_w^0 \frac{1}{D\dfrac{V_s}{V_w} + 1} \tag{7.10}$$

For *n*th successive extractions, the residual concentration of an analyte in an aqueous phase is:

$$C_w^n = C_w^0 \left[\frac{V_w}{DV_s + V_w}\right]^n = C_w^0 \left[\frac{1}{D\dfrac{V_s}{V_w} + 1}\right]^n \tag{7.11}$$

The above equations indicate that extraction efficiency increases with increasing analyte's partition coefficient (D) and with the decreasing water to solvent ratio. We use the example below to further illustrate the general principles of solvent extraction.

EXAMPLE 7.1.   Liquid–liquid Extraction. A 100-mL aqueous sample containing naphthalene is extracted using octanol as the solvent. Naphalene has a low water solubility of 31 mg/L at 20°C, and an octanol-water partition coefficient (K_{ow}) of 1950. *Calculate:*

(a) The extraction efficiency with one 100-mL octanol;

(b) The extraction efficiency with two 50-mLs of octanol.

(c) If the initial concentration of naphthalene is 25 mg/L, what is the residual concentration after four successive extractions using four 25-mLs of octanol?

SOLUTION:

(a) $V_w = V_s = 100$ mL, $D = K_{ow} = 1950$. Use Eq. 7.9, we calculate $E = 1/[1 + (100/100) \times (1/1950)] = 99.95\%$.

(b) For the first 50-mL, we have $E = 1/[1 + (100/50) \times (1/1950)] = 99.90\%$. For the second 50-mL, we have accumulative $E = 99.90\% + (100 - 99.90\%) \times 99.90\% = 99.99\%$, meaning only 0.09% is additionally extracted from the second 50-mL.

(c) $C_w^0 = 25$ mg/L, $V_s = 25$, $V_w = 100$, $n = 4$, we use Eq. 7.11 to calculate $C_w^4 = 25 \times (1/(1950 \times 25/100 + 1)^4 = 4.39 \times 10^{-10}$ mg/L.

Further generalizations about L–L extraction can be inferred from the above example: (a) Given a total volume of extracting solvent, extraction efficiency increases as the number of extraction increases. The extracted amount acquired by 2nd and subsequent extractions might be negligible if the compound is very hydrophobic (large D). (b) For one-step L–L extraction, D must be large, that is greater than 10, for a quantitative recovery ($>99\%$) of the analyte in the solvent phase. This is the consequence of the phase ratio within the practical range of $0.1 < V_s/V_w < 10$. (c) Normally two to three repeated extractions are required with fresh solvent to achieve quantitative recoveries.

7.3.2   Solid Phase Extraction

Solid phase extraction (SPE) is the technique that retains analyte from a flowing liquid sample on solid sorbent and subsequently recovers analyte by elution from the sorbent. In principle, SPE can be based on a number of solute-sorbent interaction mechanisms. SPE has various phase types, which include reverse phase, normal phase, ion exchange, and adsorption. The reverse phase C_{18} (octadecyl bonded silica) and C_8 (octyl bonded silica) are the most commonly used for the extraction of hydrophobic environmental analytes. Normal phase SPE uses cyanopropyl bonded, diol bonded, or aminopropyl bonded silica. They are used for polar contaminants such as cationic compounds and organic acids. The ionic exchange SPE is based on the electrostatic attraction of a charged group on the compound to a charged group on the sorbent's surface, which often uses quaternary amine, sulfonic acid, or carboxylic acid bonded silica. The adsorption type SPE is based on the interactions of analyte with unmodified materials, such as alumina-, Florisil-, graphitized, and resin-based packing. Figure 7.6 below shows three types of SPEs based on nonpolar, polar, and electrostatic interactions.

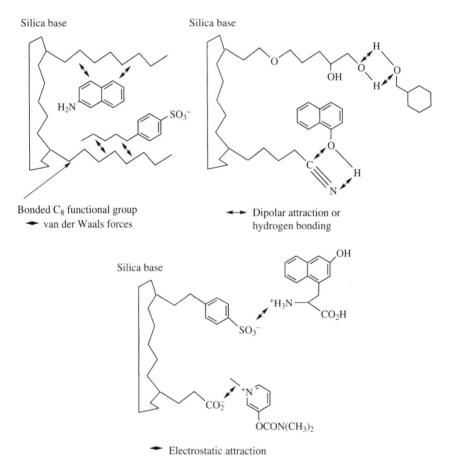

Figure 7.6   Solid-phase extraction using nonpolar, polar, and electrostatic interactions (Reprinted with permission from *American Laboratory,* 24(12):37–42. © 1992 by International Scientific Communications Inc.)

SPE is performed in a cartridge or disc, which are packed with silica particles bonded with organic coating. The SPE device can be as simple as a syringe barrel (cartridge) that can be operated manually (Fig. 7.7) with one sample at a time, or the commercial vacuum manifold with multiple cartridge units (Fig. 7.8). For all SPEs, the operation is a five-step process as shown in Fig. 7.7. SPE has the following advantage compared to the conventional L–L extraction: (a) Smaller volume of solvent used in SPE; (b) Emulsions are no longer produced;

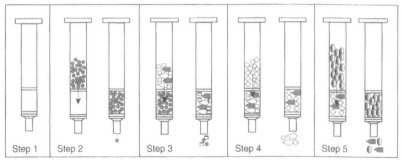

Figure 7.7   Five steps of solid phase extraction: (a) Select the proper SPE tube or disk from various commercially available SPEs, (b) Condition the SPE tube or disk, (c) Add the sample; (d) Wash the packing; (e) Elute the compounds of interest (Courtesy of Supelco)

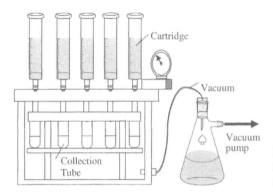

Figure 7.8   Vacuum manifold for solid phase extraction (SPE) of multiple cartridge units

(c) Automation is possible for SPE, resulting in reduced analytical cost, time, and labor.

7.3.3   Solid Phase Microextraction

Solid phase microextraction (SPME) is an "extraction" technique but, in fact, it is based on a solvent-free sorption-desorption process. SPME typically is a fused silica fiber coated with a solid adsorbent or an immobilized polymer, or a combination of the two. The fiber has a typical dimension of $1 \, cm \times 110 \, \mu m$, which is normally bonded to a stainless steel plunger and installed in a holder that looks like a modified microliter syringe. The fiber is introduced into the liquid sample or headspace, and organic analytes adsorb in the phase and establish equilibrium. The analytes are then desorbed from the fiber to a capillary GC column in the heated injection port. SPME has also been coupled with HPLC, expanding the application to nonvolatile compounds such as surfactants and pharmaceutical chemicals. SPME is relatively a new sample preparation technique, but its applications have been promising for the analysis of SVOCs as well as VOCs. The unique features of SPME make it an ideal tool by combining sampling, extraction, concentration, and injection into a single process. SPME can be used directly for both liquid and gas samples (Fig. 7.9).

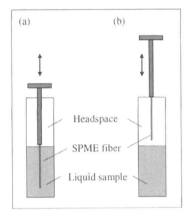

Figure 7.9   Two types of SPME applications in (a) liquid samples, and (b) headspace samples

The main principle of SPME operation is based on the partitioning of analytes between an aqueous (vapor) sample and a stationary phase on the fiber (f). The amount of analyte adsorbed by the coating at equilibrium is described by:

$$m = \frac{K_{fs} V_f C_0 V_s}{K_{fs} V_f + V_s} \tag{7.12}$$

where m is the mass of analyte adsorbed by coating, C_0 the initial analyte concentration in sample, K_{fs} the partition coefficient for analyte between coating and sample matrix, V_f the volume of coating, and V_s the volume of sample. Eq. 7.12 indicates that the adsorption is proportional to the partition coefficient (hence the hydrophobicity of analyte), and the initial analyte concentration. The adsorption is also related to the sample volume and coating volume. However, if sample volume (V_s) is very large, this equation becomes:

$$m = K_{fs} V_f C_0 \tag{7.13}$$

The above equation indicates the independence of adsorption on sample volume, which is an added advantage for SPME, because SPME can be simply exposed to air or dipped into water without measuring the sample volume.

7.3.4   Soxhlet and Automatic Soxhlet Extraction (Soxtec)

Soxhlet extraction was developed in 1879 by Franz von Soxhlet initially for the extraction of lipid from a solid test material. It can be used for any solid samples and its environmental applications are mainly for the extraction of SVOCs from solid samples, such as soil, sludge, sediment, and solid waste. The Soxhlet apparatus shown in Fig. 7.10 is a very thorough extraction system. Because of this, it is used as the benchmark to validate any new extraction technique by comparing the extraction efficiency.

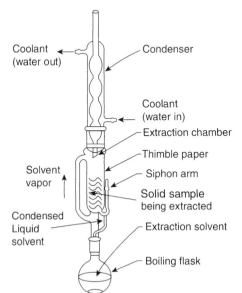

Coolant (water out)

Condenser

Coolant (water in)

Extraction chamber

Thimble paper

Solvent vapor

Siphon arm

Solid sample being extracted

Condensed Liquid solvent

Extraction solvent

Boiling flask

Figure 7.10   A schematic representation of a Soxhlet extractor

In Soxhlet extraction, dry solid sample is first placed inside a permeable cellulose "thimble", which is loaded into the Soxhlet extractor. Extraction solvent in the round bottomed flask is heated to boiling. The vapors rise through the outer chamber and into the condenser, then condense and drip down onto the extracting sample. The extraction chamber containing the sample slowly fills with warm solvent until, when it is almost full, it is emptied by a siphon arm (tube). The flushed solvent returns to the flask taking the extracted compounds with it. The solvent is redistilled from the solution in the flask and condenses in the chamber, repeating the extraction with fresh solvent. The process can be repeated as many times as necessary. Each time the process is repeated, the solution in the flask becomes more concentrated because more is being extracted from the solid mass. The Soxhlet extraction is usually completed when the solution in the Soxhlet chamber is the same color as the pure solvent. This means that nothing more is being extracted from the sample by the solvent.

Soxhlet is a rugged and well-established technique, but the major disadvantages are its time-consuming procedure (6–48 h) and a large solvent usage which often requires a concentration step to evaporate solvent. The U.S. EPA approved an automated Soxhlet extraction called *Soxtec* (Fig. 7.11) in 1994. The automated Soxtec is carried out in three stages: boiling, rinsing, and solvent recovery. In the first stage, a thimble containing the sample is immersed in the boiling solvent for about

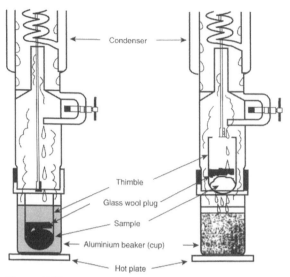

Figure 7.11   An automated Soxhlet extraction apparatus (Soxtec) (U.S. EPA, 1994)

60 min. Extraction is faster because of the rigorous contact between the boiling solvent and the sample. In the second stage, the sample-loaded thimble is lifted above the boiling solvent. The condensed solvent drips into the sample, extracts the organics, and falls back into the solvent reservoir. This takes about 60 min. The third stage is a concentration step for 10–20 min, the solvent is evaporated to 1–2 mL, as would occur in a K-D concentrator (Chapter 6). Since the concentration step is integrated in Soxtec, the extract is ready for clean-up and analysis.

7.3.5   Ultrasonic Extraction

A solid sample of 30-g or 2-g, depending on whether the concentration of an individual organic component is below or above 20 mg/kg, is mixed with anhydrous sodium sulfate (Na_2SO_4), then extracted three times with one of the variety of solvent mixtures (1:1 acetone:CH_2Cl_2; 1:1 aetone:hexane). Anhydrous Na_2SO_4 is needed to sorb the water and form the free-flowing powder. The extraction uses an ultrasonic disruptor horn of at least 300 watts power for agitation of the mixture to ensure its intimate contact between solvent and matrix. The *ultrasonic extraction* needs only a few minutes, but extraction efficiency is usually low. The U.S. EPA recommends this method (3550B) for extracting nonvolatile and semivolatile organic compounds from soils, sludges, and wastes. This method is not appropriate for applications where high extraction efficiencies of analytes at very low concentration are necessary. In addition, there are also concerns that the ultrasonic energy may lead to breakdown of some organophosphorous compounds. As a result, this extraction should not be used for some organophosphorous compounds without an extensive validation.

7.3.6 Pressured Fluid Extraction

The pressured fluid extraction (PFE), or more commonly called accelerated solvent extraction (ASE), uses elevated temperature (100–180°C) and pressure (1500–2000 psi) to achieve analyte recoveries equivalent to those from Soxhlet extraction, using less solvent and taking significantly less time than the Soxhlet procedure. The fully automated device has been commercially available since 1995 by Dionex Corp., USA. The EPA has validated a group of water insoluble or slightly water soluble SVOCs using a recommended mixture of solvents (acetone, CH_2Cl_2, hexane). A total of 10–30 g dry (air drying or by mixing with anhydrous Na_2SO_4) and ground solid samples of small particle size in the range of 100–200 mesh (150–75 μm) are loaded into the sealed extraction cell (11–33 mL). The cell is heated to the extraction temperature, pressurized with the appropriate solvent system with a pump, and extracted for five minutes, and the extract is then collected.

The mechanisms for the enhanced extraction efficiency at high temperature and pressure in the PFE system compared to extractions at or near room temperature and atmospheric pressure are suggested to be the result of the increased analyte solubility and mass transfer effects, and the disruption of surface equilibria. At high temperature, the solubility and diffusion rate of analytes are increased whereas solvent viscosity is decreased. High pressure keeps solvent liquefied above its boiling point and allows solvent to penetrate matrix readily. Elevated temperature and pressure also disrupt the analyte-matrix bonding such as van der Waals' forces, hydrogen bonding, and dipole attractions of the solute molecules and active sites on the matrix (Dean, 1998).

7.3.7 Supercritical Fluid Extraction

Unlike all other extractions using solvents, the extracting solvent in supercritical fluid extraction (SFE) is CO_2 in its supercritical fluid (SCF) state. A SCF is defined as a substance above its critical temperature (T_c) and critical pressure (P_c). The critical point represents the highest temperature and pressure at which the substance can exist as a vapor and liquid in equilibrium. This formation of SCF can be illustrated by the phase diagram of pure CO_2 (Fig. 7.12). As can be seen, CO_2 in its solid, liquid, and gas phases can coexist at the triple point. The gas-liquid coexistence curve is known as the boiling curve. If we move upward along the boiling curve, increasing both T and P, then the liquid becomes less dense as a result of thermal expansion and the gas becomes more dense as the pressure rises. Eventually, the densities of the two phases converge and become identical, the distinction between gas and liquid disappears, and the boiling curve comes to an end at the critical point ($T_c = 31.1°C$; $P_c = 74.8$ atm for CO_2).

The CO_2 in this SCF state is characterized by physical and thermal properties that are between the pure liquid and gas form of CO_2. SCF has a gas-like high mass transfer coefficients and a liquid-like high solvent property. Besides, SCF has low viscosity and almost zero surface tension. The high diffusivity of SCF makes it possible to readily penetrate porous and fibrous solids. Consequently, SCF can offer good extraction efficiency. The EPA recommends two SFE protocols; one for the extraction of total recoverable petroleum hydrocarbons (TRPHs) (Method 3560) and the other for the extraction of polynuclear aromatic hydrocarbons (PAHs) (Method 3561).

A supercritical-fluid extractor (Fig. 7.13) consists of a tank of the mobile phase (CO_2), a pump to pressurize CO_2, an oven containing the extraction vessel, a restrictor to maintain a high pressure in the extraction line, and a trapping vessel. Analytes are trapped by letting the solute-containing supercritical fluid decompress into an empty vial, through a solvent, or onto a solid sorbent material. After depressurization, CO_2 evolves as a gas.

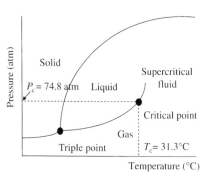

Figure 7.12   Schematic phase diagram of carbon dioxide (CO_2). P_c = critical pressure, T_c = critical temperature

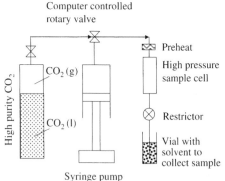

Figure 7.13   Schematic diagram of supercritical fluid extraction (SFE)

7.3.8 Comparison and Selection of Organic Extraction Methods

As described above, there are many extraction methods that appear to have some advantages over the conventional L–L extraction (for liquid samples) and the Soxhlet extraction (for solid samples). Many times, however, the extraction efficiencies are compromised by these newly developed methods–faster, greener but with a lower efficiency. A good practice is always to validate the method by performing percentage recovery with a spiked analyte in the particular sample matrix. Table 7.1 is a brief comparison among selected sample preparation techniques for SVOCs as well as VOCs. The sample preparation techniques for VOCs will be introduced in Section 7.6, but are included in Table 7.1 for a comparison purpose.

In selecting the methods listed in Table 7.1, first keep in mind what compounds (VOC vs. SVOC) and what sample matrix (liquid vs. solid) can be used for a particular method. One should note that although the classical methods such as L–L extraction and Soxhlet extraction are labor-intensive and relatively slow, they are readily available in average laboratories. The automatic or instrument methods (purge-and-trap, SFE, and PFE) are faster and greener,

Table 7.1 Comparison of sample preparation techniques for SVOCs and VOCs

Extraction method	Application	Cost	Extraction time	Solvent usage	Simplicity	EPA method
Purge & trap	VOCs (L/S)	High	30 min	None	No	5030, 5035
Headspace	VOCs (L/S)	Low	30 min	None	Yes	3810, 5021
LLE	SVOCs, NVOCs (L)	Low	1 h	500 mL	Yes	3510/3520
SPE	SVOCs, NVOCs (L)	Medium	30 min	100 mL	Yes	3535
SPME	VOCs, SVOCs, NVOCs (L)	Low	30 min	None	Yes	None
Soxhlet/ Soxtec	SVOCs, NVOCs (S)	Low	4–24 h	250/50 mL	Yes	3540/3541
Ultrasonic	SVOCs, NVOCs (S)	Medium	10 min	200 mL	Yes	3550
SFE	SVOCs, NVOCs (S/L)	High	30 min	20 mL	No	3560/3561
PFE (ASE)	SVOCs, NVOCs (S)	High	15 min	25 mL	No	3545

LLE: liquid–liquid extraction; SPE: solid phase extraction; SPME: solid phase microextraction; SFE: supercritical fluid extraction; PFE: pressured fluid extraction; ASE: accelerated solvent extraction. VOCs: volatile organic compounds; SVOCs: semivolatile organic compounds; NVOCs: nonvolatile organic compounds. L: liquid samples; S: solid samples.

but many of them come with a higher price tag. In laboratories processing a large number of samples, these automated methods may be preferred because of a reduction in labors.

7.4 POST-EXTRACTION CLEAN-UP OF ORGANIC COMPOUNDS

"Cleanup" is to remove interfering chemicals from the sample while keeping target analyte(s) in the sample matrices prior to instrumental analysis. Simply put, cleanup is to remove as much garbage as possible from complex sample matrices. Cleanup is critical to the analysis of trace analyte of interest (ppm – ppt or lower), because the analyte is often masked by high background concentration of various types of organics and the removal is essential for further analyses by instruments. Cleanup will also avoid the rapid deterioration of expensive capillary GC columns and instrument downtime caused by cleaning and rebuilding of detectors and ion sources. Many of the cleanup procedures are performed after extraction.

7.4.1 Theories and Operation Principles of Various Clean-up Methods

There are several cleanup techniques available for environmental analysis. They can be used individually or in various combinations, depending on the extent and nature of the co-extractive chemicals.

Alumina Column   Alumina is a highly porous granular aluminum oxide (Al_2O_3). It is prepared from $Al(OH)_3$, which is dehydrated and calcined at 900°C with a layer of aluminum oxycarbonate, $[Al_2(OH)_5]_2CO_3$. This allows the pH adjustment to the acidic (pH 4-5), neutral (pH 6-8), or basic (pH 9-10) grades for a particular application. The alumina solid is packed into a column topped with a water-absorbing substance over which the sample is eluted with a suitable solvent, which leaves interference on the column. Alumina is used to separate analytes from interfering compounds of a different chemical polarity.

Florisil Column   Florisil is a registered trade name of U.S. Silica Co. It is an activated form of magnesium silicate with basic properties. Florisil is more acidic and milder (with regard to compound decomposition) than alumina and silica gel. Florisil has been used for the clean-up of pesticide residues and other chlorinated hydrocarbons, for the separation of nitrogen compounds from hydrocarbons, and for the separation of aromatic compounds from aliphatic-aromatic mixtures. Florisil is also good for separations with steroids, esters, ketones, glycerides, alkaloids, and some carbohydrates.

Silica Gel Column   Silica gel is a regenerative adsorbent of silica with weak acidic amorphous silicon oxide. It is the precipitated form of silicic acid (H_2SiO_3), formed from the addition of H_2SO_4 to sodium silicate. It forms very strong hydrogen bonds to polar materials and can lead to analyte decomposition. It is somewhat soluble in methanol, which should never be used as an elution solvent. Silica gel can also remove slightly polar substances from hexane solutions and serves to isolate the strictly petroleum hydrocarbon analytes.

Gel Permeation Chromatography (GPC)   GPC is a size-based separation using a hydrophobic gel—a cross-linked divinylbenzene-styrene copolymer. Organic solvents are used to selectively pass large macromolecules, such as protein, phospholipids, resins, lignins, fuvic, and humic acids, whereas smaller molecules of target analytes are retained in the pores of gel. Methylene chloride is then used to displace the analytes. The GPC is the most universal clean-up technique to remove high molecular weight and high boiling point material from matrices for the analysis of a broad range of SVOCs and pesticides.

Acid-base Partition Clean-up   Unlike all the above methods, this is a liquid–liquid partition clean-up to separate acid analytes (e.g., organic acids and phenols) from base/neutral analytes (e.g., amines, aromatic hydrocarbons, and halogenated organic compounds), using pH adjustment in a liquid–liquid extraction. In a separatory funnel extraction, the sample is first adjusted with NaOH to pH > 12 and then extracted with CH_2Cl_2, referred to as the *base/neutral fraction*. This sample is then adjusted to pH < 2 with H_2SO_4 and extracted with CH_2Cl_2 to obtain the *acid fraction* (see also Section 2.1.2).

Sulfur Clean-up Elemental sulfur is a common contaminant that can be easily shown on chromatography as a semivolatile. It is removed by the agitation with powdered Cu, mercury metal, or tetrabutyl-ammonium sulfite $[CH_3(CH_2)_3]_4NHSO_4$.

Sulfuric Acid/Permanganate Clean-up This method uses concentrated H_2SO_4 and 5% $KMnO_4$. It is a very rigorous clean-up that will remove most interference except PCBs, chlorinated benzenes, and chlorinated naphthalenes. This method cannot be used to clean-up extracts for other target analytes, as it will destroy most organic chemicals including several pesticides. A few chlorinated hydrocarbons will survive.

7.4.2 Recommended Clean-up Method for Selected Compounds

The clean-up methods specified in the EPA SW-846 is summarized in Table 7.2. It is noted that most of the soil, sediment, and waste samples require clean-up. Four major mechanisms are also listed in this table, i.e., adsorption, size separation, partitioning, and oxidation-reduction reaction.

Table 7.2 Selection of cleanup methods and their environmental applications

Purpose of cleanup	Cleanup type (mechanism)	Method no.
Analysis of phthalate esters (neutral alumina) and nitrosamines (basic alumina)	Alumina (adsorption)	EPA 3610
Separation of petroleum wastes into aliphatic, aromatic, and polar fractions (neutral alumina)	Alumina (adsorption)	EPA 3611
Analysis of phthalate esters, nitrosamines, organochlorine pesticides, PCBs, chlorinated hydrocarbons, organophosphorus pesticides	Florisil (adsorption)	EPA 3620
Analysis of PAHs, PCBs, and derivatized phenol	Silica gel (adsorption)	EPA 3630
Analysis of phenols, phthalate esters, nitrosamines, organochlorine pesticides, PCBs, nitroaromatics, cyclic ketones, PAHs, chlorinated hydrocarbons, organophosphorus pesticides, priority pollutant SVOCs	Gel-permeation (size-separation)	EPA 3640
Analysis of phenols, priority pollutant semivolatile	Acid-base partition (partitioning)	EPA 3650
Analysis of organochlorine pesticides, PCBs, priority pollutant SVOCs	Sulfur clean-up (oxidation-reduction)	EPA 3660
Analysis of PCBs, chlorinated benzenes and chlorinated naphthalenes	H_2SO_4-permanganate (oxidation-reduction)	EPA 3665

7.5 DERIVATIZATION OF NON-VOC FOR GAS PHASE ANALYSIS

Many compounds of environmental interest, particularly high molecular weight compounds or compounds containing polar functional groups, are difficult to analyze by GC because they are not sufficiently volatile, tail badly, are strongly attracted to the stationary phase, thermally unstable or even decomposed. In all these cases, chemical derivatization is needed prior to analysis. *Derivatization* is a technique used in chemistry that transforms a chemical compound into a product of similar chemical structure, called a derivative. Derivatization in chemical analysis is used to do the following: (a) increase the volatility and decrease the polarity of compounds; (b) increase thermal stability to protect the analyte from thermal degradation; (c) increase detector response by incorporating functional groups, which lead to higher detector signals, for example, CF_3 groups for electron capture detectors; (d) improve separation and reduce tailing. Common derivatization methods can be classified into four groups depending on the type of reaction applied:

Silylation: Silylation replaces active hydrogens with a trimethylsilyl (TMS) group. The trimethylsilyl group consists of three methyl groups bonded to a silicon atom $[-Si(CH_3)_3]$. The silyl derivatives produced are more volatile and more thermally stable. The ease of reactivities of the functional group toward silylation follows the order: alcohol > phenol > carboxyl > amine > amide > hydroxyl.

$$R-AH + R'-\overset{\overset{\displaystyle OSi(CH_3)_3}{|}}{C}=N-Si(CH_3)_3 \longrightarrow R-A-Si(CH_3)_3 + R'-\overset{\overset{\displaystyle OSi(CH_3)_3}{|}}{C}=NH \quad (7.14)$$

Acylation: This derivatization adds the acryl group (RCO—) and targets highly polar, multifunctional compounds, such as carbohydrates and amino acids, and converts compounds with active hydrogens into esters, thioesters, and amides. It reduces the polarity of amino, hydroxyl, and thio groups and adds halogenated functionalities (electron-capturing groups) for GC-ECD to be detected.

Alkylation: It reduces molecular polarity by replacing active hydrogens with an alkyl group (e.g., methyl, ethyl). Through alkylation, the acidic hydrogens in compounds such as carboxylic acids and phenols form esters, ethers, and amide. Alkyl esters have excellent stability and can be isolated and stored for a long period of time.

Esterification: In the presence of a catalyst, an acid reacts with an alcohol to form an ester with a lower boiling point. An example is:

$$C_5H_{11}COOH \text{ (b.p. } 187°C) \rightarrow C_5H_{11}COOCH_3 \text{ (b.p. } 127.3°C) \quad (7.15)$$

7.6   SAMPLE PREPARATION FOR VOC, AIR AND STACK GAS EMISSION

The extraction procedures discussed in Section 7.3 do not apply to VOCs (SPME is an exception since it can be directly used to sample VOCs in headspace). A list of EPA sample preparation methods for VOCs was provided in Chapter 5 (Table 5.4) and the details are described below. As can be seen from the following discussions, methods for VOCs can be divided into two groups depending on the types of VOCs, that is, purgeable VOCs and nonpurgeable VOCs (water soluble VOCs). The later is of minor importance in routine analysis, and will be briefly introduced.

7.6.1   Dynamic Headspace Extraction (Purge-and-Trap)

In the purge-and-trap (P&T), also called *dynamic headspace extraction*, separation is carried out by purging the sample, sometimes done while heating, with an inert gas such as He or N_2 and trapping the volatile materials in a sorbent column (Tenax, silica, charcoal). The trap is designed for rapid heating so that it can be desorbed directly into a GC column (Fig. 7.14a). The purge-and-trap is applicable for compounds with boiling points below 200°C and are insoluble or slightly soluble in water. The purge efficiency can be improved for water soluble analytes, for example, ketones and alcohols, when purging at an elevated temperature of 80°C as compared to 20–40°C. Problems are often encountered with the low purging efficiency of methyl *t*-butyl ether (MTBE) and related fuel oxygenated compounds such as ethyl *t*-butyl ether (ETBE), *t*-amyl methyl ether (TAME), diisopropyl ether (DIPE), *t*-amyl ethyl ether (TAEE), *t*-amyl alcohol (TAA), and *t*-butyl alcohol (TBA). These compounds are classified as volatiles

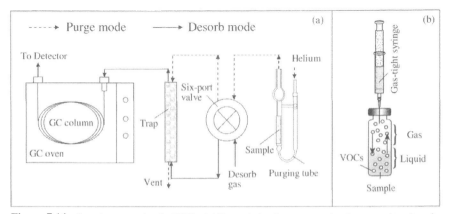

Figure 7.14   Sample preparation for VOCs: (a) Dynamic headspace extraction (purge-and-trap), and (b) Static headspace extraction (SHE)

based on their boiling points and vapor pressure, but are relatively water soluble. The EPA recommends 80°C purge conditions along with a specific capillary column (DB-WAX) for improved recoveries. As shown in Figure 7.14, P&T is often coupled with GC and programmed to operate first in the purge mode (10–12 min) and then in desorbing mode (1–2 min).

7.6.2 Static Headspace Extraction

In the static headspace extraction (SHE), two phases are in equilibrium in a sealed vial. In Figure 7.14b, the gas phase (g) is commonly referred to as the headspace and lies above the condensed sample phase (s). The sample phase (solid or liquid) initially contains the compound(s) of interest. Once the sample phase is introduced into the vial and the vial is sealed, volatile components diffuse into the gas phase until the headspace has reached a state of equilibrium as depicted by the arrows. The sample is then taken from the headspace either manually or by an automated headspace sampler. The partition coefficient (K) is defined as:

$$K = C_g/C_s \qquad (7.16)$$

where C_g is the concentration of analyte in the gas phase and C_s is the concentration of analyte in sample phase. Compounds that have high K values tend to partition less readily into the gas phase and have relatively low responses and high limits of detection. An example of this would be hexane in water; at 40°C, hexane has a K value of 7.14 in an air-water system. Compounds that have low K values will tend to partition more readily into the gas phase and have relatively high responses and low limits of detection. An example of this would be ethanol in water; at 40°C, ethanol has a K value of 0.00075 in an air-water system. Another parameter related to the partitioning is the phase ratio, defined by:

$$\beta = V_g/V_s \qquad (7.17)$$

where V_s and V_g are the volumes of sample phase and vapor phase, respectively. Lower values of β (i.e. larger sample size) will yield higher responses for volatile compounds. By using the mass balance equation (Eq. 7.18) and combining the partition coefficient (K) and phase ratio (β), the final concentration of volatile compounds (C_g) in the headspace of sample vials can be determined.

$$C_0 V_s = \frac{C_g}{K} V_s + C_g V_g \qquad (7.19)$$

$$C_g = \frac{C_0}{1/K + \beta} \qquad (7.20)$$

where C_g is the concentration of volatile analytes in the gas phase and C_0 is the original concentration of volatile analytes in the sample. Since C_g can be determined from static headspace analysis using GC, the VOCs concentration in the sample is:

$$C_0 = C_g \left(\frac{1}{K} + \beta \right) \tag{7.21}$$

Eq. 7.20 also indicates that a compound with a higher K value (hence more volatile) will result in high vapor phase concentration to be detected, and a low headspace to sample volume ratio (β) will result in higher concentrations of volatile analytes in the gas phase and therefore better sensitivity. This also implies that compound with a very low volatility cannot be analyzed by static headspace analysis.

7.6.3 Azeotropic and Vacuum Distillation

What if a volatile compound is not purgeable? For these compounds, the purge-and-trap method would not be applicable. The *azeotropic distillation* is designed for nonpurgeable, water soluble, and volatile organic compounds. These are mostly small molecules, including alcohol, aldehyde, and ketone as specified in EPA Method 5031 (Table 7.3).

An *azeotrope* is a liquid mixture of two or more substances which behaves like a single substance, in that it boils at a constant temperature and the vapors released have a constant composition. Azeotropic distillation is non-conventional distillation technique, which uses the ability of selected organic compounds to form binary azeotropes with water to facilitate the separation of the compounds from a complex matrix. The azeotropic macrodistillation system typically consists of a Vigreux column and a modified Nielson-Kryger apparatus (Fig. 7.15). The polar VOCs distill into the distillate chamber for 1 h, and are retained there. The condensate overflows back into the pot and contacts the rising steam. The VOCs are stripped by the steam and are recycled back into the distillate chamber.

Another distillation apparatus called *vacuum distillation* (EPA 5032 for details) can be used to separate organic compounds that have a boiling point below 180°C and are insoluble or slightly soluble in water (such as BTEX compounds). In vacuum distillation, sample chamber pressure is reduced using a vacuum pump and remains at approximately 10 torr (vapor pressure of water) as water is removed from the sample. The vapor is passed over a condenser coil chilled to a temperature of $-10°C$

Table 7.3 Volatile and non-purgeable compounds

Acetone	Acetonitrile	Acrylonitrile
Allyl alcohol	1-Butanol	*t*-Butyl alcohol
Crotonaldehyde	1,4-Dioxane	Ethanol
Ethyl acetate	Ethylene oxide	Isobutyl alcohol
Methanol	Methyl ethyl ketone	Methyl isobutyl ketone
n-Nitroso-di-*n*-butylamine	Paraldehyde	2-Pentanone
2-Picoline	1-Propanol	2-Propanol
Propionitrile	Pyridine	*o*-Toluidine

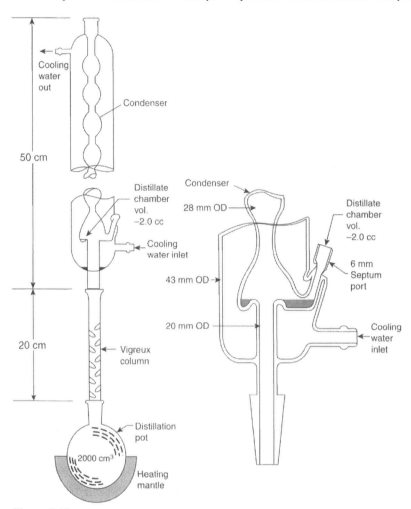

Figure 7.15   An azotropic solvent distiller for volatile and non-purgeable compounds
(U.S. EPA, 1996)

or less, which results in the condensation of water vapor. The uncondensed distillate is cryogenically trapped on a section of 1/8 inch stainless steel tubing chilled to the temperature of liquid nitrogen ($-196°C$). After an appropriate distillation period, the condensate contained in the cryotrap is thermally desorbed and transferred to the gas chromatography using helium gas.

7.6.4   Volatile Organic Sampling Train

The volatile organic sampling train (VOST) is used to collect volatile principal organic hazardous constituents (POHCs, boiling point less than $100°C$) from the gas effluents of hazardous waste incinerators. The VOST consists of mainly a sorbent cartridge containing Tenax, an empty impinger for condensate removal, a second water-cooled glass condenser, a second sorbent cartridge containing Tenax and petroleum-based charcoal, a silica gel drying tube, a calibrated rotometer, a sampling pump, and a dry gas meter. These sorbent tubes are then thermally desorbed and purged with organic-free helium through pre-purged organic-free reagent water and trapped on an analytical sorbent trap in a purge-and-trap unit (Figure 7.14a). After desorption, the analytical sorbent trap is heated rapidly and the gas flow from the analytical trap is directed to the head of a wide-bore column under subambient conditions for GC analysis. Details of VOST and the analysis of cartridge from VOST can be found in EPA 0030 and 5041.

REFERENCES

APHA (1998), *Standard Methods for the Examination of Water and Wastewater*, 20th Edition Washington, DC.

CHRISTIAN GD (2004), *Analytical Chemistry*, 6th Edition, John Wiley & Sons Hoboken, NJ.

CSUROS M, CSUROS C (2002), *Environmental Sampling and Analysis for Metals*, Lewis Publishers, Boca Raton, FL.

*DEAN JR (1998), *Extraction Methods for Environmental Analysis*, John Wiley & Sons West Sussex, English.

DEAN JR (2003), *Methods for Environmental Trace Analysis*, John Wiley & Sons West Sussex, English.

Mackay D, shiu WY (1981), Critical review of Henry's law constants' for chemicals of environmental interest, L. Phys. Chem. Ref. Data., 10:1175–1199.

Majors RE (1991), An overview of sample prepartion, LC–GC Intl., 9(1): 16–20.

MITRA S (2003), *Sample Preparation Techniques in Analytical Chemistry*, Wiley-Interscience Hoboken, NJ.

ROUESSAC F and ROUESSAC A (2000). *Chemical Analysis: Modern Instrumentation Methods and Techniques*, John Wiley & Sons, NY.

SIMPSON N (1992), Solid phase extractants utilizing nonpolar, polar, and electrostatic interactions. *Am. Lab.*, p. 37.

Supelco (1998), *Guide to Solid Phase Extraction*, Supelco Bulletin 910 Bellefonte, PA.

Supelco (1998), *Solid Phase Microextraction: Theory and Optimization of Conditions.* Supelco Bulletin 923 Bellefonte, PA.

Supelco (2001), *SPME Application Guide*, 3rd Edition. Supelco Bulletin 925B Bellefonte, PA.

Supelco, Solid phase microextraction of semivolatile compounds in US EPA method 625, Supelco 6-Pub#394006 Bellefonte, PA.

Supelco, Solid phase microextraction of volatile compounds in US EPA method 524.4, Supelco 11-Pub#394011 Bellefonte, PA.

TESSIER A, CAMPBELL PGC, BISSON M (1979). Sequential extraction procedure for the speciation of particulate trace metals. *Anal. Chem.*, 51:844–851.

US EPA (1986), Testing Methods for Evaluating Solid Wastes: Physical/Chemical Methods, EPA SW 846 on-line: http://www.epa.gov/epaoswer/hazwaste.

QUESTIONS AND PROBLEMS

1. Explain briefly: (a) why HNO_3, rather than other acids is most commonly used for acid digestion? (b) why HF digestion is used for samples containing silicates? (c) why or why not acid-preserving samples is appropriate for dissolved metals analysis?

---
*Suggested Readings

2. Explain: (a) why HCl is weaker than HNO_3 used in acid digestion? (b) why dryness should be avoided in acid digestion and why dryness is particularly harmful if $HClO_4$ is used? (c) why $HClO_4$ should not be used alone for acid digestion?

3. Describe the difference in the acid preservation procedure between total metals and dissolved metals.

4. Explain why speciation analysis is particularly important for Cr, As, and Se? What are the major valance (oxidation number) for each of these three elements? Describe their impact with regard to plant, and animal/human.

5. Describe/define briefly: (a) dissolved metals, (b) alkaline digestion for Cr(VI), (c) selenate ion vs. selenite ion, (d) arsenate and arsenite.

6. Explain: (a) why a chemical with a Henry's law constant (H) of 10^{-4} atm·m^3/mol cannot be separated using separate funnel L-L extraction; (b) why a solution containing a chemical with very high vapor pressure could be low in volatility?

7. Describe how distribution coefficient (D) and the phase ratio affect the extraction efficiency in the L–L extraction.

8. Describe the operational steps in using Soxhlet extraction.

9. Explain: (a) why acetone or acetonitrile cannot be used in solvent extraction? (b) in extracting organics from aqueous solutions using separatory funnel, which layer (top or bottom) should be collected if benzene is used? What if chloroform is used?

10. Explain: (a) why solid phase extraction (SPE) can only be used for liquid samples? (d) why anhydrous sodium sulfate (Na_2SO_4) is added during solid sample extraction? (b) why SPME can be used for both liquid and solid samples?

11. Discuss the advantage and disadvantage of the following sample preparation: (a) conventional Soxhlet extraction, (b) ultrasonic extraction, (c) solid phase extraction, (d) supercritical fluid extraction.

12. Discuss the advantage of: (a) microwave assisted acid digestion over hotplate digestion, (b) Soxtec over conventional Soxhlet, (c) pressured fluid extraction over extraction with ambient temperature and pressure.

13. Why CO_2 in its supercritical state increases the extraction efficiency? What are the major mechanisms for the enhanced extraction efficiency? Why is a minor amount of solvent used in the SFE system?

14. Draw a schematic diagram for the supercritical extraction system (SFE).

15. Draw a schematic diagram for the purge-and-trap system with GC as the analytical instrument.

16. Discuss the principles and compound applicability of the following sample cleanup procedures: (a) alumina, (b) Florisil, and (c) silica gel.

17. Derivatization is sometimes used prior to HPLC and GC analysis: (a) what are the purposes of derivatization? (b) what types of compounds (e.g., with certain functional groups) normally need to be derivatized?

18. Describe: (a) silylation, (b) acylation, (c) esterification.

19. Describe the following and indicate in general what types of contaminants (i.e., with regard to solubility, volatility, etc) are applicable: (a) dynamic headspace extraction (purge-and-trap), (b) static headspace extraction, (c) azeotropic distillation, and (d) vacuum distillation.

20. Eq. 7.10 and Eq. 7.20 are used to describe the liquid-liquid extraction and vapor-liquid equilibrium, respectively. Illustrate the similarity between these two formulas.

21. Which of the following chemical(s) need azotropic distillation for its separation from an aqueous sample: (a) hydrophobic pesticide, (b) acetone, (c) acetonitrile, (d) methyl isobutyl ketone, and (d) pyrene.

22. A clean soil sample was used as a matrix to develop an extraction method for pyrene (a hydrophobic compound), using methylene chloride (MECl) extraction. The soil has no previous contamination of this compound, so the background concentration of pyrene can be assumed zero. This analyst spiked 1.025 g soil sample with 5 mL of 10 mg/L pyrene (dissolved in acetone) stock solution, let the acetone evaporate under hood, and then extract into 10 mL MECl with a sonic extractor. The soil extract was then determined to contain 2.0 mg/L pyrene by HPLC method. Calculate the % recovery of the extraction. Is this an acceptable extraction method?

23. 100 mL of aqueous solution containing 2,4,6-trinitrotoluene (TNT) at an initial concentration of 0.05 ppm is shaken with 10 mL hexane. After phase separation, it was determined that the organic phase has 0.4 ppm. What is the partition coefficient, and what is the percent extracted? If the equilibrium concentration (rather than initial concentration) is 0.05 ppm, this would simplify the calculation. What would then be the partition coefficient and percent extracted?

24. Liquid–liquid extraction (LLE) has long been a standard method for concentrating hydrophobic organic compounds such as pesticide and polycyclic aromatic hydrocarbons (PAHs) from water. A sample of water contaminated with a pesticide is extracted with methylene chloride for determination of this pesticide.

   (a) In experiments with a standard solution of a known concentration, it is found that when a 100 mL water sample is extracted with 30 mL of methylene chloride, 75% of this pesticide is removed. What is the partition coefficient (D) of this pesticide? What fraction of this pesticide would be removed if three extractions of 10 mL of methylene chloride each were done instead?

   (b) The extraction efficiency of LLE depends on the hydrophobicity (hence D) of the chemical to be extracted as well as the volume and number of extraction. Demonstrate this by filling in the blanks of the table below regarding the total extraction efficiency (assume water sample volume = 100 mL).

	Extraction efficiency if D (unitless) =				
	10	25	50	100	1,000
One extraction with 30 mL					
Three extractions, 10 mL each					

Chapter 8

UV-Visible and Infrared Spectroscopic Methods in Environmental Analysis

This chapter and the remaining chapters of this text are the *survey* of various instrumental techniques used in environmental analysis. This *survey* is aimed at providing essential information for readers who are dealing with these instrumentations for the first time. The specific aims of this chapter are to introduce the spectroscopic principles that are fundamental to all spectroscopic analysis (namely, UV-VIS, IR, and NMR) and the applications of UV-VIS and IR in environmental measurement. Although the UV-VIS technique is a workhorse in average environmental laboratories due to its availability of standard methods, simplicity and low cost, IR technique has also found its wide applications in monitoring air pollution and in related industrial hygiene analysis.

To help the reader fully comprehend spectroscopic principles without going into much theoretical detail (e.g., quantum mechanics) is a considerable challenge. With a focus on UV-VIS and IR spectroscopy in this chapter, we start with the mechanism of radiation absorption from the molecular level (orbital theories) so that the reader will fully appreciate what environmental chemicals are subject to UV-VIS and IR analysis. Our discussions will then be focused on their use as a quantitative tool to

Fundamentals of Environmental Sampling and Analysis,   by Chunlong Zhang
Copyright © 2007 John Wiley & Sons, Inc.

determine concentrations. The discussions on structural (qualitative) analysis using IR are kept at the minimum due to its technical complexity and specialized applications only to certain professionals.

8.1 INTRODUCTION TO THE PRINCIPLES OF SPECTROSCOPY

8.1.1 Understanding the Interactions of Various Radiations with Matter

Spectroscopy is a subject that deals with the interaction of electromagnetic radiation with matter. All qualitative and quantitative determinations using spectroscopic methods need radiation to pass through the sample containing the analyte of interest (the matter). Radiation is defined by its wavelength or frequency, which is related to the energy of the radiation by the *planck's law*:

$$E = hv = hc/\lambda \tag{8.1}$$

where h = Planck constant (6.62×10^{-34} Js) and c = speed of light (3×10^8 m/s). The wavelength (λ) is the distance between two adjacent peaks of electromagnetic radiation, and may be designated in m, cm, nm, or μm (1 m = 10^2 cm = 10^6 μm = 10^9 nm). Frequency (v) is the number of wave cycles that travel past a fixed point per unit time and is usually given in cycle per second or hertz (Hz). Wavelength is inversely proportional to the frequency. The energy of radiation (E) increases with the increasing frequency and increases with decreasing wavelength (Eq. 8.1).

Figure 8.1 is an electromagnetic spectrum containing the wavelength regions with a variety of analytical applications. The energy per photon is in an increasing order: radio wave < microwave < infrared (IR) < visible light (VIS) < ultraviolet (UV) < X-ray. Our naked eyes can detect only a very limited range of wavelengths, that is, the *visible spectrum* from about 300 to 780 nm.

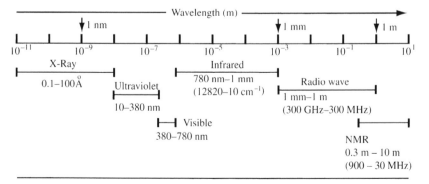

Figure 8.1 The electromagnetic spectrum (wavelength unit: 1 m = 10^2 cm = 10^3 mm = 10^9 nm = 10^{10} Å; frequency unit: 1 Hz = 1 s^{-1} = 10^{-6} MHz = 10^{-9} GHz; wavenumber has a unit of cm^{-1})

Chemicals can absorb electromagnetic radiation, but only at certain energies (wavelengths). Figure 8.2 illustrates the relationships between different energy levels within a molecule. The three groups of lines correspond to different electronic configurations. The lowest energy, termed the *ground state*, is the most stable electron configuration. Photons having certain energies in the visible and UV regions of the spectrum can cause these electrons to be excited into higher energy orbitals. Some of the possible absorption transitions are indicated by the vertical arrows. Photons that are more energetic (UV to X-ray region) may cause an electron to be ejected from the molecule (*ionization*). Photons in the *infrared (IR)* region of the spectrum have much less energy than those in the visible or UV regions of the electromagnetic spectrum. They can, however, excite vibrations in molecules. There are many possible vibrational levels within each electronic state. Transitions between the vibrational levels are indicated by the vertical arrows on the left side of the diagram.

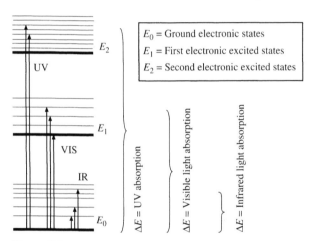

Figure 8.2   Energy levels in a molecule and three types of absorption spectrometry. ΔE is the radiation absorbed by a molecule

Figure 8.3 further illustrates molecular responses to radiation including UV, VIS, IR, and microwave. As can be seen, electron transition or ionization can occur under high energy UV radiation. The least energetic *microwave radiation* cannot excite vibrations but can only cause molecules to *rotate*. Microwave ovens are tuned to the frequency that causes molecules of water to rotate, causing heat as a result of friction of water-containing substances (Eubanks et al., 2006).

The *absorption* described in Figures 8.2 and 8.3 is only one type of the electromagnetic radiation. Absorption of radiation moves the atom to a higher energy level. Transitions among the vibrational and rotational states give rise to absorption at IR wavelengths, whereas those between electronic levels involve more energetic visible or UV radiation.

On the contrary, the energy at the higher state may also return to ground state by *emission* or may lose some of its energy as thermal energy and return to the

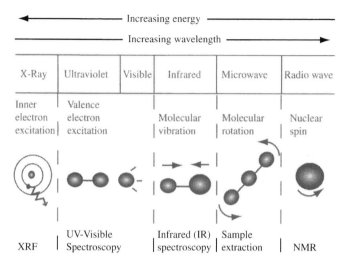

Figure 8.3   Molecular responses to radiations of various wavelength ranges (not to scale)

ground state by emitting a new and longer wavelength radiation termed as the *fluorescence* radiation (Kebbekus and Mitra, 1998) (Fig. 8.4).

8.1.2   Origins of Absorption in Relation to Molecular Orbital Theories

The above discussion as to why UV radiation causes electron excitation and why infrared radiation causes molecular vibration may not be intuitive without the

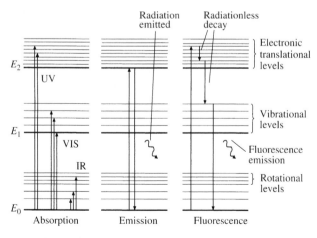

E_0 = Ground level; E_1, E_2 = Excited states

Energy spacing: vibration > rotation >> translation

Figure 8.4   Three types of spectroscopy—absorption, emission, and fluorescence

fundamental consideration of molecular orbital theories. To illustrate this, several basic concepts regarding electron configuration in an atom, bonding, and molecular orbitals are essential and will be presented below at only an introductory level. Readers who do not have fresh chemistry in mind can refer to general chemistry textbooks for a thorough review. Other readers can also skip this section for the time being if only quantitative analysis is of primary concern.

Basic Concepts of Electronic Structure

Imagine electrons are particles as well as waves. They orbit the nucleus (as particles) and their motions can be described in a volume by solving mathematical wave equations to tell us where the electron is most likely to be found. When multiple electrons are present, the electrons in an atom can be thought of as occupying a set of major shells, subshells, and orbitals that surround the nucleus. Recall from your general chemistry that

- The *major shell* determines the primary energy level of the electron. The first major shell is the one closest to the nucleus. Electrons in this first major shell has the lowest energy. Major shells are designated by a *quantum number* (n), that is, $n = 1, 2, 3, 4$ for the 1st, 2nd, 3rd, and 4th major shells.

- The *subshells* are a substructure within the major shells designated by letters s, p, d, and f. If you visualize the atom as an onion, then the large thick peels are the major shells and each onion skin or peel within each major peel is the subshell. The number of subshell within a major shell is the same as the quantum number, thus we have one subshell (s) in 1st major shell, two subshells (s and p) in the 2nd major shell, three subshells (s, p and d) in the 3rd major shell, and four subshells (s, p, d, and f) in the 4th major shell (Table 8.1).

- Each subshell is further structured into *orbitals*. The numbers of orbitals are 1 for s subshell, 3 for p subshell, 5 for d subshell, and 7 for f subshell.

- Each orbital can accommodate a maximum of two electrons (*Pauli Exclusion Principle*). As a result, the maximum numbers of electrons equal to $2n^2$ or 2, 8, 18, and 32 for 1st, 2nd, 3rd, and 4th major shell, respectively.

Table 8.1   Electron distribution in the first four shells surrounding the nucleus

Major shells (n)	1st shell	2nd shell	3rd shell	4th shell
Subshells	s	s, p	s, p, d	s, p, d, f
Number of atomic orbitals	1	1, 3	1, 3, 5	1, 3, 5, 7
Maximum number of electrons ($2n^2$)	2	8	18	32

y   y   y   y

⊕—x   ⊕—x   ⊕   x—⊕—x

z   z   z   z

2s orbital   2p$_x$ orbital   2p$_y$ orbital   2p$_z$ orbital   **Figure 8.5**   Electron in s and p orbital

Note that the closer the orbital to the nucleus the lower is the electron's energy. The relative energies of atomic orbitals are: 1s < 2s < 2p < 3s < 3p < 4s < 3d < 4p < 5s < 4d < 5p < 6s < 4f < 5d < 6p < 7s < 5f.

Types of Absorbing Electrons: σ, σ*, π, π* and n

We now look at two major types of orbitals (s and p) and how electrons in these atomic orbitals form molecular bonding. The s electron orbital has a spherical shape that is also symmetrical without directional characteristics, whereas p electron orbital is dumbbell shaped. Since there are three p atomic orbitals (Table 8.1) and they have equal energy, the three orbitals are perpendicular to each other, symmetrical about the x-, y-, and z- axis and are designated as 2p$_x$, 2p$_y$, and 2p$_z$ orbitals to emphasize their directional nature (Fig. 8.5).

When two s atomic orbitals overlap (such as two 1s hydrogen atoms combine to form hydrogen molecule H$_2$), the bond that is formed is called a *sigma (σ) bond* (Fig. 8.6, bottom). During bond formation, energy is released as the two orbitals start to overlap till maximum stability (minimum energy) is achieved when two nuclei are at a certain distance apart. This distance corresponds to the *bond length*, and the energy released corresponds to the *bond energy*. To break the bond, the same amount of energy is required (termed the bond dissociation energy).

There is another way to combine 1s atomic orbitals termed *sigma antibonding molecular orbital (σ*)* (Fig. 8.6, top). An antibonding orbital can be perceived as the detraction from formation of bond between two atoms, similar to the darkness that occurs when two light waves cancel each other or the silence that occurs when two sound waves cancel each other.

In addition to the formation from two s orbitals as described in Figure 8.6, the σ orbitals can also be formed when two 2p atomic orbitals overlap end-on (i.e., head to head), such as two 2p orbitals of fluorine (1s^2, 2s^2, 2p^5) overlap to form fluorine gas (F$_2$).

σ* antiboding molecular orbital

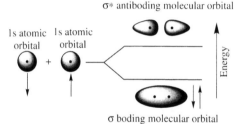

1s atomic   1s atomic
orbital   orbital

+

Energy

σ boding molecular orbital

Figure 8.6   The formation of sigma (σ) bonding molecular orbital and σ* antibonding molecular orbital. Prior to the formation of a covalent bond, each electron is in 1s atomic orbital. These two electrons are in the molecular orbital after the formation of a covalent bond

Another type of molecular orbitals (the π orbital) may be formed from two p-orbitals by a lateral overlap (i.e., shoulder-to-shoulder) as shown in Figure 8.7a. This is termed a π *bond (π-orbital)*. Since bonds consisting of occupied π-orbitals are weaker than σ bonds, π-bonding between two atoms occurs only when a σ bond has already been established. Thus, π-bonding is generally found only as a component of double and triple covalent bonds. Analogous to σ bond, a shoulder-to-shoulder overlap of two out-of-phase p orbitals will form a π *antibonding molecular orbital (π*)*.

A *hybrid orbital* forms when two or more atomic orbitals of an isolated atom mix. The shapes and orientations are different from those of atomic orbitals in isolated atoms. This is called *orbital hybridization*. For example, since carbon atoms (electron configuration: $1s^2$, $2s^2$, $2p^2$) involved in double bonds have only three bonding partners, they require only three hybrid orbitals to contribute to three sigma bonds. A mixing of the 2s-orbital with two of the 2p orbitals gives three sp^2 hybrid orbitals, leaving one of the p-orbitals unused. Two sp^2 hybridized carbon atoms are then joined together by sigma and π-bonds (a double bond) as shown in Figure 8.7b.

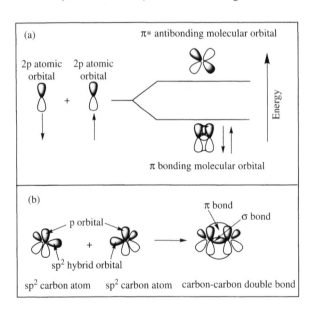

Figure 8.7   Formation of π and π* bond: (a) The formation of π orbital from two p-orbitals and (b) The formation of σ- and π-molecular orbitals from two sp^2 hybridized carbon atoms

UV Absorption and Electronic Transitions

When molecules are exposed to UV or visible radiation, three types of electronic transitions occur as a result of excitation of outer electrons in the *valence shell* (i.e., the electrons that form bonds in a molecule). These include: (1) transitions involving π, σ, and n electrons, where *n* denotes a nonbonding orbital,

(2) transitions involving charge-transfer electrons (inorganic species), and (3) transitions involving d and f electrons. In the following discussions, we only look at the possible *electronic transitions* of π, σ, and n electrons (namely, $\sigma \rightarrow \sigma^*$; $\pi \rightarrow \pi^*$; $n \rightarrow \pi^*$; $n \rightarrow \sigma^*$) (Fig. 8.8) and examine how certain structures of molecules are UV-absorbing whereas others are UV-transparent.

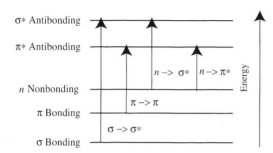

Figure 8.8 Electronic molecular energy levels

$\sigma \rightarrow \sigma^*$ *Transitions*: An electron in a σ bonding orbital is excited to the corresponding antibonding orbital. The energy required is large. In other words, a short wavelength radiation is needed for such electronic transition. For example, methane (which has only C—H single bonds and can only undergo $\sigma \rightarrow \sigma^*$ transitions) shows an absorbance maximum at 125 nm. Absorption maxima due to $\sigma \rightarrow \sigma^*$ transitions are not seen in typical UV-VIS spectra (200–700 nm). This is the reason for saturated hydrocarbon such as hexane (C_6H_{14}) or cyclohexane (C_6H_{12}) that only contain σ bonds being transparent in the conventional UV range (λ_{max} for hexane = 135 nm). Hence, hexane and cyclohexane can be used as solvents in the UV absorption spectrometry. Similarly, the transparency of water in the near UV is due to the fact that there can only be $\sigma \rightarrow \sigma^*$ and n $\rightarrow$ n* transitions in this compound.

$n \rightarrow \sigma^*$ *Transitions*: Saturated compounds containing atoms with unshared electron pairs (nonbonding electrons, n) are capable of n $\rightarrow \sigma^*$ transitions. These transitions usually need less energy than $\sigma \rightarrow \sigma^*$ transitions. They can be initiated by light whose wavelength is in the range 150–250 nm. The number of organic functional groups with n $\rightarrow \sigma^*$ peaks in this UV region is small (Table 8.2).

$n \rightarrow \pi^*$ and $\pi \rightarrow \pi^*$ *Transitions*: Most absorption spectroscopy of organic compounds is based on transitions of n or π electrons to the π^* excited state. This is because the absorption peaks for these transitions fall in an experimentally convenient region of the spectrum (200–700 nm). These transitions need an *unsaturated group* in the molecule to provide the π electrons.

Molar absorptivities (ε, see Section 8.14) from n $\rightarrow \pi^*$ transitions are relatively low and range from 10 to 100 L/mol/cm. $\pi \rightarrow \pi^*$ transitions normally give molar absorptivities between 1000 and 10,000 L/mol/cm. The n $\rightarrow \pi^*$ transition is usually observed in molecules that contain a *heteroatom* (such as C=O, N=O) as part of an unsaturated system. The most common of these bands corresponds to the carbonyl band at around 270–295 nm. Table 8.3 gives a list of UV *chromophores*

Table 8.2   Examples of absorption due to n $\rightarrow \sigma^*$ transition (Skoog et al., 1997)[a]

Compound	λ_{max} (nm)	ε_{max}
H_2O	167	1480
CH_3OH	184	150
CH_3Cl	173	200
CH_3I	258	365
$(CH_3)_2S$	229	140
$(CH_3)O$	184	2520
CH_3NH_2	215	600
$(CH_3)_3N$	227	900

[a]Sample in vapor phase except $(CH_3)_2S$ in ethanol; ε max in L/mol/cm.

(the UV-absorbing functional group in a molecule) in the UV range due to n $\rightarrow \pi^*$ and $\pi \rightarrow \pi^*$ transitions.

The UV chromophores listed in Table 8.3 reveal important information regarding which chemicals can absorb UV and, therefore, can be analyzed by UV spectroscopy skoog et al., 1997. Compounds with UV chromophores include dienes and polyenes (C=C), carbonyl compounds (C=O), and benzene derivatives. Dienes are hydrocarbons that contain two double bonds. If the double bonds are separated by two or more single bonds, they are *unconjugated dienes*. If the double bonds are separated by one single bond, they are *conjugated dienes*. The series of conjugated double bonds typically absorb strongly in the UV range of the electromagnetic spectrum.

Table 8.3   Examples of absorption due to n $\rightarrow \pi^*$ and $\pi \rightarrow \pi^*$ transitions

Chromophore	Example	Excitation	λ	ε	Solvent
C=C	Ethene	$\pi \rightarrow \pi^*$	171	15,000	Hexane
C≡C	1-Hexyne	$\pi \rightarrow \pi^*$	180	10,000	Hexane
C=O	Ethanal	$n \rightarrow \pi^*$	290	15	Hexane
		$\pi \rightarrow \pi^*$	180	10,000	Hexane
N=O	Nitromethane	$n \rightarrow \pi^*$	275	17	Ethanol
		$\pi \rightarrow \pi^*$	200	5,000	Ethanol

IR Absorption and Vibrational and Rotational Transitions

From the above discussion, we understand the origin of UV sorption from the molecular orbitals standpoint. Now let us try to understand the origin of IR absorption and how this is related to vibrational and rotational transitions (Jang, 1996; Pecsok and shield, 1968).

Molecular *vibrations* can be understood by imagining a diatomic molecule as two spheres connected by a spring. When the molecule vibrates, the atoms move toward and away from each other at a certain frequency. The energy of the system is related to how much the spring is stretched or compressed. The vibrational frequency (v) is proportional to the square root of the ratio of the spring force constant (k, N m^{-1}) to the masses on the spring (μ, kg).

$$v = \frac{1}{2\pi}\sqrt{\frac{k}{\mu}} \qquad \text{where} \qquad \mu = \frac{m_1 m_2}{m_1 + m_2} \qquad (8.2)$$

From Eq. 8.2, the lighter the masses on the spring, or the tighter (stronger) the spring, the higher will be the vibrational frequency. Similarly, vibrational frequencies for stretching bonds in molecules are related to the strength of the chemical bonds and the masses of the atoms. Molecules differ from sets of spheres-and-springs in that the vibrational frequencies are *quantized*. That is, only certain energies for the system are allowed, and only photons with certain energies will excite molecular vibrations. The symmetry of the molecule will also determine whether a photon can be absorbed.

The number of vibrational modes (different types of vibrations) in a molecule can be calculated as 3N−5 for *linear molecules* (e.g., CO_2) and 3N−6 for *nonlinear molecules* (e.g., H_2O), where N is the number of atoms. Carbon dioxide, a linear molecule, has $3 \times 3 - 5 = 4$ vibrations. These vibrational modes, shown in Figure 8.9, are responsible for the "greenhouse" effect (global warming) in which heat radiated from the earth is absorbed (trapped) by CO_2 molecules in the atmosphere. The arrows indicate the directions of motion. Vibrations labeled A and B represent the *stretching* of the chemical bonds: one in a *symmetric stretching* (a), in which both C=O bonds lengthen and contract together (in-phase) and the other in an *asymmetric stretching* (b), in which one bond shortens whereas the other lengthens. The asymmetric stretch (b) is *infrared active* because there is a change in the molecular dipole moment during this vibration. *The dipole moment* of a bond is defined as the magnitude of the charge times the distance. To be "active" means that absorption of a photon to excite the vibration is *allowed* by the rules of quantum mechanics.

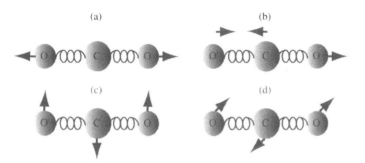

Figure 8.9   The vibration modes of CO_2: (a) symmetric stretching, (b) asymmetric stretching, (c) vertical bending, and (d) horizontal bending. The symmetric mode (a) has no change in dipole moment, hence it is not active in the IR range

For CO_2 molecule described above, infrared radiation at $2349 \, cm^{-1}$ ($4.26 \, \mu m$) excites this particular vibration. The symmetric stretch is not infrared active and so this vibration is not observed in the infrared spectrum of CO_2. The two equal-energy *bending vibrations* in CO_2 (Fig. 8.9c and d) are identical except that one bending mode is in the plane of the paper and the other is out of the plane. Infrared radiation at $667 \, cm^{-1}$ ($15 \, \mu m$) excites these vibrations. Figure 8.10 is the *IR spectrum* of CO_2 clearly showing two bands of IR absorption due to the molecular vibration—one at $4.26 \, \mu m$ due to asymmetric stretching and the other at $15 \, \mu m$ due to bending.

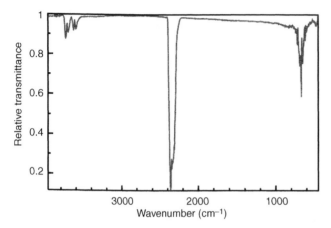

Figure 8.10   Infrared spectrum of carbon dioxide (CO_2). Source: NIST Mass Spec Data Center, S.E. Stein, director, "Infrared Spectra," in NIST Chemistry WebBook, NIST Standard Reference Database Number 69, Eds. P.J. Linstrom and W.G. Mallard, June 2005, National Institute of Standards and Technology, Gaithersburg MD, 20899 (http://webbook.nist.gov)

Besides stretching and bending of a chemical bond, more complicated molecules vibrate in *rocking, twisting, scissoring, and wagging* modes, which arise from combinations of bond bending in adjacent portions of a molecule. Without revealing the details of molecular vibrations, some general trends can be stated. That is, the stronger the bond the more energy will be required to excite the stretching vibration. Thus, stretches of organic compounds with triple bonds such as $C\equiv C$ and $C\equiv N$ occur at higher frequencies than stretches for double bonds ($C=C$, $C=N$, $C=O$), which are in turn at higher frequencies than single bonds ($C-C$, $C-N$, $C-H$, $O-H$, or $N-H$). The heavier an atom the lower the frequencies for vibrations that involve that atom.

8.1.3 Molecular Structure and UV-Visible/Infrared Spectra

A *spectrum* is a plot of absorption vs. wavelength. Spectra of various types are the basis to deduce structures for chemicals of unknown identities to confirm chemical identity with a spectral library established by the analytical community. In the 1940s, before the advent of more powerful identification techniques, UV-VIS spectroscopy was the main tool used for structural identification. Structural analysis from electronic spectra of UV radiation yields little structural information because of their relative simplicity. In comparison, IR radiations are associated with molecule's rotational and vibrational sublevels. In a molecule, there are numerous possible transitions that are quite close in energy. As a result, it produces a continuous broad absorbance/emission band. The IR spectra are richer than UV-VIS in structural information. Even though, IR spectrum has diminished its major role as other recent techniques become available, such as MS and NMR (Chapter 12). With all available structural identification techniques, the overall utility is in the increasing order: UV < IR < MS < NMR (Field et al., 2003).

Figure 8.11 compares the IR spectrum vs. UV spectrum for the same compound benzene. It is noted that UV-visible spectrum typically is a plot of absorbance vs. wavelength, whereas IR spectrum is more conveniently plotted as percentage transmittance vs. wave number (cm^{-1}). The concept of absorbance and transmittance will be introduced in the next section.

Wave number ($\bar{v}$) is related to wavelength (λ) as follows:

$$\bar{v} \, [cm^{-1}] = \frac{10,000}{\lambda \, [\mu m]} \tag{8.3}$$

The wave number has a unit of cm^{-1}. It is directly proportional to the frequency (v) and hence the energy of IR. Also note that the wave number on the axis is in the descending order and the scale is nonlinear on a typical IR spectrum. There are two regions of the IR spectra: (a) *a group frequency region*, which encompasses radiation from about $3600 \, cm^{-1}$ to approximately $1200 \, cm^{-1}$ (3–8 μm); (b) *fingerprint region* in the range of $1200–600 \, cm^{-1}$ (8–14 μm), which is particularly useful because differences in the structure and constitution of a molecule result in significant changes in the appearance and distribution of absorption peaks in this region.

From the standpoint of both application and instrumentation, the infrared spectrum is conveniently divided into near-, mid-, and far-infrared radiations.

- *Near-infrared (NIR):* 0.78–2.5 μm ($12,800–4,000 \, cm^{-1}$). The photometers and spectrometers in this range are similar in design and components to UV-VIS spectrometry. They are less useful in structural identification, but found in most applications of the quantitative analysis of compounds containing functional groups that are made up of H bonded to C, N, and O. Applications include water, CO_2, S, low molecular weight hydrocarbons, amine nitrogen, and many other simple compounds in industrial and agricultural materials.
- *Mid-infrared (MIR):* 2.5–50 μm ($4,000–200 \, cm^{-1}$). The mid-infrared range is the most widely used IR as it has found use in measuring the

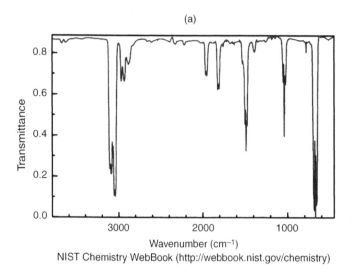

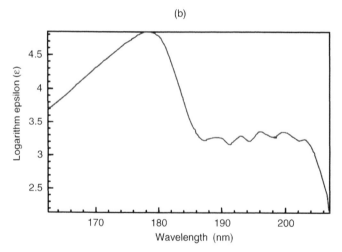

Figure 8.11   A comparison of (a) IR spectrum (top) and (b) UV spectrum (bottom) of benzene.
Logarithm epsilon (ε) denotes concentration–corrected absorbance, where ε is the molar absorptivity
(see sec 8.1.4). Source: (a) Coblentz Society, Inc., "Evaluated Infrared Reference Spectra" and
(b) V. Talrose, E.B. Stern, A.A. Goncharova, N.A. Messineva, N.V. Trusova, M.V. Efimkina, "UV-Visible
Spectra," in NIST Chemistry WebBook, NIST Standard Reference Database Number 69, Eds.
P.J. Linstrom and W.G. Mallard, June 2005, National Institute of Standards and Technology, Gaithersburg
MD, 20899 (http://webbook.nist.gov)

composition of gases and atmospheric contaminants with potentially
sensitive, rapid, and highly specific methods.

- *Far-infrared (FIR):* 50–1,000 μm (200–10 cm^{-1}). The far-infrared range
  has limited use because of experimental difficulties. FIR is mostly for the

structural determination of inorganic and metal-organic species because absorption due to stretching and bending vibrations of bonds between metal atoms and both inorganic and organic ligands generally occur at frequencies lower than $650\,cm^{-1}$ ($>15\,\mu m$).

The interpretation of IR spectrum is not a straightforward task. To simplify, Figure 8.12 can be used as a general guide for the common stretching and bending vibrations. The reader can relate the IR spectrum of benzene (Fig. 8.11) for an example use of Figure 8.12. In benzene, every carbon has a single bond to hydrogen and each carbon is bonded to two other carbons and the carbon-carbon bonds are alike for all six carbons. The aromatic C—H stretch appears at $3100–3000\,cm^{-1}$. There are aromatic C—C stretch bands (for the carbon-carbon bonds in the aromatic ring) at about $1500\,cm^{-1}$. Two bands are caused by bending motions involving C—H bonds. The bands for C—H bends appear at approximately $1000\,cm^{-1}$ for the in-plane bends and at about $675\,cm^{-1}$ for the out-of-plane bend. As shown, the IR spectrum for benzene is relatively simple because benzene has only a few prominent bands owing to its symmetric and planar molecular structure.

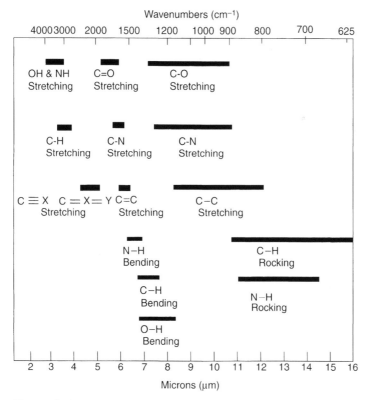

Figure 8.12 Chart of characteristic vibrations (http://www.wag.caltech.edu/home/jang/genchem/infrared.htm)

The structural interpretation of more complex molecules may be frustrating to beginners, but keep in mind that it takes time and practice for anyone to get to a comfortable level. Fortunately, the structural elucidation is only the routine work for a skilled organic chemist. For most environmental work, qualitative analysis for structural analysis is of less importance. Most environmental applications are done at the quantitative level for concentration measurement, which does not require the skilled techniques in spectrum interpretation. The quantitative principles are discussed in the following section.

8.1.4   Quantitative Analysis with Beer-Lambert's Law

In spectroscopy, the Beer-Lambert's law, also known as Beer's law is an empirical relationship in relating the absorption of light to the amount of chemical when light is traveling through the sample. The *Beer's law* states that *absorbance* (*A*) is proportional to the concentration (*C*) of the light-absorbing chemicals in the sample according to the following:

$$A = \varepsilon l C \tag{8.4}$$

where ε = molar absorptivity (L/mol/cm), l = length of light path in a sample cell, and C = concentration of the chemical (mol/L). *Molar absorptivity* is an intrinsic property of the chemical species, which is very large for strongly absorbing compounds ($\varepsilon > 10,000$) and very small if absorption is weak ($\varepsilon = 10 - 100$). The Beer's law (Eq. 8.4) is also additive if there are several absorbing species (1, 2, ..., *n*) in the solution, thus we have

$$A = \varepsilon_1 l C_1 + \varepsilon_2 l C_2 + \varepsilon_3 l C_3 + \cdots + \varepsilon_n l C_n \tag{8.5}$$

Occasionally, *transmittance* (*T*) is used, which is defined as the fraction of the original light that has passed through the sample:

$$T(\%) = P/P_0 \times 100\% \tag{8.6}$$

where P_0 and P are the power of incident and transmitted light, respectively (Fig. 8.13). Therefore, absorbance (*A*) and transmittance are related by

$$A = -\log_{10}T = \log P_0/P \tag{8.7}$$

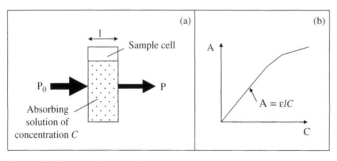

Figure 8.13   UV-VIS light absorption and the Beer's law: (a) Light absorption in a sample cell. (b) The Beer's law as shown in a calibration curve

For example, a blank sample without a light-absorbing chemical should be 100% transparent, that is, $T = 100\%$ or $A = 0$.

When a calibration curve, that is, a plot of A vs. known concentrations (C) of standard solutions ($A = a\,C + b$) is acquired, a linear range is always sought for the measurement of analyte concentration in samples (Chapter 2). Care should be excised on the limitations of the Beer's law. The linearity of the Beer's law is sometimes deviated by chemical and instrumental factors, and the analyst should be aware of the solutions. The nonlinearity may be caused by: (a) electrostatic interactions between molecules in close proximity, (b) scattering of light caused by particulates in the sample, (c) fluorescence or phosphorescence of the sample, (d) changes in refractive index at high analyte concentration, (e) shifts in chemical equilibria as a function of concentration, (f) nonmonochromatic radiation (this deviation can be minimized by using a relatively flat part of the absorption spectrum such as the maximum of an absorption band), (g) stray light that reaches the detector without passing through the sample.

With IR, instrumental deviations from the Beer's law are more common than with UV and visible wavelengths. This is because infrared absorption bands are relatively narrow. For dispersive IR (Section 8.3.2), the low intensity of sources and low sensitivities of detectors in this region require the use of relatively wide monochromator slit widths, leading to nonlinear relationship between absorbance and concentration.

EXAMPLE 8.1.   A spectrometric method was developed to determine nitrite (NO_2^-, MW $= 46$ g/mol) in human saliva and rain water (Helaleh and Korenaga, *J. AOAC Int.,* 84:53–58, 2001). The method was reported to have an absorption maximum of 546 nm and molar absorptivity of 4.6×10^4 L/(mol cm). What range of concentrations can be measured, so the absorbance remains within the range 0.05–0.80, using a 1.0 cm cell?

SOLUTION: For $A = 0.05$,

$$C = \frac{A}{\varepsilon\, l} = \frac{0.05}{(4.61 \times 10^4 \text{ L/cm/mol}) \times 1 \text{ cm}} = 1.08 \times 10^{-6} \text{ mol/L} = 0.050 \text{ µg/mL}$$

For $A = 0.80$,

$$C = \frac{A}{\varepsilon\, l} = \frac{0.80}{(4.61 \times 10^4 \text{ L/cm/mol}) \times 1 \text{ cm}} = 1.74 \times 10^{-5} \text{ mol/L} = 0.798 \text{ µg/mL}$$

Note that a conversion factor of 46×1000 is used to convert mol/L into µg/mL. In the absorbance range 0.05–0.80, the above calculated concentration range compares favorably with the linear range (0.054–0.816 µg/mL) reported in this paper.

EXAMPLE 8.2.   Beer's law for multiple components. Permanganate ($KMnO_4$; $\lambda = 525$ nm) and chromate ($K_2Cr_2O_7$; $\lambda = 440$ nm) can be determined in the same sample by visible light spectroscopy. The absorbance readings of two standard solutions (0.01 M $KMnO_4$ and 0.02 M $K_2Cr_2O_7$) and a sample containing two analytes obtained from a UV-VIS spectrometer are given below. What is the concentration of $KMnO_4$ and $K_2Cr_2O_7$ in the sample?

Solution	Measured at $\lambda = 440$ nm	Measured at $\lambda = 525$ nm
0.02 M $KMnO_4$ standard	$A = 0.021$	$A = 0.880$
0.04 M $K_2Cr_2O_7$ standard	$A = 0.645$	$A = 0.010$
Unknown sample	$A = 0.450$	$A = 0.576$

SOLUTION: From Beer's law (Eqs 8.4 and 8.5), we can use a lump parameter k to account for the product of two constants (ε and l), that is, $A = k\,C$, where k has a unit of L/mol. Recall that k is also the slope of the UV-VIS calibration curve. For $KMnO_4$ standard solution, we have

$$0.021 = k_{440,\,Mn} \times 0.02, \; k_{440,\,Mn} = 1.05; \text{ and } 0.880 = k_{525,\,Mn} \times 0.02, \; k_{525,\,Mn} = 44.0$$

For $K_2Cr_2O_7$ standard solution, we can similarly calculate the k values as

$$0.645 = k_{440,\,Cr} \times 0.04, \; k_{440,\,Cr} = 16.13; \text{ and } 0.010 = k_{525,\,Cr} \times 0.04, \; k_{525,\,Cr} = 0.250$$

For the unknown sample, two simultaneous equations can be generated at two different wavelengths as

$$0.450 = 1.05 \times C_{Mn} + 16.13 \times C_{Cr}$$
$$0.576 = 44.0 \times C_{Mn} + 0.250 \times C_{Cr}$$

Solving these two equations will give $C_{Mn} = 0.0129$ M and $C_{Cr} = 0.0268$ M.

8.2   UV-VISIBLE SPECTROSCOPY

8.2.1   UV-Visible Instrumentation

Spectroscopic instruments differ considerably depending on the wavelength regions (i.e., visible, UV, IR). Regardless of the variations, however, all spectrometers (spectrophotometers) require five major components: (1) a source of continuous radiation over the wavelengths of interest, (2) a monochromator for selecting a narrow band of wavelength from the source spectrum, (3) a sample cell, (4) a detector for converting radiant energy into an electrical signal, and (5) a device to read out the response of the detector (absorbance, transmittance, etc.) (Fig. 8.14).

A student version spectrometer in the visible wavelength range is shown in Figure 8.15. A brief description of each component is as follows.

- *Radiation source*: Usually a low-pressure hydrogen or *deuterium lamp* is used for UV (185–375 nm) and a tungsten filament *incandescent lamp* for visible measurements. The tungsten filament emits approximately 325 nm to 3 μm, so it can also be used in the near-UV and near-IR regions.

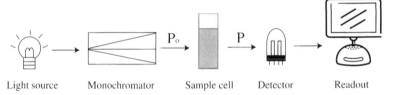

Light source   Monochromator   Sample cell   Detector   Readout

Figure 8.14   Schematic diagram of a typical single beam UV-visible light spectrometer

- *Monochromator (wavelength selector)*: It consists of an entrance slit that narrows the bandwidth of radiation, a collimator that makes the radiation parallel, a grating or prism that disperses unwanted radiation, and an exit slit that isolates the desired wavelength of radiation from the grating. Glass prisms and lens can be used in the visible range, but quartz or fused silica materials must be used in the UV region.

- *Cuvette (sample cell)*: It is used to hold sample. Quartz cuvettes are transparent to UV and visible radiation and, therefore, are more commonly used for UV-VIS spectroscopy.

- *Detector (transducer)*: It changes the radiation transmitted from the UV-VIS spectrometer into a current or voltage for the readout device to use. Solid-state photodiodes, photoemissive tubes, and photomultiplier tubers (PMT) are commonly used as single element detectors, whereas solid-state array detectors are used as multielement detectors. The operational principles are complicated (Albert Einstein received the Nobel Prize in 1921 for his discovery of the photoelectric effect, not for his well-known relativity theory), but fortunately, practical users can consider it as a black box and they rarely need to know the theory behind these detectors in examining the electronic and mechanical details of these optical instruments.

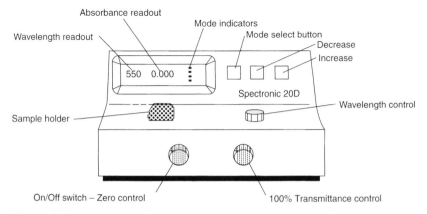

Figure 8.15   A simple visible wavelength light spectrometer (http://www.chemistry.nmsu.edu/Instrumentation/Spectronic–20.html)

8.2.2 UV-VIS as a Workhorse in Environmental Analysis

The U.S. EPA has adopted a number of spectrophotometric methods for the determination of pollutants in both air and water (Manahan, 2005). In the following discussions, we first summarize EPA methods (Tables 8.4 and 8.5), then briefly introduce the chemical principles (with an emphasis on visible colorimetric methods). We will focus on the chemical principles since operation details can always be found in cookbooks of these standard methods.

Table 8.4 Spectrometric (colorimetric) methods for the analysis of pollutants in air

Pollutant	Method	EPA method no.
SO_2	Pararosaniline (colorimetric)	40 CFR 50 Appendix A
O_3	Potassium iodide reaction (colorimetric)	40 CFR 50 Appendix E
NO_2	Azo dye reaction (colorimetric)	40 CFR 50 Appendix F

UV-VIS Methods for Atmospheric Pollutants

The spectrometric methods in Table 8.4 specify reference analytical methods by the U.S. EPA. These methods are published in CFR. Although they are not necessarily technically advanced, for regulatory and legal purposes, they provide reliable measurement proven to be valid.

Sulfur dioxide (SO_2): This cumbersome EPA reference method uses an impinger (Fig. 4.9c) to scrub air samples through a collection solution containing tetrachloromercurate or formaldehyde. The reaction with SO_2 results in the oxidation of sulfite to bisulfite:

$$[HgCl_4]^{2-} + 2\,SO_2 + 2\,H_2O \rightarrow [Hg(SO_3)_2]^{2-} + 4\,Cl^- + 4\,H^+ \qquad (8.8)$$

$$SO_2 + CH_2O + H_2O \rightarrow HOCH_2SO_3H \qquad (8.9)$$

The red-purple pararosaniline methylsulfonic acid is formed when acid-bleached pararosaniline and formaldehyde scrubbed the solution containing a stable, nonvolatile dichlorosulfitomercurate (West and Gaeke, 1956).

p-Rosaniline

$$(8.10)$$

Table 8.5 Spectrometric (colorimetric) methods for the analysis of pollutants in water

Pollutant	Method description	EPA method
Ammonia	Alkaline Hg(II) iodide reacts with ammonia, producing colloidal orange-brown $NH_2Hg_2I_3$, which absorbs light at λ 400 and 500 nm	350.2
Arsenic	Reaction of arsine, AsH_3, with silver diethylthiocarbamate in pyridine, forming a red complex ($\lambda = 535$ nm)	206.4
Boron	Reaction with curcumin, forming red rosocyanine ($\lambda = 540$ nm)	212.3
Chloride	Cl^- liberates thiocyante (SCN^-) from mercuric thiocyanate. The SCN^- forms highly colored ferric thiocyanate ($\lambda = 480$ nm)	325.1–2
Chromium	Reacts with 1,5-diphenylcarbohydrazide to form a purple color complex ($\lambda = 540$ nm)	7196
Cyanide	Formation of a blue dye from reaction of cyanogen chloride, CNCl, with pyridine-pyrazolone reagent ($\lambda = 620$ nm)	335.2
Fluoride	Decolorization of a zirconium-dye colloidal precipitate ($\lambda = 570$ nm) by formation of colorless zirconium fluoride and free dye	340.1
Nitrate and nitrite	NO_3^- is reduced to NO_2^-, which is diazotized with sulfanilamide and coupled with N-(1-naphthyl)-ethylenediamine dihydrochloride to produce a highly colored azo dye ($\lambda = 540$ nm)	353.1
Phenol	Reaction with 4-aminoantipyrine (pH 10) in the presence of $K_3Fe(CN)_6$ forming an antipyrine dye that is extracted into pyridine ($\lambda = 460$ nm)	420.1
Phosphate	Reaction with molybdate ion to form a phosphomolybdate that is selectively reduced to intensely colored molybdenum blue ($\lambda = 690$ nm)	365.1–3
Silica	Formation of molybdosilicic acid with molybdate followed by reaction to a heteropoly blue ($\lambda = 650$ or 815 nm).	370.1
Sulfide	Sulfide reacts with dimethyl-p-penylenediamine in the presence of ferric chloride to produce methylene blue ($\lambda = 625$ nm)	376.2
Surfactants	Reaction with methylene blue to form blue salt ($\lambda = 652$ nm)	425.1
Total Kjeldahl nitrogen	Digestion in sulfuric acid to NH_4^+ followed by treatment with alkaline phenol and sodium hypochlorite to form blue indophenol ($\lambda = 630$ nm)	351.1

Ozone (O_3): Potassium iodide (KI) is used to scrub O_3 from air. The resulting I_2 can be measured by a spectrometric method.

$$2KI + O_3 + H_2O \rightarrow 2KOH + O_2 + I_2 \qquad (8.11)$$

$$I_2 + starch \rightarrow blue/purple \ complex \qquad (8.12)$$

Nitric oxide (NO) and nitrogen dioxide (NO$_2$): This is a classical colorimetric method developed by Saltzman (1954). The process begins with bubbling air through an impinge containing aqueous solution of a number of ingredients. When NO$_2$ is bubbled through water, it forms HNO$_2$, which will react with an organic base (amine). The reaction mixture also contains acetic acid and sufficient free protons to protonate the oxygen on the nitrosamine compound. Water is then eliminated from the structure resulting in a diazonium ion, as shown below.

$$2NO_2 \text{ (g)} + H_2O \text{ (l)} \rightarrow HNO_2 \text{ (aq)} + H^+ \text{ (aq)} + NO_3^- \text{ (aq)} \tag{8.13}$$

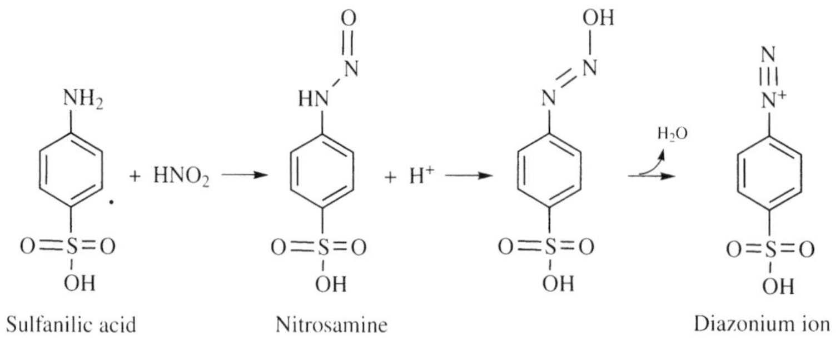

$$(8.14)$$

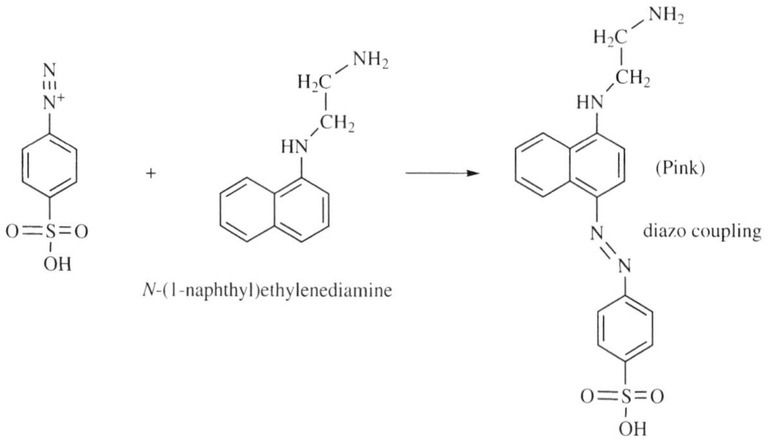

Sulfanilic acid   Nitrosamine   Diazonium ion

N-(1-naphthyl)ethylenediamine   (Pink)

diazo coupling (8.15)

The diazonium ions (characterized by N≡N) are electrophiles, which will seek out electron-containing species to share with. This species, known as *N*-(1-naphthyl) ethylenediamine (N-NED), forms a diazo coupling (—N=N—) by a coupling reaction at the position opposite to the amine group on the N-NED. Diazo coupling always results in a strong colored compound. In this instance, the final compound is pink with maximum absorption at 550 nm. Since nitrite solution is more conveniently used as a standard than gaseous NO_2, Saltzman (1954) also examined stoichiometry of nitrite and concluded that 1 mole of NO_2 (gas) produces the same color intensity as 0.72 mole of sodium nitrite (NO_2^-). This conversion must be applied to the final determination in order to compare with the standard nitrite solution.

UV-VIS Methods for Pollutants in Water

Table 8.5 is a list of common colorimetric methods (UV methods not included) that comprise many of the methods employed in average labs. Key reactions are also described for selected analytes. It is noted that although there are always other alternative methods for the measurement, for special applications, some of the UV-VIS methods are more widely used than their counterparts due to their simplicity. An example is the colorimetric method for Cr(VI) analysis. At concentration higher than 0.5 mg/L, this method is superior to other instrumental methods for a reliable and quick analysis.

To illustrate the principles of colorimetric methods, selected chemical reactions are listed in Figure 8.16.

8.3 INFRARED SPECTROSCOPY

Three types of IR instruments are described below: (a) Fourier transform infrared (FTIR) spectrometers, (b) Dispersive infrared (DIR) instruments, (c) Nondispersive infrared (NDIR) instruments. Nowadays, the interferometric spectrometer coupled with Fourier transform (FTIR) is the most common of all and has largely displaced dispersive IR instruments. For compound-specific applications such as in environmental analysis, the nondispersive IR is also used due to its low cost.

8.3.1 Fourier Transform Infrared Spectrometers (FTIR)

The FTIR uses the Michelson interferometer to produce the interference signals of the sample source light, with two perpendicularly arranged mirrors, one movable and the other fixed. The source light is split into two half beams and reflected back by each mirror. As there are phase-lags caused by the movable mirror, an interference pattern is produced after the combination of two beams, which is composed of all frequency signals of sample in a *time-domain*. This signal is then Fourier-transformed into *frequency-domain*, giving the FTIR spectrum. The *Fourier transform*, named after a French mathematician and physicist Joseph Fourier (1768–1830), is an integral transform that reexpresses a function in terms of sinusoidal basis functions, that is, as a sum or integral of sinusoidal functions multiplied by some coefficients (amplitude).

Major instrument components shown in Figure 8.17 are similar to UV-VIS (Fig. 8.14), but the materials are made very differently as described below.

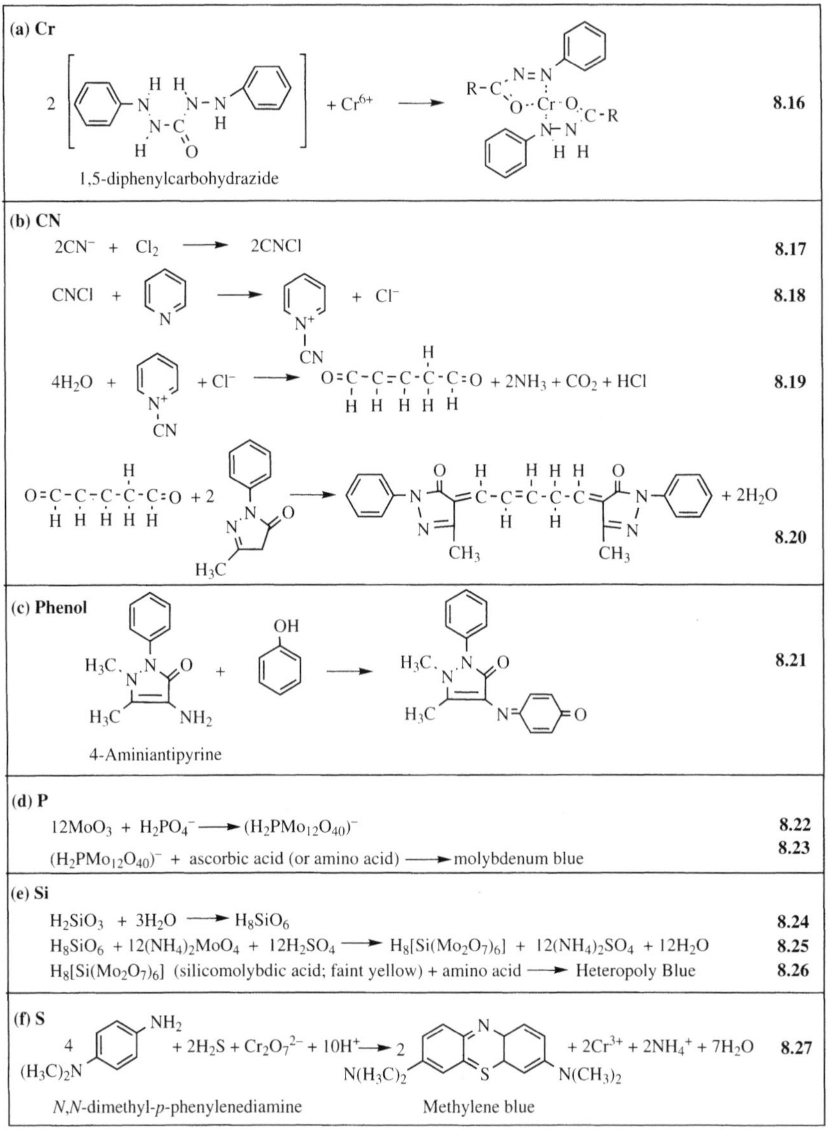

Figure 8.16 Reaction mechanisms for the spectrometric measurement of (a) chromium (Cr): Eq. 8.16, (b) cyanide (CN⁻): Eqs 8.17 to 8.20, (c) phenol: Eq. 8.21, (d) phosphate: Eqs 8.22 to 23, (e) silicon (Si): Eqs 24 to 26, (f) sulfide (S^{2-}): Eqs 8.27

- *Light source*: A typical infrared source is the Nernst glower. This is a rod consisting of a mixture of rare-earth oxides, which is heated to 1500–2000°C to emit the radiation. Another infrared source is the Globar, consisting of a rod made of sintered silicon carbide (SiC), which is heated to about 1300–1700°C.

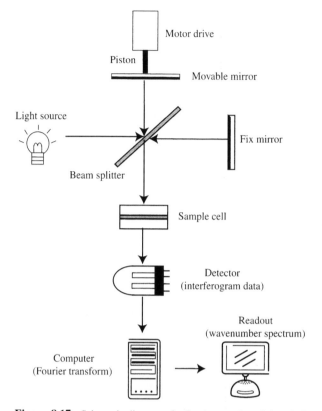

Figure 8.17 Schematic diagram of a Fourier transform infrared (FTIR) spectrometer

- *Michelson interferometer*: This includes a beam splitter (germanium film on a KBr plate support), one fixed mirror, and one movable mirror. The beam splitter is semitransparent, allowing the generation of two beams, one of which falls on a fixed mirror and the other on the mobile mirror. The two beams travel through the sample before hitting the detector. The moving mirror oscillates in time between two extreme positions. The transmitted signal, which varies with time, is recorded as an *interferogram*.
- *Detector*: It is similar to dispersive spectrometer, but employs pyroelectric bolometer as detector, and has fast response time required by the fast scan. Deuterated triglycine sulfate (DTGS) is an example of pyroelectric material.

8.3.2 Dispersive Infrared Instruments (DIR)

The optical designs for the dispersive IR instruments do not differ greatly from the double-beam UV-VIS spectrophotometers except that the sample and reference compartments are always located between the infrared sources and the monochromator in infrared instruments. Major components of the dispersive spectrometer are in between the UV-VIS and FTIR instruments. The infrared source is the same as FTIR, which is essentially the heat; so hot wires, a light bulb, or glowing ceramics are used as infrared light sources. However, instead of Michelson interferometer used in FTIR, a monochromator similar to those used in UV-VIS is used to *disperse* infrared radiations of various wavelengths. The materials of monochromator are different. Since glass and fused silica transmit very little infrared light, prisms and other optics must be made from large crystals of alkaline earth halides, which are transparent to IR radiation. NaCl can be used in the entire region from 2.5 to 15.4 µm (4000–650 cm^{-1}). For longer wavelengths, KBr (19–25 µm) or CsI (10–38 µm) can be used. The monochromator compartment must be kept dry.

8.3.3 Nondispersive Infrared Instruments (NDIR)

This type of IR instrument is designed for target application, that is, for the quantification of one or several compounds. The nondispersive infrared (NDIR) spectrometers incorporate interchangeable filters to isolate a particular wavelength for measurement. Therefore, the nondispersive infrared instruments do not record a spectrum. Major environmental applications include analysis of CO and CO_2 from car exhausts by recording absorbance at 2170 and 2350 cm^{-1}, respectively. The commercial total organic carbon (TOC) analyzer also uses nondispersive IR.

8.3.4 Applications in Industrial Hygiene and Air Pollution Monitoring

IR has a variety of applications as various inorganic and organic chemicals respond to infrared light. It is estimated that, of more than 400 chemicals for which maximum tolerable limits have been set by the OSHA, more than half appear to have absorption characteristics suitable for determination by means of midinfrared filter photometers or spectrophotometers. Out of 189 hazardous air pollutants or HAPs (Table 2.7) listed by the U.S. EPA, more than 100 HAPs have their established IR spectra (http://www.epa.gov/ttn/emc/ftir/welcome.html).

One major disadvantage of IR, however, is its limitation in measuring compounds at low concentrations. This explains why many IR applications are limited to high concentration detection, such as, auto exhaust emission testing, occupational exposure testing in industrial hygiene, and measurement of aggregate organic contaminants in environmental analysis (oil and petroleum). An exhaustive list of such applications is impossible, but some of the major IR-related EPA methods are summarized in Table 8.6 (air methods) and Table 8.7 (water methods). Methods with IR coupled with GC have also been included in EPA's SW-846. For a list of OSHA and NIOSH approved IR methods, the reader should refer to the web sites given in Chapter 5. IR has also found special applications in tropospheric studies on the global distribution of CO_2, O_3, H_2O, HNO_3, and NO_2, and even in stratospheric monitoring of Freon that causes ozone layer depletion.

Table 8.6   IR methods for the analysis of pollutants in air

Pollutant	IR method description	EPA method
CO	Absorption at $2165–2183 \, cm^{-1}$; Measured by nondispersive IR	40 CFR 50, Appendix C
CO_2	Absorption at $2342 \, cm^{-1}$	–
O_3	Absorption at $1045 \, cm^{-1}$ (2 ppb DL, 100 m path length)	–
NO	Absorption at $1920–1870 \, cm^{-1}$ (4 ppb DL, 100 m path length)	–
SO_2	Absorption at $1361 \, cm^{-1}$ (2 ppb DL, 100 m path length)	–
Formic acid	Absorption at $1105 \, cm^{-1}$	
VOC	VOC measurement real-time by FTIR	TO 16

Table 8.7   IR methods for the analysis of pollutants in water and other samples

Pollutant	IR method description	EPA method
Total recoverable oil and grease	Oil and grease from waters and wastewater. Applicable to the determination of hydrocarbons, vegetable oils, animal fats, waxes, grease, and related matter.	413.2
Semivolatile organics	GC-FTIR analysis of SVOCs. Particularly for the identification of specific isomers that are not differentiated using GC/MS.	8410
Ethers and alcohols	Bis(2-chloroethy) ether and its hydrolysis compounds in aqueous matrices by direct aqueous injection (DAI) and GC-FTIR.	8430
TRPH[a]	Sediment, soil, and sludge are extracted by supercritical CO_2 and then analyzed for TRPHs. Petroleum hydrocarbons in water and wastewater.	8440 418.1

[a]TRPH, total recoverable petroleum hydrocarbons.

8.4   PRACTICAL ASPECTS OF UV-VISIBLE AND INFRARED SPECTROMETRY

8.4.1   Common Tips for UV-Visible Spectroscopic Analysis

The use of UV-VIS spectrometers is pretty straightforward. The needed training is minimal compared to all other instruments. Few helpful tips to the beginners are listed below.

- Use only quartz cuvettes (sample cell) in the UV range (200–380 nm). Plastic and disposable cuvettes are fine in the visible range (380–800 nm), but will be melted if samples are in certain solvents such as hexane (Fig. 8.18 a, b).

- Watch out different sides of the cuvette and the light path; do not hold (scratch) the smooth sides of the cuvette. Make sure that the light passes through the cuvette with the smooth sides.

- Always select λ_{max} for a maximum sensitivity. Some UV-visible spectrophotometers have filters in the light paths. Make sure that the right filter is selected for that particular wavelength.

- Unlike IR, only liquid samples are allowed for UV-VIS. Any solid or colloidal particles must be removed before UV-VIS chemical analysis.

- Always keep in mind the linear analytical range of the particular methods. For the visible range spectrophotometer, when absorbance (A) is greater than 0.8–0.9, it is always better to dilute the sample first. If A is too low, such as less than 0.05, a large sample amount is needed whenever possible.

8.4.2 Sample Preparation for Infrared Spectroscopic Analysis

Sample preparation for IR analysis is unique and is the key to acquire a good spectrum. Always keep in mind the useful IR range (infrared transmitting properties) of the materials used in sample preparations.

- IR can analyze samples in various forms—gas, liquid, or solid (Fig. 8.18c):
- *Gas samples*: Usually a long path cell is needed to maximize the sensitivity (Beer's law), ranging from 10 cm (high concentration) to 20–100 m (low concentration). The standard window material for gas cell is KBr.

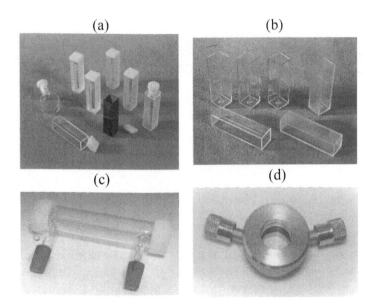

(a)   (b)

(c)   (d)

Figure 8.18   Sample cells for UV, visible and IR: (a) UV quartz cuvettes, (b) Disposable cuvettes for visible light, (c) IR cell for gas samples, and (d) IR cell for liquid samples. Courtesy of Sigma-Aldrich Co. (a and b); Courtesy of Bruker Optics Inc (c and d)

- *Pure liquids*: Sandwiched between two halide crystal disks made of NaCl or KBr. A clean crystal window can be used as a reference background.

- *Diluted solutions*: Put in cells (NaCl, KBr) or IR cards; Carbon tetrachloride (CCl_4) and carbon disulfide (CS_2) are the common solvents used for dilution. A longer path length than pure liquid is needed.

- *Solid samples*: Grinding sample with a mineral oil (Nujol) to form liquid slurry (mull). Otherwise, make a KBr pellet by grinding solid sample with solid KBr and apply high pressure to the dry mixture to form a thin transparent film.

- Recommended pathlength for near-IR: 5–10 cm for 800–1100 nm or 0.1–2 cm for 1100–3000 nm (0.1–1 cm for UV-VIS). In low concentration air pollution monitoring, the cell length can be very long.

- Sodium chloride crystal cells must be placed in desiccators to protect from atmospheric moisture. They require periodic polishing to remove "fogging" due to moisture contamination. For the same reason, maintaining IR instrument at moisture-free condition is essential. Silver chloride windows are often used for wet samples or aqueous solutions. They are soft and will gradually darken due to exposure to visible light.

REFERENCES

BRUICE PY (2001), *Organic Chemistry*, 3rd Edition, Prentice Hall, Upper Saddle River, NJ pp. 1–32.

EUBANKS LP, Middlecamp C, Pienta N, Heltzel C, Weaver G (2006), *Chemistry in Context: Applying Chemistry to Society*, 5th Edition, American Chemical Society, Washington, DC.

Field LD, Sternhell S, Kalman JR (2003), *Organic Structures from Spectra*, 3rd Edition, John Wiley & Sons West Sussex, English, pp. 1–19.

GAUGLITZ G, Vo-Dinh T (2003), Handbook of Spectroscopy, Vol. 1, Wiley-VCH Weinheim.

GIRARD JE (2005), Principles of Environmental Chemistry, Johns and Bartlett Publishers Sudbury, MA.

GÜNZLER H, Gremlich H-U (2002), IR Spectroscopy: An Introduction, Wiley-VCH Weinheim.

HACH (1992), Water Analysis Handbook, 2nd Edition, HACH Company, Loveland, Colorado pp. 754–802.

HOLLAS JM (2004), *Modern Spectroscopy*, 4th Edition, John Wiley & Sons, NY.

*JANG YH (1996), Lecture Notes on Infrared Spectroscopy, http://www.wag.caltech.edu/home/jang/genchem/infrared.htm.

KEBBEKUS BB, Mitra S (1998), *Environmental Chemical Analysis*, Blackie Academic & Professional London, UK pp. 54–102.

MANAHAN S.E. 2005. *Environmental Chemistry*, 8th Ed, 2005, CRC press, pp. 681–741.

PECSOK RL, Shield LD (1968), *Modern Methods of Chemical Analysis*, Wiley, New York, NY.

ROUESSAC F, Rouessac A (2000), Chemical Analysis: Modern Instrumentation Methods and Techniques, John Wiley & Sons, New York, NY.

SALTZMAN BE (1954), Colorimetric microdetermination of nitrogen dioxide in the atmosphere, *Anal. Chem.*, 26 (12):1949–1955.

SKOOG DA, Holler FJ, Nieman TA (1997), *Principles of Instrumental Analysis*, 5th Edition, Saunders College Publishing, New York pp. 299–354.

*Suggested readings.

Stuart B (2004), *Infrared Spectroscopy: Fundamentals and Applications*, John Wiley & Sons West Sussex, England, pp.224.

US EPA (1983), Methods for Chemical Analysis of Water and Wastes, EPA/600/4-79/020.

West PW, Gaeke GC (1956), Fixation of sulfur dioxide as disulfitomercurate (II) and subsequent colorimetric estimation. *Anal. Chem.* 28(12):1816–1819.

QUESTIONS AND PROBLEMS

1. Describe the differences between absorption, emission, and fluorescence?

2. Calculate (a) the energy (in eV) of a 5.3-Å X-ray photon and (b) the energy (kJ/mol) of a 530-nm photon of visible radiation. Given the following constants and unit conversion factors: $1\ \text{Å} = 10^{-10}$ m, 1 nm $= 10^{-9}$ m, 1 nm $= 10$ Å, $1\ \text{J} = 6.26 \times 10^{18}$ eV, Planck constant $(h) = 6.63 \times 10^{-34}$ Js, speed of light $(c) = 3 \times 10^{8}$ m/s.

3. Use molecular orbital theories to explain: (a) Why hexane and water can be used as a solvent for UV absorption spectrometry? (b) Why CO_2 has two sorption bands in the mid-IR range? (c) Why the frequency of bonds is in the increasing order of: C–C < C=C < C≡C ?

4. Explain: (a) How two s orbitals form a σ bond? and (b) How two p orbitals form a σ bond?

5. Explain in general: (a) Why wavelength of less than approximately 180 nm is hardly used in UV spectrophotometer? (b) Why IR is less sensitive than UV but more useful than UV to deduce the structure of unknown chemicals? (c) Why CO_2 and water vapor always present problem when measuring atmospheric contaminants using IR?

6. Define/describe the following terms: (a) the difference between absorption spectroscopy and emission spectroscopy, (b) π electrons and n electrons, (c) sigma bonding and sigma antibonding molecular orbital, (d) dispersive IR and nondispersive IR?

7. Which of the following electronic transition(s) is usually concerned with UV-VIS spectroscopy of organic compounds: (a) $\sigma \rightarrow \sigma^*$, (b) n $\rightarrow \pi$ and $\pi \rightarrow \pi^*$, (c) n $\rightarrow \sigma^*$?

8. Draw electron configuration $(1s^2, 2s^2, \ldots)$ for (a) C and (b) O. The atomic number for C and O are 6 and 8, respectively.

9. Draw the electron dot formula for: (a) O_2, (b) H_2O, and (c) CH_4.

10. Explain: (a) Why do you think it takes more energy (shorter wavelengths, higher frequencies) to excite the stretching vibration than the bending vibration? (Refer to Fig. 8.9 for CO_2), (b) Why C≡C triple bonds have higher stretching frequencies than C=C double bonds, which in turn have higher stretching frequencies than C–C single bonds?

11. Refer to Figure 8.12 and explain why the IR absorption wave number for C=N is greater than that for C–N?

12. (a) Explain what the linear molecules are and what nonlinear molecules are? (b) Calculate how many vibrational modes are there for CO and H_2S?

13. Describe the difference between a typical UV spectrum and an IR spectrum?

14. Explain why the nonlinear deviation in IR is more severe than UV-VIS?

15. Which one of the following is true regarding Beer's law: (a) Absorbance is proportional to both path length and concentration of absorbing species? (b) Absorbance is proportional to the log of the concentration of absorbing species? (c) Absorbance is equal to P_0/P?

16. Perform the following calculation: (a) $A = 0.012$, $\% T = ?$; (b) $T = 95\%$, $A = ?$; (c) wavelength $= 350\,nm$, wave number $= ?\,cm^{-1}$; (d) Wave number $= 4000\,cm^{-1}$, frequency $= ?$ hertz.

17. For the following compound formaldehyde:

$$\begin{matrix} H \\ \\ H \end{matrix}\!\!\!>\!\!C\!\!=\!\!O$$

    (a) Draw the electronic structure that shows all the sharing and nonsharing bonds,
    (b) Indicate the types of molecular orbitals in formaldehyde?

18. The absorbance of a 2-cm sample cell of a 10 ppm solution is 0.43, what would be the absorbance of a 1-cm cell of 15 ppm solution of the same chemical under the same condition?

19. A 2-cm cell of 0.003% (w/v, i.e., 0.003 g/100 mL) solution of $C_{12}H_{17}NO_3$ (MW $= 223.3$) has a λ_{max} of 280 nm with an absorbance of 0.92. What is the absorbance of the same chemical when the concentration is 0.005% using 1/CM cell under the same wavelength? Report ε in 1/CM cm).

20. The molar absorptivities (ε) of two compounds A and B were measured with pure samples at two wavelengths. From these data in the following table, determine the concentration (in M) of A and B in the mixture. The absorbance of a mixture of A and B in a 1.0 cm cell was determined to be 0.886 at 277 nm and 0.552 at 437 nm.

	ε at 277 nm (1/M/cm)	ε at 437 nm (M^{-1}/cm^{-1})
Chemical A	14,780	5112
Chemical B	2377	10,996

21. Draw a schematic diagram of the following: (a) UV-VIS spectrometer and (b) FTIR spectrometer.

22. From the following methods, do a literature/internet search (*Hint:* NEMI database and APHA's standard method book), give a list of methods that use only "UV" spectrometry (i.e., not to include visible spectrometry). (a) EPA 100–600 series methods, (b) APHA's 4000 series methods for inorganic nonmetallic constituents

23. Which one of the following IR instruments cannot record spectrum: (a) FTIR, (b) DIR, or (c) NDIR? Explain why?

24. Explain: (a) Why in situ atmospheric CO_2 can be monitored by IR? (b) Why in situ atmospheric (stratospheric) O_3 can be measured by UV?

25. Explain: (a) Why O_2 and N_2 will not affect the monitoring of CO in auto emission using IR? (b) Why a long cell is needed to monitor trace organic compounds using IR?

Chapter 9

Atomic Spectroscopy for Metal Analysis

Chapter 8 deals with the *molecular* spectroscopy for the determination of inorganic or organic molecules in solution (for UV–VIS) or in other forms (solution, gas, or solid for IR). The spectrum and the amount of energy (UV, visible, or IR radiation) absorbed are the basis for qualitative and quantitative determination of these molecules, respectively. In Chapter 9, *atomic* spectroscopy is used to quantify the elemental concentration based on the amount of energy absorbed or emitted and sometimes to qualify (determine) which element is present.

The primary objective of this chapter is to introduce the principles and environmental applications of four atomic spectroscopic techniques (FAA, GFAA, ICP–OES, and XRF). One particular aspect is to understand the chemistry of molecules and atoms under various radiation sources with energy higher than UV–VIS and IR, including high temperature flame, graphite furnace, plasma, and X ray. A solid understanding of these fundamental principles will help the analyst to address interference problems encountered in atomic spectroscopy and help the analyst to make the appropriate adjustments needed to meet the applicable analytical needs. We will continue our discussions on instrumental components and the

Fundamentals of Environmental Sampling and Analysis,   by Chunlong Zhang
Copyright © 2007 John Wiley & Sons, Inc.

comparisons among various elemental analysis methods. Many of these methods are complementary to each other with regard to sensitivity, linear range, cost, sample throughput, and skills required for operation. This chapter will be concluded with materials on some of the practical tips helpful to beginners in this field.

9.1 INTRODUCTION TO THE PRINCIPLES OF ATOMIC SPECTROSCOPY

To begin our discussions on atomic spectroscopy, we first need to appreciate what differs it from molecular spectroscopy described in Chapter 8. In molecular spectroscopy, low-energy radiation (IR, visible, UV) causes a molecule to vibrate/ rotate or *outer electron* (valence electron) to transit from low- to high-energy state. In atomic spectroscopy, radiation of a much higher energy is used. As atoms are the simplest and purest form of matter and cannot rotate or vibrate as a molecule does, the high-energy radiation causes *inner electrons* to transit within the atom. This high-energy radiation is commonly provided by (a) flame in flame atomic absorption spectroscopy (FAA); (b) electrothermal furnace in flameless graphite furnace atomic absorption spectroscopy (GFAA); (c) plasma in inductively coupled plasma–optical emission spectroscopy (ICP–OES); or (d) X ray in X-ray fluorescence spectroscopy (XRF).

The discussion in this section details the principles of FAA, GFAA, ICP–OES, and XRF. Like molecular spectroscopy, these atomic spectroscopic techniques belong to one of the three major types of atomic spectroscopy described in Figure 8.4, namely absorption, emission, and fluorescene. In *atomic absorption spectroscopy* (FAA and GFAA), light at a wavelength characteristic of the element of interest radiates through the atom vapor. The atoms of that element then absorb some of the light. The amount of light absorbed is measured. In *atomic emission spectrometry* (ICP–OES), the sample is subjected to temperatures high enough to cause excitation and/or ionization of the sample atoms. These excited and ionized atoms are then decayed to a lower energy state through emission. The intensity of the light emitted at a wavelength specific to the element of interest is measured. In *atomic fluorescence spectrometry* (XRF), a short wavelength is absorbed by the sample atoms, whereas a longer wavelength (lower energy) radiation characteristic of the element is emitted and measured.

9.1.1 Flame and Flameless Atomic Absorption

Fundamental to atomic spectroscopy, let us now look at how elevated temperature in flame or furnace changes the chemistry of molecules and atoms and how temperature affects the ratio of excited and unexcited (ground state) atoms. We will also examine how a liquid sample is aspirated to become aerosols of fine particles (nebulization) and then vapor atoms (atomization). Finally, we will discuss Beer's law, introduced in Chapter 8, and how it is also applicable to the quantitative measurement of elements in atomic spectroscopy.

Processes Occurring in Flame and Flameless Furnace

The thermal chemistry of molecules and atoms under high-temperature flame or flameless furnace is the heart of atomic absorption techniques as well as the flame emission techniques. When a solution of acid digestate (see Section 7.2) is introduced into a high-temperature flame or furnace (direct introduction of solid sample is possible only through graphite furnace), molecules containing the elemental atoms will eventually become gaseous atom through a series of reactions. Figure 9.1 is an illustration of these reactions when calcium (Ca) is analyzed.

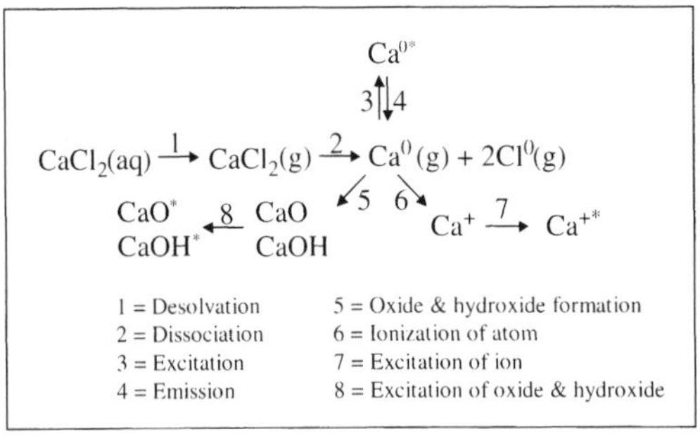

$$Ca^{0*}$$
$$3 \uparrow \downarrow 4$$
$$CaCl_2(aq) \xrightarrow{1} CaCl_2(g) \xrightarrow{2} Ca^0(g) + 2Cl^0(g)$$
$$\nearrow 5 \quad 6 \searrow$$
$$CaO^* \xleftarrow{8} CaO \qquad\qquad Ca^+ \xrightarrow{7} Ca^{+*}$$
$$CaOH^* \qquad CaOH$$

1 = Desolvation	5 = Oxide & hydroxide formation
2 = Dissociation	6 = Ionization of atom
3 = Excitation	7 = Excitation of ion
4 = Emission	8 = Excitation of oxide & hydroxide

Figure 9.1 Process occurring in flame for a solution containing $CaCl_2$ (Christian, 2003, Reproduced with permission from John Wiley & Sons, Ltd.)

As can be seen from Figure 9.1, Ca is initially present in its salt form such as $CaCl_2$ in the aqueous sample. After the removal of water (desolvation), a gaseous $CaCl_2$ is produced, which is further dissociated into gaseous atom (Ca^0). At an elevated temperature, Ca can have other electronic configurations, including Ca^{0*} (excited vapor Ca atom), Ca^+ (ionic Ca), and Ca^{+*} (ionic Ca with an excited electron).

If Ca is analyzed by atomic absorption spectroscopy, we measure the radiation absorption of gaseous atom (Ca^0) based on the excitation reaction in Figure 9.1. Ca can also be analyzed by flame emission spectroscopy. In this case, we measure the emission of Ca^{0*} based on the emission process shown in Figure 9.1. If M is used to denote the vapor form of other atoms (metal elements), the general reactions are:

$$M + h\nu \rightarrow M^* \text{ (for flame atomic absorption spectrometry)} \qquad (9.1)$$

$$M^* \rightarrow M + h\nu \text{ (for flame atomic emission spectrometry)} \qquad (9.2)$$

Note that only the reactions leading to the formation of gaseous atoms (Ca^0) are the desired reactions for atomic absorption spectroscopy. All other reactions (Reactions 4–8) are undesired and will result in interferences. For flame emission spectroscopy, reactions leading to the formation of excited gaseous atom (Ca^{0*}) and the subsequent emission to the ground state Ca^0 are the desired reactions. It is important to note that formation of metal oxide/hydroxide (Reaction 5) and the ionization of gaseous atom (Reaction 6) are common sources of interference that must be minimized.

Flame and flameless graphite furnaces are two common radiation sources used in atomic spectroscopy (see Section 9.2.1 for a schematic diagram). The flame used in atomic absorption or emission spectroscopy is a relatively low energy excitation source, because temperature is often around 2000–3000 K, depending on the type and velocity of the fuel gas. For example, the maximum temperatures that can be achieved for air–acetylene and nitrous oxide–acetylene are 2250 and 2955°C, respectively. The second common type of excitation source used in atomic absorption is *flameless graphite furnace*. When a potential difference of a few volts is applied on a graphite boat that holds a sample, the graphite rod behaves as an electric resistor and is heated. The heating process is accomplished through a four-stage cycle of a programmed temperature profile: (a) drying (100°C); (b) decomposition (400°C); (c) atomization (2000°C); and (d) pyrolysis (cleaning, 2300°C). Drying is needed to avoid splashing of sample when a sudden temperature increase is employed. Note that the atomization reactions in furnace are similar to the processes in flame as described in Figure 9.1.

The Boltzmann Equation: Understanding the Temperature Effect and the Sensitivity Difference Between Absorption and Emission Spectrometry

Temperature has a profound effect on the ratio between the excited atoms and unexcited (ground state) atoms in the flame or flameless atomizer. The quantitative description of this effect is the *Boltzmann equation*:

$$\frac{N^*}{N} = \frac{P^*}{P} e^{-(E^* - E)/kT} \tag{9.3}$$

where N^* and N are the number of atoms in an excited state and the ground state, respectively, k is the Boltzmann constant (1.38×10^{-23} J/K), T is the temperature in kelvin, and $E^* - E$ is the energy difference between the excited state and the ground state. The P^* and P are the statistical weights of the excited and ground states, respectively, which can be determined by the number of states having equal energy at each quantum level n (Section 8.1.2). In a hydrogen atom, for example, there are $P = 2$ ways that an atom can exist at the $n = 1$ energy level ($1s^2$), and $P = 8$ ways that an atom can arrange itself at the $n = 2$ energy level ($2s^2$, $2p^6$). The example below illustrates the use and implications of Boltzmann equation.

EXAMPLE 9.1. Determine the ratio of sodium (Na) atoms in the 3p (excited state) to the number in the 3s (ground state) at 2600 and 2610 K. Assume an average wavelength of 5893 Å for the 3p → 3s transition. There are two quantum states in the 3s level and six in the 3p, hence $P^*/P = 6/2 = 3$.

SOLUTION: Recall from Chapter 8 that the wavelength (λ) and energy are related by: $\Delta E = hc/\lambda$ where h = Planck constant (6.62×10^{-34} Js) and c = speed of light (3×10^8 m/s). The wavelength needs to be converted into metric units in meters. Since $1m = 10^{10}$ Å, 5893Å $= 5.893 \times 10^{-7}$ m.

$$\Delta E = E^* - E = hc/\lambda = 6.62 \times 10^{-34} \times 3 \times 10^8/(5.893 \times 10^{-7})$$
$$= 3.37 \times 10^{-19} \text{J}$$

$$\frac{N^*}{N} = \frac{6}{2}e^{-3.37 \times 10^{-19}/(1.38 \times 10^{-23} \times 2600)} = 2.50 \times 10^{-4}(\text{at } 2600 \text{ K})$$

For the N^*/N ratio at 2610 K, we replace 2600 K with 2610 K in the above equation, which yields $N^*/N = 2.59 \times 10^{-4}$, a 3.7% increase in the number of excited sodium atoms due to the temperature increase from 2600 to 2610 K.

The above example has several important implications: (a) The ratio of excited vs. ground state is very low (e.g., 2.50×10^{-4} at 2500 K), hence most of the atom at such temperature is in the ground state (e.g., Ca^0). (b) From earlier discussions, we know that the atomic absorption measures the absorption of atom in the ground state. This means that the atomic absorption is less dependent on the temperature variation. On the contrary, atomic emission relies on the number of the excited atoms. An increase in 10 K (2600 to 2610 K) results in a 3.7% increase in excited atoms (Ca^{0*}). This means that temperature variation needs to be better controlled in atomic emission to assure reproducible data. (c) Since atomic absorption methods are based on a much larger population of ground state atoms, they are expected to be more sensitive than the flame emission methods.

Understanding "Nebulization" and "Atomization" Process: Why Higher Sensitivity is Achieved in Flameless GFAA Than Flame FAA

In flame absorption spectrometry, a liquid sample (normally several milliliters) is pumped into a mixing chamber where a process called "nebulization" is taking place. *Nebulization* is to aspirate liquid sample into small liquid particles (aerosols). It is estimated that only 10% of the fine aerosols will reach the burner, and all the remaining larger droplets of the samples will condense and will be drained out of the chamber. This results in lower efficiency in flame FAA as compared with graphite furnace GFAA, because the entire sample is atomized for the latter.

It is critical not to be confused between "nebulization" and "atomization" which are taking place in flame or furnace. *Atomization* is a process that converts sample elements into atomic vapor (gaseous atoms such as Ca^0). In flame atomization (FAA), the overall efficiency of atomic conversion and measurement of ions present in aspirated solutions has been estimated to be as little as 0.1%. This is in significant contrast with almost 100% in electrothermal atomization in graphite furnace (GFAA). This difference simply implies that, in theory, the detection limits in GFAA will be 100–1000 times improved over the FAA.

Quantitation and Qualification of Atomic Spectroscopy

Performing the qualitative and quantitative measurement is generally straightforward in atomic spectroscopy. Qualitative information, such as which elements are present in the sample, is normally obtained for atomic emission spectroscopy rather than atomic absorption. This is because in emission, a series of spectral lines characteristic of the element of interest is identified. For quantitation, the concentration of an element present in the sample is described by *Beer's law*, which was discussed in Chapter 8. In atomic absorption spectroscopy (FAA and GFAA), the absorption depends on the number of ground state atoms N in the optical path. Measurements are made by comparing the unknown to standard solutions.

$$A = k \cdot l \cdot C \tag{9.4}$$

where A is the absorbance, C is the concentration of the element, l is path length of the flame, and k is a coefficient unique to each element at a given wavelength.

For flame emission spectroscopy, the emitted light intensity I of a population of n excited atoms depends on the number of atoms dn that return to the ground state during an interval time dt ($dn/dt = k\, n$). As n is proportional to the concentration of the element, the emitted light intensity I, which varies as dn/dt, is also proportional to the concentration:

$$I = k \cdot l \cdot C \tag{9.5}$$

9.1.2   Inductively Coupled Plasma Atomic Emission

Flame emission atomic spectrometry and plasma atomic emission are the two atomic emission techniques. However, only the latter is critically important and is becoming a predominant tool in environmental metal analysis. The principles of flame emission have been discussed in the previous section, because the same processes occur with the flame atomic absorption. The plasma atomic emission has similar processes as shown in Figure 9.1, but the principles are distinctly different.

Inductively coupled plasma (ICP), or *plasma* in short, is an ionized gas at an extremely high temperature. It is the fourth state of matter besides gas, liquid, and solid. At a typical plasma temperature of approximately 9000 K (recall that flame temperature is only 2000–3000 K), the commonly used argon gas (Ar) is ionized according to the following reaction:

$$Ar \rightarrow Ar^+ + e^- \tag{9.6}$$

The mean energy of argon ions (Ar^+) thus produced is 15.76 electron volts (eV). This energy in the plasma is transferred by collision of Ar^+ with the atoms of interest from a sample. As shown in Table 9.1, such energy is enough high to ionize many metals with a typical ionization energy of approximately 7–8 eV. The alkali metals are particularly vulnerable to ionization because of the low energy of ionization ($\sim$4 eV). As a result, the alkali metals emit the lowest energy or the longest wavelength visible light. Most metals are ionizable little more readily than metalloids followed by nonmetals. Most metals will emit in the UV range and therefore can be measured using a UV spectrometer. For nonmetals, since their typical ionization energies of about 12 eV are just below the ionization of Ar, nonmetals will not readily form ions in the ICP. Because of the difficulty in ionizing nonmetals and the high energy needed, vacuum UV (150–160 nm) is required for nonmetals such

Table 9.1 Ionization energy for selected elements and fraction of the ionized species in the argon plasma

Element	Ionization energy (eV)	Degree of ionization (%)
Alkali metals		
Li	5.39	99.99
Na	5.14	99.9
K	4.34	99.9
Metals		
Cu	7.73	90
Zn	9.39	75
Pb	7.42	97
Cd	8.99	65
Ni	7.64	91
Hg	10.44	38
Metalloid		
As	9.79	52
Se	9.75	33
Nonmetals		
P	10.49	33
S	10.36	14
Cl	12.97	0.9

as P, S, and C to have emission lines. This explains why ICP–OES is not commonly used for nonmetal analysis.

ICP can theoretically analyze almost all elements. This is in significant contrast to only approximately 70 elements (metals and nonmetals) that can be analyzed by atomic absorption and flame emission spectroscopy. Another significant advantage of ICP is its ability to analyze multiple elements simultaneously as compared with the absorption-based atomic spectroscopy, which can analyze only one element at a time. This ability is the result of the simultaneous emission of multielements and the ability of modern optical devices to resolve various wavelengths from complex emission spectra.

We interchangeably used ICP or ICP–OES for the above discussion. Note that various names have been used for the plasma emission spectroscopy, including ICP, ICP–AES, and ICP–OES. The use of ICP should be avoided because this represents only the method of energy transfer. ICP–AES is more commonly used, likely because users and manufacturers who in many cases have worked with atomic emission spectrometry. However, the use of the term "atomic" is probably misleading because in ICP, most particles in the plasma are ions rather than atoms— unlike what we discussed in other atomic absorption and emission techniques. With this regard, ICP–OES (inductively coupled plasma–optical emission spectrometry) is technically a more proper term.

9.1.3   Atomic X-ray Fluorescence

The atomic X-ray fluorescence for elemental analysis is a two-step process. The first step is excitation of electrons at the lower shell such as K electron (i.e., $n = 1$, the electron that is located in the inner orbitals rather than outer shells). The energy needed for such excitation comes from X ray or other alternative sources such as radioactive decay. Because the inner K shell electrons are stripped off, it creates a vacancy in this shell. The second step involves the "jumps in" of the electrons from L shell ($n = 2$) or M shell ($n = 3$) to fill the vacancy so that the ionized atom will be stabilized. In this second process, it emits a characteristic X ray unique to this element and in turn, produces a vacancy in the L or M shell.

Figure 9.2 illustrates the X-ray fluorescence process of a titanium atom. The figure on the left shows how an electron in the K shell is ejected from the atom by an external primary excitation X ray, creating a vacancy. The figure on the right shows how an electron from the L or M shell "jumps in" to fill the vacancy. In the process, it emits a characteristic X-ray fluorescence unique to this element.

The XRF instrument measures the photon energy from the X-ray fluorescence to identify which element is present and the intensity of the photon to measure the amount of the element in the sample.

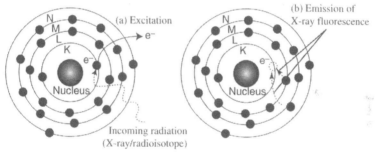

Figure 9.2   An example of X-ray fluorescence from a titanium atom (atomic number is 22). The distribution of electrons in a titanium atom is: $1s^2\ 2s^2\ 2p^6\ 3s^2\ 3d^2\ 4s^2$

9.2   INSTRUMENTS FOR ATOMIC SPECTROSCOPY

9.2.1   Flame and Flameless Atomic Absorption

Although variations exist, the basic instrument components of an atomic absorption spectrometer can be depicted in Figure 9.3 for a single-beam system (In a double-beam system, the light from the source lamp is divided into a sample beam, which is focused through the sample cell, and a reference beam, which is directed around the sample cell.) The instrument components for graphite furnace atomic absorption spectrometry are similar to the flame system except for the method of atomization. Details of these components are described below. Practical users need

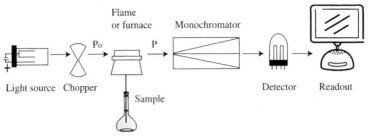

Figure 9.3   Instrument components for a single-beam atomic absorption spectrometer

to be familiar with the light source and the atomizer rather than the optical and electrical detection components. The latter can be effectively considered a "black box" since their repair and maintenance are generally beyond the ability of most analytical chemists.

1. Light source: The light source is usually a *hollow cathode lamp* (HCL) of the element being measured. Its purpose is to provide the spectral line for the element of interest. The HCL uses a cathode made of the element of interest with a low internal pressure of an inert gas. A low electrical current (~10 mA) is imposed in such a way that the metal is excited and emits a few spectral lines characteristic of that element (For instance, Cu has 324.7 nm and a couple of other lines; Se has 196.0 nm and other lines.) The light is emitted directionally through the lamp's window, which is made of a glass transparent in the UV and visible wavelengths.

2. Nebulizer and atomizer: In a flame system (Figure 9.4a), the *nebulizer* is to suck up liquid sample at a controlled rate, create a fine aerosol, and mix

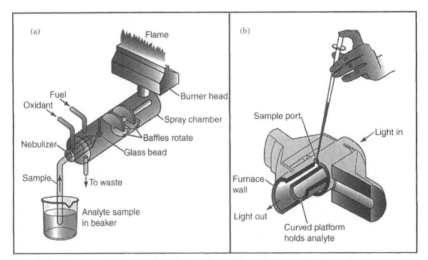

Figure 9.4   Diagram of (a) a premix nebulizer burner for a flame atomic absorption spectrometer, and (b) a graphite furnace for a flameless atomic absorption spectrometer (Girard, JE, Principles of Environmental chemistry, 2005, Jones and Bartlett publishers, Sudbury, MA. WWW.jbpub.com. Reprinted with permission.)

the aerosol with fuel oxidant thoroughly for introduction into the flame. The flame then destroys any analyte ions and breakdown complexes, and creates atoms (the elemental form) of the element of interest into vapor form such as Fe^0, Cu^0, Zn^0 (Fig. 9.1). As shown in Figure 9.4a, the rapid flow of the reactant gases, along with the spinning baffles in the chamber, acts to reduce the droplet size of the sample to a fine aerosol. Excess liquid that is no longer in the gas stream collects on the bottom of the premix chamber and flows out as waste. In a flameless *graphite furnace* system (Fig. 9.4b), instead of using an aspiration device, both liquid and solid samples (a few microgram or microliter sample) can be deposited directly to a graphite boat using a syringe inserted through a cavity. The graphite furnace can hold an atomized sample in the optical path for several seconds, compared with only a fraction of a second in the flame system. This results in a significantly higher sensitivity of the GFAA as compared with FAA.

3. Monochromator: The role of a *monochromator* is to isolate photons of various wavelengths that pass through the flame or furnace and remove scattered light of other wavelengths from the light source. In doing this, only a narrow spectral line impinges on the photomultiplier tubes (PMT) detector. The monochromator of atomic absorption uses UV–VIS radiation, which is governed by the same principles as those described in Section 8.2.1 for a UV–Visible spectrometer. These two devices are similar in mechanical construction in the sense that they employ slits, lenses, mirrors, windows, and either gratings or prisms.

4. Detector: The *PMT detector* determines the intensity of photons in the analytical line exiting the monochromator. Since the basis for both the flame and flameless system is atomic absorption, the monochromator seeks to allow only the light not absorbed by the analyte atoms in the flame or furnace to reach the PMT. That is, before an analyte is atomized, a measured signal is generated by the PMT as light from the HCL passes through the flame or furnace. When analyte atoms are present in the flame or furnace—while the sample is atomized—some part of that light is absorbed by those atoms. This causes a decrease in PMT signal that is proportional to the amount of analyte. In a double-beam system, the readout represents the ratio of the sample to reference beams. Therefore, fluctuations in source intensity do not become fluctuations in instrument readout, and signal stability is enhanced.

9.2.2 Cold Vapor and Hydride Generation Atomic Absorption

The two atomic absorption spectrometric methods using flame and graphite furnace can measure most elements with satisfactory detection limits. There are notable exceptions to these techniques, including several elements (mercury, selenium, and arsenic) that are of environmental importance. Mercury (Hg) is a volatile metal;

it does not need to be heated in a flame or furnace during spectroscopic analysis. It can be analyzed by a "cold vapor" atomic absorption (CVAA) technique. Arsenic (As) and selenium (Se) are not enough volatile at room temperature to be analyzed by the "cold vapor" techniques, but they are too volatile to be analyzed by FAA or GFAA. The hydride generation atomic absorption (HGAA) technique, introduced in Chapter 7 as a sample preparation method (Section 7.2.3), is applicable in atomic absorption spectroscopy for the analysis of As, Se, and several other elements (Bi, Ge, Pb, Sb, Sn, Te).

Cold Vapor Atomic Absorption (CVAA) Spectroscopy for Hg

Free mercury (Hg) atoms can exist at room temperature and, therefore, Hg can be measured by atomic absorption without a heated sample cell. In the CVAA technique, Hg is chemically reduced to the free atomic state by reacting the sample with a strong reducing agent like stannous chloride ($SnCl_2$) or sodium borohydride ($NaBH_4$) in a closed reaction system (Eq. 7.7). The volatile-free mercury is then driven from the reaction flask by bubbling air or argon through the solution. Mercury atoms are carried in the gas stream through tubing connected to an absorption cell, which is placed in the light path of the AA spectrometer (Fig. 9.5). Sometimes the cell is heated slightly to avoid water condensation but otherwise the cell is completely unheated. As the mercury atoms pass into the sampling cell, measured absorbance rises indicating the increasing concentration of mercury atoms in the light path.

The sensitivity of the cold vapor technique is far greater than can be achieved by conventional flame AA. This improved sensitivity is achieved, first of all, through a 100% sampling efficiency. All of the mercury in the sample solution placed in the

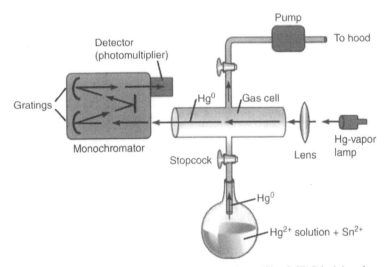

Figure 9.5 Schematic diagram of cold vapor mercury analyzer (Girard, JE, Principles of Environmental chemistry, 2005, Jones and Bartlett publishers, Sudbury, MA. WWW.jbpub.com. Reprinted with permission.)

reaction flask is chemically atomized and transported to the sample cell for measurement. The sensitivity can be further increased by using very large sample volumes. Since all of the mercury contained in the sample is released for measurement, increasing the sample volume means that more mercury atoms are available to be transported to the sample cell and measured. The detection limit for mercury by this cold vapor technique is approximately 0.02 mg/L. Where the need exists to measure even lower mercury concentrations, some systems offer an *amalgamation* option. Mercury vapor liberated from one or more sample aliquots in the reduction step is trapped on a gold or gold alloy gauze. The gauze is then heated to drive off the trapped mercury, and the vapor is directed into the sample cell. The only theoretical limit to this technique would be that imposed by background or contamination levels of mercury in the reagents or system hardware.

Although the cold vapor system is the most sensitive and reliable technique for determining very low concentrations of mercury by atomic absorption, this technique requires a specialized instrument that can be used to only analyze mercury since no other element offers the possibility of chemical reduction to a volatile-free atomic state at room temperature.

Hydride Generation Atomic Absorption (HGAA) Spectroscopy for As and Se

In Chapter 7 (Section 7.2.3), we discussed the chemical principle of how to generate volatile hydride (AsH_3 and SeH_2) from samples containing various species of As or Se. This hydride generation module can be incorporated into an atomic absorption instrument, namely the HGAA spectrometer. As depicted in Figure 9.6, many of the main parts of the HGAA system are identical to that of FAA including a hollow cathode lamp, air–acetylene flame, and optical system. The nebulizer required in FAA is not used in HGAA, and the only required addition is the hydride generation module. Consequently, HGAA spectrometer is available through an

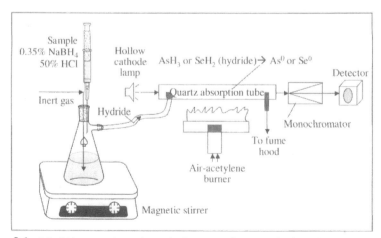

Figure 9.6   A hydride generation and atomization system in an atomic absorption spectrometer

option for many modern FAA instruments. The HGAA system shown in Figure 9.6, for illustration, is a batch flow system, but continuous flow systems are more commonly used. Less-expensive systems use manual operation and a flame-heated cell. The most advanced systems combine automation of the sample chemistries and hydride separation using flow injection techniques with decomposition of the hydride in an electrically heated, temperature-controlled quartz cell.

Similar to the cold vapor mercury systems, samples in a hydride generation module are reacted in an external system with a reducing agent (usually $NaBH_4$). Gaseous reaction products are then carried to a sampling cell in the light path of the AA spectrometer. Unlike the mercury technique, the gaseous reaction products are not free analyte atoms but the volatile hydrides (SeH_3 and SeH_2). These molecular species are not capable of causing atomic absorption. To dissociate the hydride gas into free atoms, the sample cell must be heated. In some hydride systems, the absorption cell is mounted over the burner head of the AA spectrometer, and the cell is heated by an air–acetylene flame. In other systems, the cell is heated electrically. In either case, the hydride gas is dissociated in the heated cell into free atoms (Se^0 and Se^0), and the atomic absorption rises and falls as the atoms are created and then escaped from the absorption cell. The maximum absorption reading, or peak height, or the integrated peak area is taken as the analytical signal.

The elements determinable by hydride generation techniques can not only measure As and Se, but also measure several other metals and metalloids such as Bi, Ge, Pb, Sb, Sn, and Te. For these elements, detection limits well below the mg/L range are achievable. Like cold vapor mercury, the extremely low detection limits result from a much higher sampling efficiency. In addition, separation of the analyte element from the sample matrix by hydride generation is commonly used to eliminate matrix-related interferences.

The major limitation to the hydride generation technique is that it is restricted primarily to several elements listed above. Results depend heavily on a variety of parameters, including the valence state of the analyte, reaction time, gas pressures, acid concentration, and cell temperature. Therefore, the success of the hydride generation technique will vary with the care taken by the operator in attending to the required detail. The formation of the analyte hydrides is also suppressed by a number of common matrix components, leaving the technique subject to certain types of chemical interference.

9.2.3 Inductively Coupled Plasma Atomic Emission

Figure 9.7 is a representation of the layout of a typical ICP–OES. The sample is nebulized and entrained in the flow of plasma support gas, which is typically argon (Ar). The plasma torch consists of concentric quartz tubes. The inner tube contains the sample aerosol and Ar support gas and the outer tube contains flowing gas to keep the tubes cool. A *radio frequency* (RF) *generator* (typically 1–5 kW at 27 MHz) produces an oscillating current in an induction coil that wraps around the tubes. The induction coil creates an oscillating magnetic field, which in turn sets up an

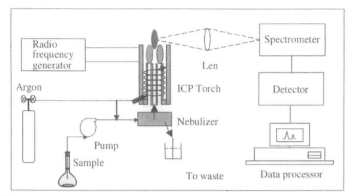

Figure 9.7 Schematic diagram of inductively coupled plasma-optical emission spectrometry (ICP–OES)

oscillating current in the ions and electrons of the support gas (Ar). The energy transferred to a stream of argon through an induction coil produces temperatures up to 10,000 K. The high temperature of the plasma atomizes the sample and promotes atomic and ionic transitions, which are observable at UV and visible wavelength. The excited atoms and ions emit their characteristic radiation, which are collected by a device that sorts the radiation by wavelength. The intensity of the emission for the analyte is detected and turned into electronic signals that are output as concentration information.

There are two ways of viewing the light emitted from the ICP. In the classical radial ICP–OES configuration, the light across the plasma is viewed radially, resulting in the highest upper linear ranges. By viewing the light emitted by the sample looking down the center of torch or axially, the continuum background from the ICP itself is reduced and the sample path is maximized. The axial ICP–OES provides better detection limits—by as much as a factor of 10, than those obtained by radial ICP–OES.

9.2.4 Atomic X-ray Fluorescence

There are three main types of X-ray fluorescence (XRF) instruments in use today: wavelength dispersive, energy dispersive, and nondispersive. In *wavelength dispersive* XRF, the fluorescence radiation is separated according to wavelength by diffraction on an analyzer crystal. This is typically a very expensive type of XRF not in common uses. The XRF with current environmental applications is the energy dispersive. As shown in Figure 9.8, the *energy dispersive* XRF consists of a polychromatic source (either an X-ray tube or a radioactive material), a sample holder, a semiconductor detector, and the various electronic components required for energy discrimination. Note that the emitted photons released from the sample are observed from the sample at a 90° angle to the incident X-ray beam so that the incident light will not interfere the detector. In the detector, each photon strikes a silicon wafer that

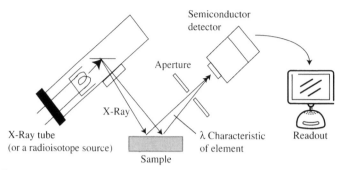

Figure 9.8   Schematic diagram of an energy-dispersive XRF elemental analyzer

has been treated with lithium and generates an electrical pulse that is proportional to the energy of the photon. The concentration of the element is determined by counting the number of pulses.

The energy dispersive XRF has no moving parts and is smaller and cheaper than the wavelength dispersive XRF. As the semiconductor detector is much closer to the sample, this results in strong signal and high sensitivity. However, the resolution of energy dispersive XRF is lower than wavelength dispersive XRF. For the details of various XRF techniques, readers should consult the book by Jenkins (1988).

Under the U.S. EPA's SW-846 method (EPA 6200), inorganic analytes of interest are identified and quantitated using a field portable energy dispersive X-ray fluorescence spectrometer. Radiation from one or more radioisotope sources (Fe-55, Cd-109, Am-241, Cm-244) or an electrically excited X-ray tube (Cu, Mo, Ag in anode) is used to generate characteristic X-ray emissions from elements in a sample. Up to three sources may be used to irradiate a sample. Each source emits a specific set of primary X rays that excite a corresponding range of elements in a sample. When more than one source can excite the element of interest, the source is selected according to its excitation efficiency for the element of interest.

XRF is unique among all atomic spectroscopic techniques in that it provides a nondestructive way to analyze samples. This method is well adapted to qualitative analysis; however, for quantitative analysis, serious problems can be encountered and reference standards have to be analyzed in matrices that are almost identical in composition to that of the sample. XRF has been the method of choice for field applications and industrial process control for the measurement of elemental composition of materials. Detection limits are poor, in the range of 0.01–10%. It does not work for light elements such as C, B, because they do not have L or M shells. XRF as a tool for field screening of heavily contaminated media has become increasingly important, particularly when hand-held XRF instruments have been commercially available. Reported uses include brownfield site investigation, testing metallic soil contamination, contaminated land remediation, industrial hygiene testing, analyzing bulk soil and powders, elemental dust-wipe analysis, and air quality-air filter analysis. The detection limits with hand-held XRF are in the ppm (mg/kg) range.

9.3 SELECTION OF THE PROPER ATOMIC SPECTROSCOPIC TECHNIQUES

It is not always intuitive for a beginner to choose appropriate atomic spectroscopic techniques for a specific type of elemental (metal) analysis. In the relatively small settings of analytical and environmental labs, analysts really do not have many options and are forced to use whatever instruments are available. In other cases, such as in commercial labs, the analysts may have various instrument options to choose by deciding which technique is optimum to meet a particular analytical need. A clear understanding of the analytical problem and the capability provided by different techniques are therefore necessary. Important factors include detection limits, analytical working range, sample throughput, cost, interferences, ease of use, and availability of proven methodology. These are summarized below in Sections 9.3.1 and 9.3.2 for the comparison of FAA, GFAA, ICP–OES, and ICP–MS. Technical details of ICP–MS will be described in Chapter 12. As a semiquantitative tool, XRF is excluded in the following comparison; its environmental applications and limitations have been briefly discussed in the previous section.

9.3.1 Comparison of Detection Limits and Working Range

A low detection limit is essential for trace analysis of metals. Without adequate detection limit capabilities, lengthy sample preparation (concentration, extraction, and so forth) for the manipulation of the analytes may be required prior to analysis. In general, the detection limits are in a decreasing order of: FAA > ICP–OES (radial) > ICP–OES (axial) > hydride generation AA > GFAA > ICP–MS (Fig. 9.9). FAA is the least sensitive, and ICP–MS is the most sensitive.

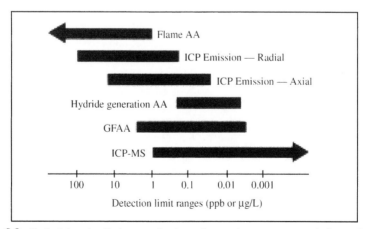

Figure 9.9 Typical detection limit ranges for the major atomic spectroscopy techniques (Courtesy of Perkin-Elmer, Inc.)

For mercury and those elements that form hydrides, the cold vapor mercury or hydride generation techniques offer exceptional (low) detection limits.

The *analytical range* is the concentration range over which quantitative results can be obtained without having to recalibrate the system. An ideal working range minimizes analytical effort (e.g., dilution, concentration) and time by allowing many samples with analyte concentrations within that range to be analyzed together. Figure 9.10 shows the typical analytical working ranges with a single set of instrumental conditions. The wideness of the working range increases in the order of: GFAA < FLAA < ICP–OES < ICP-MS. For instance, the ICP–MS has 8 orders of magnitude of signal intensity.

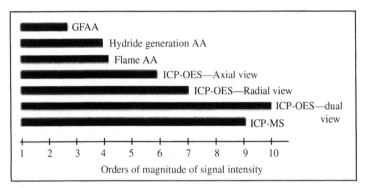

Figure 9.10   Typical analytical working ranges for the major atomic spectroscopy techniques (Courtesy of Perkin-Elmer, Inc.)

9.3.2   Comparison of Interferences and Other Considerations

Interference

Interference is one of the major issues to deal with in atomic spectroscopy. There are three types of interferences: spectral, chemical, and physical. Table 9.2 shows the types of interference for various atomic spectroscopic analyses and the methods of compensation. Basic terminology is provided below. For specific details, the reader should consult the excellent textbook by Skoog et al. (1997).

Spectral interference: Spectral interference occurs in both atomic emission and atomic absorption spectroscopy. In atomic emission, the spectral interference occurs when either another emission line or a molecular emission band is close to the emitted line of the test element and is not resolved from it by the monochromator. Such molecular emission could come from the oxides of other elements in the sample (refer to Fig. 9.1). In atomic absorption spectroscopy, solid particles, unvaporized solvent droplets, or molecular species in the flame can cause a positive spectral interference.

The interelement effects and background absorption are more pronounced in the GFAA than in the FAA system. Such spectral interference can be corrected by a

Table 9.2 Atomic spectroscopy interferences

Technique	Type of interferences	Method of compensation[a]
FAA	Ionization	Ionization buffer
	Chemical	Releasing agent or nitrous-acetylene flame
	Physical (self-absorption)	Dilution, matrix matching, or method of additions
GFAA	Physical and chemical	Stabilized temperature platform furnace (STPF) condition
	Molecular absorption	Zeeman or continuum source background correction
	Spectral	Zeeman background correction
ICP–OES	Spectral	Background correction or the use of alternative analytical lines, IECs or MSF
	Matrix	Internal standardization
ICP–MS	Mass overlap	Interelement correction, use of dynamic reaction cell (DRC) technology, use of alternate mass values or higher mass resolution
	Matrix	Internal standardization

[a]Details of these compensation methods can be found in Skoog et al. (1997). IEC = Interfering element correction; MSF = Multicomponent spectral fitting. *Zeeman background correction* = A method to correct for background absorption in furnace AA that uses a magnetic field around the atomizer. The field splits the energy levels of the absorbing atoms and allows discrimination of atomic absorption from other sources of absorption. (Courtesy pf Perkin–Elmer, Inc.)

standard additions method for calibration (see Example 9.2 in Section 9.4) or background correction. With background correction, the standard is added to a separate aliquot of the sample and the increase in the measured signal is proportional to the concentration added. In this manner, the standard is subjected to the same matrix as the sample. The background absorption interference, particularly serious in complex biological and environmental samples, can be readily eliminated by an automatic background subtraction (the resonance line absorption from the hollow-cathode lamp minus the broadband absorption of a continuum source).

Chemical interference: The chemical interference is a result of the formation of undesired chemical species during the atomization process, such as ions and refractory compounds (Fig. 9.1). Their effects are more common than spectral ones, but can be frequently minimized by selecting a proper operating condition. Chemical interferences are more common in low-temperature systems such as FAA and GFAA than in high-temperature ICP systems.

When easily ionized elements (such as alkali and alkaline earth elements) are measured, chemical ionization interference occurs as we are measuring the non-ionized atoms. The ionization of these easily ionizable elements will decrease the signals for both absorption and emission, hence creating negative interference. When other elements are measured, the presence of such elements will also cause positive interference. This is because the free electrons added to the flame from alkali or alkaline elements will suppress the ionization of the test element.

Another source of chemical interference is the formation of refractory metal compounds, which prevents the test element to form atomic atoms thereby causing negative interference. Refractory metal phosphates (e.g., $Ca_2P_2O_7$) can be avoided by adding *releasing agents* such as strontium ion or lanthanum ion, which will preferentially combine with phosphate and prevent its reaction with calcium. The formation of refractory metal oxides (e.g., CaO) can be avoided by the use of high-temperature flame (nitrous oxide–acetylene rather than air–acetylene).

Physical interference: Physical interference is caused by the variation of instrumental parameters that affect the rate of sample uptake in the burner and atomization efficiency. This includes variations in the gas flow rates, variation in sample viscosity due to temperature or solvent, high solids content, and changes in flame temperature. Physical interference can be corrected by the use of internal standards.

Other Considerations

Other considerations include sample throughput, cost, ease of use, and availability of proved methodology. The analytical time for metal analysis using atomic spectroscopy is of less concern compared with the analysis of organic compounds using chromatographic instruments. The time required for elemental analysis ranges from a few seconds to several minutes. However, one fundamental consideration is the single-element technique (FAA and GFAA) vs. multi-element technique (ICP–OES and ICP–MS).

The *single-element technique* requires specific light source and optical parameters for each element to be determined. Therefore, the lamp must be changed between each element. Although FAA or GFAA now has automated multi-element systems, it is still considered to be a single element method. A typical run for a single element takes only 3–10 s, whereas GFAA takes much longer, approximately 2–3 min for each element, because the system needs to remove solvent and matrix components prior to atomization. The *multi-element technique* (ICP) can determine 10–40 elements per minute in individual samples. ICP–MS is typically 20–30 element determinations per minute depending on such factors as the concentration levels and required precision. For both ICP–OES and ICP–MS methods, the time is only limited by the equilibrium time needed between new sample and the plasma (typically 15–30 s).

Capital costs for atomic spectroscopic equipment in the increasing order are FAA < GFAA < ICP–OES ≪ ICP-MS (Fig. 9.11). There is a considerable variation in cost between instrumentation for the same technique depending on the features (e.g., manual vs. automation).

In terms of skill required and ease of operation, FAA is the easiest to use, followed by GFAA, where skills to optimize operational parameters are needed. Skill requirements for ICP are intermediate between FAA and GFAA. ICP–MS requires operator skills similar to ICP–OES and GFAA.

As a general selection guide to summarize the above discussions, FAA and ICP–OES are favored for moderate to high concentration, while GFAA and ICP–MS are

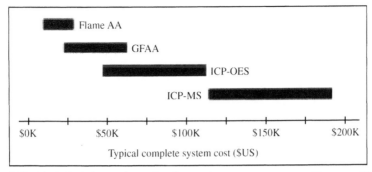

Figure 9.11 Typical relative purchase prices for atomic spectroscopy systems (Courtesy of Perkin-Elmer, Inc.)

favored for lower concentrations. ICP–OES and ICP–MS are multi-element techniques, favored where large numbers of samples are to be analyzed and cost is not the primary concern (Fig. 9.12).

Whereas the cost of ICP–MS is still prohibitive for many labs, ICP–OES appears to have become the dominant instrument for routine environmental metal analysis and the role of other atomic absorption techniques has decreased. Compared with FAA, ICP–OES offers several advantages: (a) Lower interelement interference, which is a direct consequence of their higher temperatures; (b) Better spectra for many elements under the same excitation conditions, and hence spectra for many elements can be recorded simultaneously. This is why ICP can perform sequential multielemental analysis; (c) Higher temperature allowing refractory compounds (e.g., metal oxides) to be measurable; (d) Determination of nonmetals such as Cl, Br, I, and S; and (e) A wider linear dynamic range of 4–6 orders of magnitude.

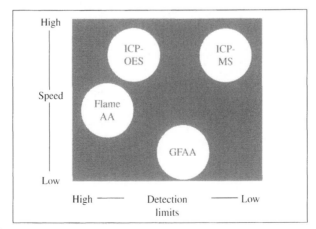

Figure 9.12 General selection guide for atomic spectroscopy instrumentation based on sample throughput and concentration range (Courtesy of Perkin-Elmer, Inc.)

There are disadvantages of ICP–OES, which make other atomic absorption spectrometers still competitive under certain circumstances. These include: (a) ICP spectra from plasma are often made up of hundred, or even thousands, of lines. This requires higher resolution and more sophisticated optical equipment than is needed for atomic absorption methods. This explains why ICP has much higher instrument capital cost. (b) Higher operating cost due to the consumption of significant amount of argon gas (5 ~20 L/min). (c) Need for skilled operators to develop methods. (d) Often less precise than with atomic absorption. (e) For B, P, N, S, and C, a vacuum UV spectrometer is necessary since emission of these elements lie at wavelengths below 180 nm where components of the atmosphere absorb. (f) Limited use for alkali metals (Li, K, Rb, and Cs) since the most prominent lines are located at near-infrared wavelengths which lead to the detection problems.

9.4   PRACTICAL TIPS TO SAMPLING, SAMPLE PREPARATION, AND METAL ANALYSIS

In Chapter 7 (Section 7.2), we have discussed the principles and some practical tips of sample preparation for metal analysis. In the following discussions, we further introduce several practical guides that are essential to beginners during sample running, calibration, and results calculation for metal analysis. For procedure details, the reader can refer to Csuros and Csuros (2002).

Correction for Sample Moisture and Dilution

Oftentimes, beginners forget to keep a subsample for the measurement of moisture content (w, %) of solid samples and improperly report the analytical results. It is essential to record the mass m (to an accurate 0.0001 g) of a solid sample used for digestion and the final volume after digestate is diluted (V, mL). If sample is a liquid, record the original sample volume (V_1) and the final volume (V_2) if digestion or dilution has been performed. In both cases, if the concentration from the instrument analysis is c mg/L, then the concentration of the original sample can be calculated as follows:

$$C(\text{mg/kg}) = \frac{c(\text{mg/L}) \times V(\text{mL}) \times 100}{m(\text{g}) \times (100 - w)} \text{(solid sample)} \tag{9.7}$$

$$C(\text{mg/L}) = c(\text{mg/L}) \times \frac{V_2(\text{mL})}{V_1(\text{mL})} \text{(liquid sample)} \tag{9.8}$$

Instrumental Drift and Run Sequence QA/QC

Instrumental drifting is a common problem in atomic spectrometry. Inexperienced analysts are often frustrated by the broad variations of results for the same sample tested in different batches. The bottom line is that you will very unlikely get the exact same results when the same sample is run several times. As long as proper QA/QC protocols are followed, variations should be minimal and a margin of error is allowed with a certain degree of confidence. A good practice is therefore to follow the QA/QC guidelines, such as the one recommended by the U.S. EPA in SW-846 Method 6010B, as described in Table 9.3 (Tatro, 2000). The run sequence shown in this table can be easily set up in the sequence table of many types of instrumental software.

It is critical to have an externally prepared standard in all metal analysis. The initial calibration verification (ICV) in Table 9.3 should be prepared from a source different from the source used for the calibration standard solutions. The ICV should be run immediately after the standard solution. As shown in Table 9.3, a 10% error is allowed. If greater error is received, the problem should be resolved before running further samples. The method blank (MB), as described in Section 5.4.2, should be carried through the entire sample preparation (digestion) process. If the MB is greater than three times the instrument detection limit ($3 \times$ IDL), this is an indication of contamination, and all samples must be redigested. The continuing calibration verification (CCV) standard and blank, placed at a certain interval between samples, are used to check any instrumental drift and potential carryover between samples. The sample log shown in this table appears to be tedious, but it eventually will save the analyst time by showing when and what problems occurred.

Table 9.3 SW-846 Method 6010B ICP–OES run log

Sample sequence	QC error limits
Calibration blank	
Calibration standard (e.g., five-point standard solution)	
Initial calibration verification (ICV) standard	$\pm 10\%$
Initial calibration verification (ICV) blank	$<3 \times$ IDL
Interference check standard (ICS)	$\pm 20\%$
Method blank (MB)	$<3 \times$ IDL
Sample 1	
Sample 1 + matrix spike (MS)	$\pm 25\%$
Sample 1 + matrix spike duplicate (MSD)	$\pm 20\%$ RPD
Sample 2 through Sample 6	
Continuing calibration verification (CCV) standard	$\pm 10\%$
Continuing calibration verification (CCV) blank	$<3 \times$ IDL[a]
Sample 7 through 10	
Continuing calibration verification (CCV) standard	$\pm 10\%$
Continuing calibration verification (CCV) blank	$<3 \times$ IDL[a]

[a]IDL=Instrument detection limit; RPD=Relative percent difference (see Section 2.1).
(Tatro, 2000, Reproduced with permission from John Wiley & Sons Ltd.)

Erroneous Data and Methods of Compensation

Sometimes erroneous results are obtained even when proper QA/QC protocols are strictly followed. The analyst should then consider one of the sources of analytical interference as discussed above in Table 9.2. Think about the chemistry introduced at the beginning of this chapter, and try various compensation methods as appropriate, such as background correction, use of higher temperature flame, addition of releasing agents, use of an alternative wavelength, matrix spike, internal standard, and standard addition calibration.

For the *internal standard method*, an internal standard undergoes similar interference as the analyte. Measurement of the ratio of the analyte to internal standard signals (e.g., K/Li, Na/Li) cancels the interference. The internal standard element should be chemically similar to the analyte elements (such as Li as the internal standard for K and Na), and their wavelengths should also be similar. In *standard addition calibration*, the standard is added to the sample, and so it experiences the same matrix effects as the analyte. This method is used to account for the sample matrix effect that may be caused by high viscosity or chemical reaction with the analyte. The example below demonstrates how the standard addition method works to eliminate the interference and how results should be properly reported.

EXAMPLE 9.2.   An acid digestate of sludge sample is analyzed for zinc by flame emission spectrometry using the method of single standard addition. Two portions of digested aliquots (1 mL each) are added to 25 mL portion of deionized water. To one portion is added 5 μL standard $ZnCl_2$ solution containing 100 mg/L (as Zn). The instrumental signals of the flame emission spectrometry in arbitrary units are 23.5 and 95.6. (a) What is the concentration of Zn in the sludge digestate? (b) If 0.5 g of wet sludge was used, and the total digestate volume is 50 mL, what is the Zn concentration in sludge (The sludge moisture was determined to be 35%?)

SOLUTION: (a) Assume there is x mg Zn present in 1 mL of the digestate, then the total milligrams of Zn in the aliquot with added standard is: x (mg) + 100 mg/L *5 μL = $x + 5*10^{-4}$ mg (Note 1 L = 10^6 μL). Therefore:

$$\frac{x}{23.5} = \frac{x + 5 \times 10^{-4}}{95.6} \quad \text{Solve for } x = \frac{23.5 \times 5 \times 10^{-4}}{95.6 - 23.5} = 1.63 \times 10^{-4} \text{ mg}$$

This is the mass of Zn in 1 mL of the digestate, therefore Zn concentration in the acid digestate is: 1.63×10^{-4} mg/1 mL = 0.163 mg/L

(b) Note the total volume of the digestate from 0.5-g wet sludge is 50 mL. The concentration of Zn in sludge is:

$$\text{Zn(mg/kg, wet basis)} = \frac{0.163 \frac{mg}{L} \times 50 \text{ mL} \times \frac{L}{1000 \text{ mL}}}{0.5 \text{ g}} = 0.0163 \text{ mg/g} = 16.3 \text{ mg/kg}$$

$$\text{Zn(mg/kg, dry basis)} = \frac{16.3 \frac{mg}{kg}}{(100 - 35)/100} = 25.08 \text{ mg/kg}$$

Since the moisture content in this sludge sample is quite high, the concentration on a dry basis is considerably higher than the concentration reported on a wet basis.

REFERENCES

BEATY R.D., KERBER J.D. 1993. *Concepts, Instrumentation and Techniques in Atomic Absorption Spectrometry*, 2nd edition. The Perkin-Elmer Corporation. Norwalk, CT.

BROEKAERT J.A.C. 2002. *Analytical Atomic Spectrometry with Flames and Plasmas*. Wiley-VCH Weinheim.

BOSS C.B., FREDEEN K.J. 1989. *Concepts, Instrumentation, and Techniques in Inductively Coupled Plasma Atomic Emission Spectrometry*, Perkin Elmer Shelton, CT.

CHRISTIAN G.D. 2004. *Analytical Chemistry*, 6th edition, John Wiley & Sons Hoboken, NJ.

*CSUROS M., CSUROS C. 2002. *Environmental Sampling and Analysis for Metals*, CRC Press, Boca Raton, FL.

GIRARD J.E. 2005. *Principles of Environmental Chemistry*. Jones and Bartlett Publishers Sudbury, MA.

HARRIS D.C. 2003. *Quantitative Chemical Analysis*, 6th edition, W.H. Freenman, New York.

JENKINS R. 1988. *X-Ray Fluorescence*, Wiley, New York, NY.

NÖLTE J. 2003. *ICP Emission Spectrometry: A Practical Guide*, Wiley-VCH Weinheion.

Perkin Elmer Instruments (2000), *An Overview of Atomic Spectroscopy*. PerkinElmer Instruments, Shelton, CT.

*Perkin Elmer Instruments (2004), *Guide to Inorganic Analysis*, PerkinElmer Instruments, Shelton, CT (http://labs.perkinelmer.com/Content/RelatedMaterials/005139C_01_Inorganic_Guide_web.pdf)

*ROBISON K.A., and ROBISON J.F. 2000. *Contemporary Instrumental Analysis*. Prentice Hall, Upper Saddle River, NJ.

SCHWEDT G. 1999. *The Essential Guide to Analytical Chemistry*, John Wiley & Sons, New York, NY.

SKOOG D.A., HOLLER F.J., NIEMAN T.A. 1997. *Principles of Instrumental Analysis*, 5th edition, Saunders College Publishing, New York.

TATRO M.E. 2000. *Optical Emission Inductively Coupled Plasma in Environmental Analysis*. *Encyclopedia of Analytical Chemistry*, Edited by Meyers RA. John Wiley & Sons West Sussex, UK.

QUESTIONS AND PROBLEMS

1. Explain: (a) The difference in the electronic configuration among various species of calcium, that is, Ca in $CaCl_2$, Ca^0, Ca^{0*}, Ca^+, and Ca^{+*}; (b) Of these species, which one(s) are the desired for AAS measurement of Ca and which one(s) are unwanted species that may cause interference for AES measurement of Ca?

2. Explain: (a) What fractions of atoms in the flame are typically in the excited states? (b) Why flame absorption based spectrometers are normally more sensitive than flame emission spectrometer?

3. Explain the difference in light source used in atomic absorption vs. the light source used in UV–VIS spectrometer.

4. For the following atomic absorption spectrometers—FAA, GFAA, and CVAA: (a) Sketch the schematic diagram; (b) Describe the principles (functions) of major instrumental components.

5. For the following atomic emission spectrometers—flame atomic emission, ICP–OES, and XRF: (a) Sketch the schematic diagram; (b) Describe the principles (functions) of major instrumental components.

*Suggested Readings

6. Briefly answer the following questions: (a) mechanism of hollow cathode lamp; (b) atomization procedure in graphite furnace; (c) principles of cold vapor mercury analyzer; (d) principles for the formation of plasma; (d) how X-ray fluorescence is produced in XRF.

7. Describe (a) nebulizer and atomizer in the FAA; (b) is there a nebulizer in GFAA?

8. Briefly describe: (a) advantage of GFAA over FAA; (b) why CVAA rather than FAA should be particularly useful for Hg analysis; (c) advantage of ICP over flame AA; (d) principal advantage and drawback of XRF and why XRF is unique among all atomic spectroscopic techniques?

9. In comparing the flame atomic absorption (FAA) and the electrothermal (graphite furnace) atomic absorption (GFAA): (a) Explain why atomization efficiency is very low for FAA; (b) Explain why GFAA is much more sensitive than FAA; (c) Explain why GFAA analysis usually is slower (2–3 min per sample) than FAA (few seconds); (d) What is the major problem with GFAA.

10. Briefly define what are (a) spectral interference; (b) chemical interference; and (c) physical interference.

11. The standard method 3500-Li B determines lithium (Li) at a wavelength of 670.8 nm using a flame emission photometric method. In the Interference section, it states: "A molecular band of strontium hydroxide with an absorption maximum at 671.0 nm interferes in the flame photometric determination of lithium. Ionization of lithium can be significant in both the air–acetylene and nitrous oxide–acetylene flames and can be suppressed by adding potassium." (a) Describe the spectral interference of this method; (b) Describe why lithium ionization is a problem and how the addition of potassium can mitigate such a problem.

12. Sulfate (SO_4^{2-}) and phosphate (PO_4^{3-}) ions hinder the atomization of Ca^{2+} by the formation of nonvolatile salts. Explain why EDTA (ethylenediaminetetraacetic acid) can be added to protect Ca measurement from SO_4^{2-} and PO_4^{3-} interference.

13. Explain why cesium (Cs) salt should be used to minimize the ionization interference when analyzing K and Na using FAA.

14. Explain how Zeeman background correction is used to minimize spectral interference.

15. Explain why NH_4NO_3 is added to seawater when Pb and Ca are analyzed by GFAA (*Hint:* $NH_4NO_3 + NaCl \rightarrow NH_4Cl + NaNO_3$).

16. Explain why (a) HCl cannot be used as a sample matrix when Cu, Pb, Zn, Ca are analyzed using GFAA; (b) why moisture is needed to be removed in cold vapor gas cell.

17. What are the major factors (criteria) in selecting which atomic spectroscopic instruments for metal analysis?

18. Generally speaking, which of the following has the lowest (best) detection limits, the widest linear range, the fastest sample throughput, or the lowest purchasing price: (a) FAA; (b) GFAA; (c) ICP–OES; (d) ICP–MS?

19. Explain: (a) Why XRF is particularly useful as a tool in screening test of contaminated site; (b) What problem occurs if flame atomization is used for mercury analysis.

20. ICP–OES has become increasingly popular in environmental labs for metal analysis and it seems that the flame and flameless AA techniques are diminishing. (a) Explain why; (b) List major advantages and disadvantages of ICP–OES for elemental (metal) analysis.

21. (a) Sketch and briefly describe the hydride generation atomic absorption (HGAA) technique for the analysis of arsenic and selenium. (b) What are the operational parameters that affect the sensitivities?

22. In the U.S. EPA's SW-846 methods: (a) Describe two multielement atomic spectrometric methods; (b) Describe two most common single element atomic AA; (c) What else atomization techniques are used for several other elements such as Hg, As, and Sn?

23. A groundwater sample is analyzed for its K by FAA using the method of standard additions. Two 500 µL of this groundwater sample are added to 10.0-mL DI water. To one portion, it was added 10.0 µL of 10 mM KCl solution. The net emission signals in arbitrary units are 20.2 and 75.1. What is the concentration of K in this groundwater in mg/L and mM?

24. A 5-point calibration curve was made for the determination of Pb in FAA. The regression equation was: $y = 0.155x + 0.0016$, where y is the signal output as absorbance, and x is the Pb concentration in mg/L. (a) A contaminated groundwater sample was collected, diluted from 10 to 50 mL, and analyzed without digestion. The absorbance reading was 0.203 for the sample. Calculate the concentration of Pb in this groundwater sample; (b) A sediment sample suspected of Pb contamination was collected. After decanting the overlying water, a 1-g wet sediment sample was digested and diluted to 50 mL. FAA measurement gave an absorbance reading of 0.350. A subsample of this wet sediment was taken to measure the moisture content with oven drying overnight. The moisture was 35%. Report the Pb concentration in sediment sample on a dry basis.

25. In flameless graphite furnace method, which quantitative method is more appropriate: (a) Internal standard method; (b) Standard addition methods? Explain.

26. What precautions regarding QA/QC should be made in running a batch of samples in the automatic sample log (sequence table)?

Chapter 10

Chromatographic Methods for Environmental Analysis

The purpose of this chapter is to introduce the fundamentals of chromatography used primarily for the analysis of environmental organic compounds. We will first introduce the *principles of separation* employed in gas chromatography (GC) and high performance liquid chromatography (HPLC). Separation is the heart for all chromatography, and the selection of a separation column from various commercially available sources is always the key to developing a successful chromatographic method. Further details are provided on the *nomenclature of chromatograms* and the *qualitative/quantitative basis* common to all chromatographic analyses. *Instrumental components* are then described for common chromatographic techniques, including GC, HPLC, and ion chromatography (IC), as well as less common, supercritical fluid chromatography (SFC). This is followed by the discussion of *common detectors* used in GC, HPLC, and IC.

The readers are further introduced to the *environmental applications* of chromatographic methods for the analysis of pollutants in air, water, and wastes. These pollutants are organized into three categories, related to each of the three major chromatographic methods: gas, volatile, and semivolatile by GC; semivolatile and nonvolatile by HPLC; and ionic species by IC. At the end of this chapter, some *practical*

Fundamentals of Environmental Sampling and Analysis, by Chunlong Zhang
Copyright © 2007 John Wiley & Sons, Inc.

tips including method development, maintenance, and troubleshooting are provided to help the beginner of chromatographic analysis. Suggested readings are provided for the experienced reader and for those who may need further details.

10.1 INTRODUCTION TO CHROMATOGRAPHY

10.1.1 Types of Chromatography and Separation Columns

Chromatography is the method of separation in which several chemicals to be separated for subsequent analyses are distributed between two phases. The first one is stationary (*stationary phase*), whereas the other moves in one direction (*mobile phase*). Chromatographic separation has many different forms, depending on the type of mobile and stationary phase, the type of equilibrium involved between mobile and stationary phase, and the physical means by which these two phases are brought into contact. Despite various forms, all chromatographic analyses are based on the establishment of equilibrium between a stationary phase and a mobile phase. Chemicals are separated according to their different affinities to the stationary phase.

Three general forms of chromatography are GC, *liquid chromatography* (LC), and SFC. As the names imply, their mobile phases are gas, liquid, and supercritical fluid, respectively. The liquid chromatography discussed in this text refers exclusively to the *high-performance liquid chromatography* or *high-pressure liquid chromatography* (HPLC), which markedly differs from the classical low-pressure and gravity-flow liquid chromatography.

In GC, separation is based mainly on the partitioning between a gas mobile phase and a liquid stationary phase. Sometimes separation is based on a gaseous or volatile compounds' adsorption ability on a solid stationary phase. These are *gas–liquid chromatography* (GLC) or the *gas–solid chromatography* (GSC), respectively. GLC is typically correlated to the volatility (or boiling point) of the compound to be separated. A more volatile compound with a lower boiling point will be eluted first.

In LC, separation is often achieved on the basis of molecular polarity via partitioning between a liquid mobile phase and a liquid film adsorbed on a solid support material (*liquid–liquid chromatography*, LLC), and in some cases through adsorption on a solid stationary phase (*liquid–solid chromatograph*, LSC). Compounds can also be separated on the basis of molecular size (*size exclusion chromatography*, or *gel permeation chromatography*), or molecular charge (IC).

Separation in most modern chromatographic analyses, including LC and GC, is performed in a column, namely *column chromatography*, in which the stationary phase is held in a narrow tube through which the mobile phase flows under pressure. A less common separation technique is *planar chromatography*, in which a stationary phase is supported on a flat plate such as a glass, plastic, or metal surface, and the mobile phase moves through the stationary phase by capillary force or by gravity.

With column chromatography, the physical appearance of a *LC column* is often a straight, stainless steel tube between 3 cm and 25 cm in length with packing

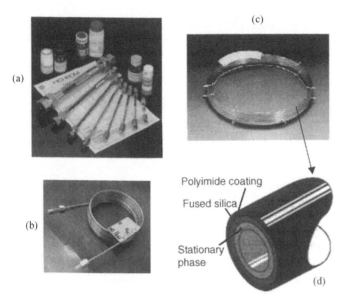

Figure 10.1 Separation columns used in LC and GC. (a) LC packed columns (Courtesy of Richard Scientific, Inc.); (b) GC packed column (© 2003 varian, Inc. Reprinted with permission of varian, Inc.); (c) GC capillary column or "wall-coated open tubular" (WCOT column) (Courtesy of SGE Analytical Science); and (d) A cross-sectional view of WCOT column (© Agilent technologies, Inc. 2006, Reproduced with permission, Courtesy of Agilent Technologies, Inc.)

materials inside the tube. *GC columns* are of two design types, packed column, and capillary column (Fig. 10.1). A *packed column* is typically a glass or stainless steel coil (1–5 m total length and 5 mm i.d.) that is filled with the stationary phase, or a packing coated with the stationary phase. A *capillary column* is a thin fused-silica (purified silicate glass SiO_2) tube (typically 10–100 m in length and 250 µ i.d.) that has the stationary phase inside. Capillary columns have an open tubular structure without packing materials, and are coated by one of the three ways.

The first type of capillary column, named "wall-coated open-tubular" or *WCOT column* has a thin liquid film stationary phase coated on the walls of the capillary. The second type, "support coated open-tubular" or *SCOT column* has microparticles attached to the wall of the capillary, which in turn is coated with the liquid stationary phase. The third type, "porous layer open-tubular" or *PLOT column* has solid-phase microparticles attached to the capillary walls. Unlike WCOT or SCOT columns, the PLOT columns are used in adsorption-based gas–solid chromatography rather than partition-based gas–liquid chromatography.

Because the open tubular structure has very low resistance to the gas flow, capillary columns can be made much longer, up to 100 m, than packed columns. Such long lengths permit very efficient separations of samples with a complex mixture. Therefore, capillary columns have become predominately used in approximately 80% of all trace environmental analysis using GC. However, packed columns will continue to be employed primarily for fixed gas analysis and separation where high resolution is not required or not always desirable.

10.1.2 Common Stationary Phases: The Key to Separation

The stationary phases mentioned above have many forms and thousands are commercially available from various manufacturers. Since it is the stationary phase that has the greatest influence on the separations obtained and the key to good separation, choosing a good column with the right stationary phase becomes a critical step in developing a chromatographic method. The discussions below are intended to help the reader understand the chemistry and the major characteristics of column stationary phases of GC and HPLC, which are the two most commonly used nowadays. The stationary phases used in IC will be briefly discussed in Section 10.2.3, and size exclusion chromatography is omitted because it has limited utility for most environmental contaminants with small molecular weight.

GC Column Stationary Phase

The gas-solid chromatography (GSC) has a stationary phase made of alumina (Al_2O_3) or porous polymers in the PLOT columns. The GSC is based on the *adsorption* of gaseous chemicals on the stationary phase. Owing to the semi-permanent and the nonlinear sorption nature, the applications of GSC are limited to certain low-molecular-weight gaseous species, such as components in air, H_2S, CS_2, CO, CO_2, and rare gases.

The *partition*-based gas–liquid chromatography (GLC) is commonly shortened to gas chromatography (GC). GLC is based on the partition of an analyte between a gaseous mobile phase and a liquid stationary phase, immobilized on the surface of an inert solid. It is reported that the common stationary phases in GLC that are listed in Table 10.1 can provide separation of almost 90% or more of the sample commonly encountered.

Five of the six liquid stationary phases listed in Table 10.1 have the general structure of polydimethyl siloxane, and the fifth entry in the table is a polyethylene glycol. Their basic structures are shown below.

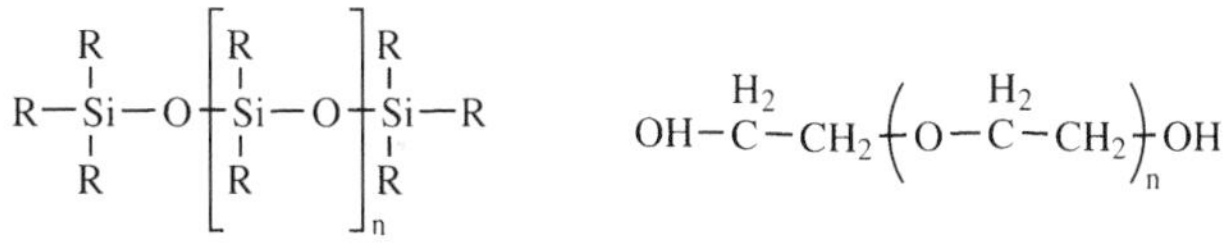

(a) Polydimethyl siloxane   (b) Polyethylene glycol

For polydimethyl siloxane, the —R groups are all hydrophobic —CH_3, giving the liquid the least polarity. For other polysiloxanes shown in the table, a fraction of the

Table 10.1   Common stationary phases for gas-liquid chromatography in gas chromatography[a]

Stationary phase	Polarity	Maximum temperature	Environmental applications
Polydimethyl siloxane	Nonpolar	350°C	General-purpose nonpolar phase, hydrocarbons, PAHs, PCBs, drugs, steroids
Poly(phenylmethyldimethyl) siloxane (10% phenyl)	Intermediate	350°C	Halogenated compounds, fatty acid methyl esters, alkaloids, drugs
Poly(phenylmethyl) siloxane (50% phenyl)	Intermediate	250°C	Pesticides, drugs, steroids, glycols
Poly(trifluoropropyldimethyl) siloxane	Intermediate	200°C	Chlorinated aromatics, nitroaromatics, alkyl-substituted benzenes
Polyethylene glycol	Very polar	250°C	Alcohols, aldehydes, ketones, and separation of aromatic isomers, e.g., xylenes
Poly(dicyanoallydimethyl) siloxane	Very polar	240°C	Polyunsaturated fatty acids, free acids, alcohols

[a]Stationary phases are arranged in order of increasing polarity. Adapted from Principles of Instrumental Analysis, 5th edition by Skoog et al. (1998). Reprinted with permission of Brooks/Cole, a division of Thomson Learning: http://www.thomsonrights.com.

methyl groups is replaced by functional groups such as phenyl ($-C_6H_5$), cyanopropyl ($-C_3H_6CN$), and trifluoropropyl ($-C_3H_6CF_3$). The increase in the percentage of substitution of these relatively polar groups increases the polarity of the liquids to various degrees.

In selecting a column from various stationary phases, one common rule is "*like dissolves like.*" In other words, one chooses a nonpolar column for a nonpolar mixture and a polar column for a polar mixture. Using a less polar column will provide the best resolution for the most difficult separation, whereas a more polar column will benefit when difficult isomer separations are required. An exception to this generalization occurs when one attempts to separate similar solutes, such as isomers (For instance, isomers of xylene that are more or less nonpolar and have similar boiling points). A nonpolar stationary phase will not be satisfactory for their separation because they do not vary much in either boiling point or in polarity. Instead, a polar stationary phase, like DB-WAX, is required to separate compounds with small differences in polarity (McNair and Miller, 1998).

HPLC Column Stationary Phase

Of the four types of liquid chromatography introduced, we will only describe the stationary phases for adsorption-based LSC and the partition-based LLC.

LSC has limited choices of stationary phases, using either silica or alumina, and is best suited to samples that are soluble in nonpolar solvents (i.e., low solubility in aqueous solvents). One particular application of LSC is its ability to differentiate isomeric mixtures.

Most of the liquid chromatography are performed by LLC, which is based on the partition between liquid mobile phase and a liquid stationary film attached on a support surface (packing material). The liquid stationary film can be retained on surface through physical adsorption or by chemical bonding (the *bonded phase*). There are two types of bonded phase packing, the normal phase and the reverse phase. Normal phase HPLC, as the name implies, was considered to be normal during the early stage of HPLC development, whereas the reverse phase HPLC has become the most popular. Approximately three quarters of HPLC instruments are operated in the reverse phase mode.

In *normal phase HPLC* (NP-HPLC), the stationary phase is highly polar, whereas the mobile phase is a relatively nonpolar solvent, such as hexane, methylene chloride, or chloroform. Polar stationary phases for normal-phase chromatography in increasing order of polarity include cyano $[(CH_2)_3CN]$, diol $[(CH_2)_2]CH_2(OH)$-$CH_2OH]$, amino $[(CH_2)_3NH_2]$, and dimethylamino $[(CH_2)_3N(CH_3)_2]$. In *reverse phase HPLC* (RP-HPLC), the stationary phase is a nonpolar liquid, whereas the mobile phase is relatively polar solvent such as water, methanol, acetonitrile, or a mixture of water with one of the organic solvents. The most common nonpolar bonded phases are hydrocarbons such as *n*-decyl $(C_8 = C_8H_{17})$ or *n*-octyldecyl $(C_{18} = C_{18}H_{37})$. C_8 is intermediate in hydrophobicity and C_{18} is very nonpolar.

One should bear in mind different elution sequences of chemicals between the two types of liquid chromatography. In normal phase, the polar stationary phase and nonpolar mobile phase favor the retention of polar compounds so that the least polar component elutes first. The least polar component is also the most soluble component in the hydrophobic mobile phase solvent. Increase in the polarity of the mobile phase will decrease the elution time. By contrast, in the reverse phase mode, the most polar component elutes first, and increasing mobile phase polarity increases the elution time.

10.1.3   Other Parameters Important to Compound Separation

Up to this point, we have learned that an appropriate column with the right chemistry is essential for the separation of a given set of compounds. In the worst case scenario, separation of compounds may be impossible when the compounds have the same affinity with the selected column stationary phase. This is where chemistry can play a critical role, and understanding the chemistry fundamentals is crucial to the success of chromatographic analysis.

There are, however, other factors that are also important to separation. Even though they may not be as critical as the stationary phase, it is likely that separation can be greatly improved if these parameters are selected at the optimal values. In the discussions below, separation is referred to as the capacity, speed of analysis, and the resolution. We will define these concepts in Section 10.1.3, but let us first examine how qualitatively these factors affect separation. Both column-related parameters (column length, internal diameter (i.d.), stationary film thickness, particle size of column packing materials) and operational parameters are included. The latter includes the flow velocity of mobile phase, and more importantly, the control of temperature for GC and selection of mobile phase for HPLC. A summary of such effects is listed in Table 10.2. The reader should be aware that exceptions to these generations do exist as a result of a variety of chromatographic conditions.

Table 10.2   Major factors affecting separation in column chromatography

Effects on separation if	Capacity	Speed	Resolution
Optimal stationary phase	Optimal	Optimal	Optimal
Increase in column length	$+^a$	−	+
Increase in column internal diameter (i.d.)	+	−	+
Increase in stationary phase film thickness	+	−	$\pm^d$
Increase in particle size	$-^b$	+	−
Increase in mobile phase flow velocity	$?^c$	+	$\pm^e$
Increase in column temperature (GC)	?	+	−
Increase in solvent polarity (RP-HPLC)	?	−	+

[a]Increase; [b]Decrease; [c]Minor or negligible effect; [d]Increase or decrease depending on whether the compound(s) are the low or high boiling points in GC; [e]Increase or decrease depending on whether the flow velocity is below or above the optimal flow velocity.

From Table 10.2, the capacity, analytical speed, and resolution respond to various column parameters differently, and oftentimes a compromise has to be made with regard to the optimal conditions. Sample capacity can be increased by the use of a longer or wider column with a thicker film of stationary phase, but doing so will increase the analytical time. Resolution is proportional to the square root of column length and the square of column internal diameter, so the resolution can be improved with the use of a longer column or a column with a larger diameter. The use of longer column in HPLC, however, may be limited by the pressure constrain. The effect of film thickness on separation is dependent upon the type of chemicals. For example, thin-film columns offer a higher resolution for high-boiling point chemicals but a lower resolution for more volatile components in GC (Grob and Barry, 2004).

The effect of mobile phase flow velocity on separation is not straightforward. A higher flow velocity will certainly reduce the analytical time, but its effect on resolution depends on the range of flow velocity. A quantitative examination on how flow affects chemical diffusion in various types of columns is beyond the scope of this text. The interested reader should consult the references listed in this chapter for a full account of the *van Deemter equation*, originally developed for packed columns and the *Golay equation* for capillary columns. With these equations, an optimal flow velocity can be justified. Understanding these quantitative mathematical expressions is helpful for the chromatographer to effectively manipulate the chromatography analysis, but it should be realized that these modeling equations are not employed on a daily basis.

A last, but not least, important parameter in separation is column temperature control for GC and mobile phase selection in HPLC. The column temperature is a compromise between speed, sensitivity, and resolution. At high column temperatures, analytes spend most of their time in the gas mobile phase so they are eluted from the column quickly (short retention time), but resolution is poor and sensitivity is increased because of decreased spreading of the peaks. Temperature can be easily programmed in operating software and, therefore, is considered less critical than the selection of stationary phase. In HPLC, commonly employed in environmental analysis, temperature has little effect on separation and hence room temperature is well served. The equivalent operational parameter in HPLC is the types and composition of mobile phases (solvent), or polarity of the mobile phase. This is because polarity is the basis for the separation in HPLC. The selection of mobile phase solvents in HPLC will be further discussed in Section 10.2.2.

Practical tips

- *Column internal diameter:* GC capillary columns of 0.25 or 0.32 mm i.d. represent the best compromise between resolution, speed, sample capacity, and ease of operation. These are the reference columns against which all other internal diameters are measured. The 0.10 mm fused silica columns have limited sample capacity, and are not well suited for trace analysis. The 0.53 mm, called *megabore* or *widebore* columns, have lower resolution but offer increased capacity and ease of operation (i.e., direct on-column syringe injection).

- *Column length:* Increased column length will increase resolution, but will also proportionally increase the retention time resulting in slow analysis. Medium length GC capillary columns (25 or 30 meters) are recommended for most environmental applications. If more than 50 components are to be analyzed (such as for petroleum analysis), then 60 meters or longer column length is needed.

- *Film thickness:* For most analysis, a standard film thickness of 0.25 μm is a reasonable starting point. It represents a compromise between the high resolution attainable with thin film and the high capacity available with thick films. High capacity means that larger sample quantities be accommodated while keeping usually the injection technique simple.

- *HPLC column parameters* (i.d., column length, particle size, and pore size): HPLC analytical methods usually are best developed with 0.46- or 0.3 cm i.d. columns having particles in the range of 3–10 μm. Columns with 5-μm particles give the best compromise of efficiency, reproducibility, and reliability. Columns of 0.3 cm i.d. reduce solvent consumption to one-half of widely used 0.46-cm units. HPLC columns of these types should allow mobile phase flow not to exceed a few mL/min. Smaller i.d. columns are especially useful (and often necessary) when interfacing an HPLC instrument with mass spectrometers and other instruments requiring small solvent input volumes. The narrow bore column can be used for a flow of a few μL/min such as those used in LC/MS. Since most of the environmental organic contaminants are hydrophobic (i.e., solvent soluble compounds), a reverse phase C_8 or C_{18} column is recommended. C_8 is an excellent starting point for method development because of its moderate hydrophobicity.

10.1.4 Terms and Theories of Chromatogram

A *chromatogram* is the instrumental output of all chromatographic analyses. It is a plot of detector signal vs. time after sample introduction, that is, the time from sample injection to the time when detector responds to the compound (Fig. 10.2). The chromatograms are the same regardless of GC, HPLC, or any other chromatographic methods. They vary only with the types of signals. In this section, we define several important terms regarding chromatograms and introduce several quantitative equations helpful to further understand chromatographic separation. Detailed mathematical derivations are avoided to simplify the illustration.

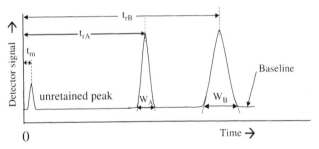

Figure 10.2 A typical chromatogram showing an un-retained peak and two analyte peaks

Shown in Figure 10.2 are a peak of an unretained compound at time $t = t_m$, the first eluted compound (A) at $t = t_{rA}$, and the second eluted compound (B) at $t = t_{rB}$. Here, t_m is the *mobile-phase holdup time*, defined as the time required for an average molecule of the mobile phase to pass through the column; t_r is the *retention time*, corresponding to the time it takes after sample injection for the analyte to reach the detector.

The peak of an unretained compound may be due to the response to air or methanol in GC (air peak) or the solvent in HPLC (solvent peak), which may not be present at all times, depending on whether or not the detector has a response to such. Another feature of chromatogram is the *broadening* of peaks as the time increases. This broadening is typical in any chromatography because of the random diffusion processes in the column. This is to say that various molecules of the same compound elute from the column at slightly varied times. A third feature of a typical chromatogram is the symmetrical nature of all eluting peaks. The symmetry can be mathematically described by the Normal (Gaussian) distribution introduced in Chapter 2.

All peaks in an ideal chromatogram should be narrow, symmetrical, well-spaced (one compound per peak) but not too far apart from each other (for a short run time). Unfortunately, this is not always the case in practice. This is where the skilled chromatographer who can examine the chromatogram from a trial run, and then effectively adjust the chromatographic conditions for a better separation, is paid off. The following terms and concepts related to separation principles are introduced.

Distribution Constant

The *distribution constant* (K_c) is defined as the concentration of a solute in the stationary phase (C_s) divided by its concentration in the mobile phase (C_m):

$$K_c = C_s/C_m \tag{10.1}$$

The larger the K_c, the more the solute will be sorbed or partitioned in the stationary phase, and the longer it is retained on the column. K_c is dependent upon the types of stationary phase and mobile phase. Separation will be impossible if two compounds have the same K_c values. This explains why the selection of stationary phase described in Section 10.1.2 is critical. In addition, since K_c is a temperature-dependent thermodynamic constant, temperature plays a role in separation, particularly in GC.

Retention Factor

Retention factor (k) is defined as the ratio of the amount (not concentration) of a solute in the stationary phase (M_s) to the amount in the mobile phase (M_m).

$$k = M_s/M_m \tag{10.2}$$

The larger this value, the greater the amount of a given solute in the stationary phase, and the longer it will be retained on the column. As such, it resembles distribution constant (K_c) and depends on both mobile-stationary phases and the temperature. Unlike K_c, however, retention factor can be readily measured from a chromatogram. Its working definition is the ratio of the time an analyte spends in the stationary phase ($t_r - t_m$) to the time it spends in the mobile phase (t_m).

$$k = \frac{t_r - t_m}{t_m} \tag{10.3}$$

An ideal separation of solutes in a mixture can be achieved if the k values lie in the range of 2–10. However, the analytical time will be too long if k values are larger than the range 20–30.

Separation Factor

Separation factor (α), defined as the ratio of distribution constants of two solutes, can be derived in terms of the ratio of the retention factors (k) of two solutes (Eq. 10.2) or can be measured from a chromatogram with the retention times of two solutes

(Eq. 10.3). In the equation below, subscript A denotes a rapidly eluted chemical with a weak retention factor (refer to Fig. 10.2) so that α is always greater than 1.0.

$$\alpha = \frac{k_B}{k_A} = \frac{t_{rB} - t_m}{t_{rA} - t_m} \tag{10.4}$$

The separation factor represents the relative interaction between each of the solutes and the stationary phase, and can be used to express the relative intermolecular forces and the magnitude of their similarities or differences. In practice, it tells us how difficult it is to separate these two solutes. The larger the value of α, the easier the separation. Two compounds can be separated only if α is higher than 1.0 in the selected phase system. For HPLC separation α should be 1.05 or higher.

Resolution

Resolution (R) is the measure of a column's ability to separate two peaks. The column *resolution* of two adjacent peaks is defined as:

$$R = \frac{\text{Peak separation}}{\text{Average peak width}} = \frac{t_{rB} - t_{rA}}{(w_A + w_B)/2} \tag{10.5}$$

where t_{rA} and t_{rB} are the retention times of the two peaks (compound A elutes first), and w_A and w_B are the baseline width of two peaks. Note that tangents are drawn to the inflection points in order to determine the widths of peaks at their bases (Fig. 10.2). The unit for w is the same as the time unit, hence resolution is a unitless parameter.

When a resolution (R) is equal to 1.0, two peaks of equal widths will have a 2.3% overlap and is considered to be the minimum for quantitative separation. A resolution value of 1.5 results in only 0.1% overlap, and can be considered sufficient for a complete baseline resolution. Strictly, Eq. 10.5 is valid only when two peaks have the same heights. Typically an R value of greater than 1.5 is needed, depending on the height ratio of the peaks. If a smaller peak is adjacent to its large neighbor, then a higher resolution is necessary to achieve a good separation.

Number of Theoretical Plates

From the classical work of Martin and Synge (1941; 1952 Nobel Prize), a chromatographic column with a length L is divided into equal length N theoretical plates numbered from 1 to N. For each of these plates, the concentration of analyte in the mobile phase reaches equilibrium with the concentration in the stationary phase. The *number of theoretical plates* (N) can be calculated from a chromatogram on the basis of the retention time (t_r) and the peak width (w).

$$N = 16\left(\frac{t_r}{w}\right)^2 \tag{10.6}$$

Since one theoretical plate represents a single equilibrium step, a column with more theoretical plates will have greater resolving power. Therefore, the plate number (N) is a measure of the separation performance of a column, or the *column efficiency*. In

GC, an efficient packed column should have several thousand theoretical plates, and an efficient capillary column should have more than 10,000 theoretical plates.

Since the ratio in Eq. 10.6 compares the retention time with the peak width, one can see that N is also a measure of the band broadening per unit time. The narrower the peak (w) the greater the number of plates (N). The plate number can then be used to calculate the height (H) of a theoretical plate using column length (L).

$$H = L/N \qquad (10.7)$$

A good column will have a large N and small H. Note that the theoretical plate model, adapted from distillation, has no physical significance, but is universally used to explain the separation in the column. One final equation relates resolution (R) to three dependent variables, that is, number of plates (N), separation factor (α), and the retention factor (k):

$$R = \frac{1}{4} \sqrt{N} \left(1 - \frac{1}{\alpha}\right) \left(\frac{k_B}{1+k}\right) \quad (\text{where } \bar{k} = (k_A + k_B)/2) \qquad (10.8)$$

From Eq. 10.8, it is important to realize that the resolution can be optimized with combined values of α, N, and k. The separation factor (α) has the largest effect. If a separation needs to be improved, it is well worth the effort of increasing α although it is impossible to give a general proposal on how to do this. If the plate number (N) is increased, the effect is only by the factor $\sqrt{N}$. This means that if the column length is doubled, the resolution will improve only $\sqrt{2} = 1.4$. Increasing the retention factor (k) only has a notable influence on resolution if k is small to start with.

EXAMPLE 10.1. Two chemicals (A and B) are separated on a GC capillary column with retention times of 14.6 and 18.3 min, and peak widths (at base) 0.80 and 0.93 min for A and B, respectively. An unretained air peak occurs at 1.1 min. Calculate: (a) retention factor, (b) separation factor, (c) resolution, (d) average plate number.

SOLUTION: By substituting the given values of t_m (1.1 min), t_{rA} (14.6 min), and t_{rB} (18.3 min) into Eq. 10.3 to 10.6, we have the following:

(a) $k_A = \dfrac{t_{rA} - t_m}{t_m} = \dfrac{14.6 - 1.1}{1.1} = 12.3;$ and $k_B = \dfrac{t_{rB} - t_m}{t_m} = \dfrac{18.3 - 1.1}{1.1} = 15.6$

(b) $\alpha = \dfrac{k_B}{k_A} = \dfrac{t_{rB} - t_m}{t_{rA} - t_m} = \dfrac{15.6}{12.3} = \dfrac{18.3 - 1.1}{14.6 - 1.1} = 1.3$

(c) $R = \dfrac{t_{rB} - t_{rA}}{(w_A + w_B)/2} = \dfrac{18.3 - 14.6}{(0.80 + 0.93)/2} = 4.35$

(d) $N_A = 16 \left(\dfrac{t_{rA}}{w_A}\right)^2 = 16 \left(\dfrac{14.6}{0.8}\right)^2 = 5329;$ and $N_B = 16 \left(\dfrac{t_{rB}}{w_B}\right)^2 = 16 \left(\dfrac{18.3}{0.93}\right)^2 = 6195$

$$N(\text{average}) = (5329 + 6195)/2 = 5762.$$

Recall that a good range of k is in the range of 2–10. The k values for both compounds in this example exceed this range, implying improvement is needed by making both chemicals less retentive. This will save analytical time. The α value is okay (indicative of possible separation), and the resolution ($R = 4.35$) is large enough for a good separation of two chemicals. It is perhaps unnecessarily too large from the analytical time standpoint, which is consistent with the k values calculated above.

10.1.5 Use of Chromatograms for Qualitative and Quantitative Analysis

In a chromatogram, qualitative information of all chromatographic techniques can be obtained by comparing the retention time of an analyte to that of a pure standard. This will provide *tentative identification* of unknown compounds if the analyst has a short list of compounds in mind. Further confirmative identification of the compounds, if needed, can be obtained by other techniques, such as GC-MS, LC-MS, and NMR (Chapter 12). It should also be noted that, although chromatograms may not lead to positive identification of chemical species, they can often provide sure evidence of the *absence* of certain compounds. That is, if the sample does not produce the peak with the same retention time as the standard, then it can be assumed that the chemical in question is not present at the given detection limit.

Quantitative chromatographic analyses are achieved by determining instrumental signals from the chromatograms. Signals are normally measured by *peak area* or less commonly by *peak height*, either one of which should be in direct proportion with the change in analyte concentrations. Peak areas are better because areas are independent of broadening effects. Peak heights suffer from many variations, such as column temperature, flow rate, and sample introduction rate. A standard calibration curve (i.e., a regression between area and concentration, see Chapter 2) must be performed first. In some cases, an internal standard method (area ratio of analyte to internal standard) can be used as an alternative (Section 9.4). In many areas of environmental research, authentic standards are not commercially available and the analyst would have to synthesize a standard on their own. If this is impossible, definite concentrations cannot be reported and relative amounts of the compound based on instrumental signal may be permissible.

10.2 INSTRUMENTS OF CHROMATOGRAPHIC METHODS

This section describes the operating principles of GC, HPLC, and IC. SFC, a hybrid of GC and HPLC, is only briefly discussed since the popularity of SFC is still limited at the present time.

10.2.1 Gas Chromatography

A schematic diagram of a typical GC system is given in Figure 10.3. The basic components of a simple gas chromatography include the following: (a) carrier gas

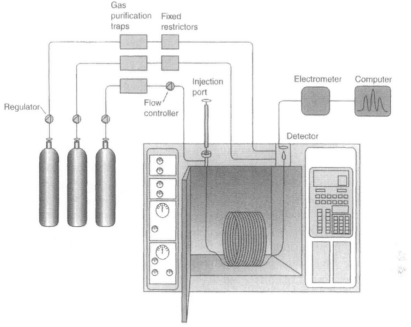

Figure 10.3 Schematic diagram of a typical GC system (© Agilent Technologies, Inc. 2006, Reproduced with permission, courtesy of Agilent Technologies, Inc.)

supply, (b) flow control, (c) sample introduction and splitter, (d) separation column, (e) temperature control zones (ovens), (f) detector, and (g) data-acquisition system. As can be seen from the graph, the analytes are carried by an inert carrier gas through the injection port, the column, and the detector. The detector measures the quantities of the analytes and generates an electrical signal. The signal then goes to a data system/integrator that generates a chromatogram (signal vs. run time). The purpose of each component, its positioning within the system, and its functioning are further described below.

Carrier gases are the mobile phase in GC, and as the name implies, their purpose is to carry analytes through the GC system. Helium and nitrogen are the most common carrier gases, but other *auxiliary gases*, such as air and hydrogen shown in Figure 10.3, may be used in certain detectors. Carrier gases should be inert and high in purity, because impurities, such as oxygen and moisture, may chemically attack the liquid stationary phase (polyester, polyglycol, and polyamide columns are particularly susceptible). These impurities should be removed using gas purification traps placed between the gas cylinder and the instrument. The high pressure in the gas cylinder is also reduced to approximately 20–60 psig through a *two-stage regulator* before it is connected to the GC. The flow is carefully controlled to ensure reproducible retention times and to minimize detector drift and noise.

Samples can be introduced either by manual injection or by an autosampler. By piercing a microsyringe through a *septum* (a polymeric silicone with high temperature stability), gaseous or liquid samples can be directly introduced into a heated region at the head of the column. An *inlet* is the device on the GC that accepts the sample and transfers it to the column. For packed columns, a sample is introduced directly into the column (*on-column injection*), or introduced into the heated region where it is vaporized and carried into the column (*flash vaporizer*). Four inlets are in common use in capillary GC: split, splitless, on-column, and programmed-temperature vaporization. A *split/splitless injector* is used when sample quantity should be reduced at the split mode. Care should be exercised not to *overload* the column, because sample size (liquid) is typically very small in GC (e.g., 0.2–20 μL for packed columns and 0.01–3 μL for capillary columns).

The column is held in an *oven* between two other separate heaters for an injector port and a detector. These three components have independent programmed temperature control. Since the key role of columns in chromatographic separation has been discussed in the previous section, we will now focus on the appropriate temperature of these three components, which is the key variable in GC. For the injection port, a general rule is to have the temperature at 50 °C higher than the boiling point of the sample. This temperature should be hot enough to vaporize the sample rapidly, but low enough not to thermally decompose the analytes. Column temperature is the easiest and most efficient way to optimize separation. The control of column temperature within ± 0.5 °C is essential. Raising the column temperature speeds both the elution (higher vapor pressure) and the rate of approach to equilibrium between the mobile and stationary phases. In GC, the change in temperature as the separation proceeds is called a *temperature gradient*, which can range from the ambient to 360°C. Higher column temperature should be avoided, since it will cause *column bleeding* where the stationary phase itself can be vaporized and/or decomposed and the material is then passed along the column and eluted.

The detector temperature depends on the type of detector used in the GC. As a general rule, however, detector temperature from the column exit must be hot enough to prevent condensation of the sample and/or liquid phase. The *detector* is where the chemical signal is sensed in the form of a chromatogram. GC detectors commonly used in environmental analysis will be described in Section 10.3.1.

10.2.2 High Performance Liquid Chromatography (HPLC)

Although variations exist, a typical HPLC system with the basic components is schematically shown in Figure 10.4. From the beginning of the solvent flow to the end of waste effluent, an HPLC system includes the following basic components: (a) solvent reservoirs and a degassing unit, (b) a solvent pump, (c) a sample injection system, (d) a column (guard column and analytical column), (e) a detector, and (f) a data acquisition system. The purpose, positioning, and functioning of the major components are described below. Since mobile phases (solvents) are crucial in

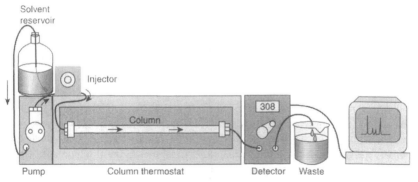

Figure 10.4   Schematic diagram of a typical HPLC system (Girard, JE, Principles of Environmental Chemistry, 2005, Jones and Bartlett Publishers, Sudbury, MA. WWW.jbpub.com. Reprinted with permission.)

HPLC, solvent properties that are fundamental to the HPLC method development are also introduced.

Solvent reservoirs contain either neat liquids, mixtures of liquids, or liquids with modifiers such as buffers. These solvents must have a purity of "HPLC" grade or higher, and must be free of particles and dissolved gases. Particles must be removed by filtering solvents through 0.25-μm filters. Doing so will extend the pump life by preventing scoring and reducing contamination or plugging of the column. A replaceable *inlet filter* inside each solvent container is also commonly used to further protect particular matter or residual salt from entering into the system. Dissolved gases in solvents are detrimental to the HPLC system because they result in bubbles, low pressure, noise, and inaccurate flow. Dissolved gases are removed by a *degassing unit*, which can be a sintered glass connected to a helium gas cylinder (sparging), or an online vacuum degassing system in which solvent passes through a thin-walled porous polymer tube in a vacuum chamber.

Solvents in the reservoirs are drawn by a corrosion resistant *pump* capable of delivering high pressure, pulse free, accurate flow rate of solvents. The mobile phase solvents are then forced through an *analytical column* for compound separation, where high pressures are developed. In HPLC system, the pressure between the pump and detector are high. Typically, pressures of 1000–3000 psi (1000 psi ≈ 70 bar) are required to provide flow rates of 1–2 mL/min in columns of 3- to 5-mm i.d. and 10–30 cm long. The usual type of HPLC pump is the so-called short-stroke piston pump. Pumps should never run dry. *Purging* or *priming* may be done when needed to remove gas bubbles in HPLC system with higher flow rates of mobile solvent. Flush the pump tubing with solvent when the pump chamber and the inlet lines from the solvent reservoir are empty. This usually occurs after the pump has been idle for a while.

Liquid samples are introduced into a *sample injection system* located between the pump and column by either manually filling the sample loop with a syringe, or with an autosampler having 6-port values. In addition to the components shown in Figure 10.4, a *pressure transducer*, an *in-line filter*, and a *guard column* are always present. These components are critical in, respectively, monitoring system pressure, removing particulates generated by mobile phase or piston seals, and protecting expensive analytical columns from clogging. The guard column is a very short replicate (same packing material) of the analytical column and is used to protect the sample-born materials (Note that the inline filter only removes the particulate matter from the mobile phases).

When compounds enter into the detector, their signals are detected and the data are analyzed by the data processing unit. HPLC *detectors* include ultraviolet detectors (most popular), the differential refractometers, fluorescence detectors, and conductivity detectors. Details on these HPLC detectors will be described in Section 10.3.2.

Since solvent mobile phase is the most often changed component in the HPLC once the separation conditions are established, we will further examine more details on solvent (mobile phase) selection in this section. The question here is, given a set of compounds and the selected column, what solvent or mixtures of solvents should be chosen to maximize the separation efficiency while keeping other problems at a minimum. Solvent regime can be *isocratic* (solvent or mixed solvents at constant composition over time), or *gradient* (solvent or mixed solvents composition changes over time). Note that the gradient elution is used to improve separation, just like the temperature programming is used in GC to improve the column separation of the analytes.

In selecting solvents as the mobile phase, several properties must be examined, including viscosity, UV cutoff, refractive index, boiling point, and polarity (Table 10.3). Solvents with a higher viscosity will cause a larger pressure drop and, therefore, should be cautioned. If a UV detector is used, the solvents used in

Table 10.3 Properties of selected solvents used as HPLC mobile phase

Solvent	UV cutoff (nm)	Refractive index (20°C)	Viscosity (cP)	Boiling point (°C)	Polarity (P)
Pentane	190	1.3575	0.23	36.07	0.0
Hexane	195	1.3749	0.31	68.7	0.1
Cyclohexane	200	1.4242	1.0	80.72	0.2
Toluene	284	1.4969	0.59	110.62	2.4
Methyl *tert*-butyl ether	210	1.3689	0.27	55.2	2.5
Methylene chloride	233	1.4241	0.44	39.75	3.1
Isopropyl alcohol	205	1.3772	2.40	82.26	3.9
Tetrahydrofuran	212	1.4072	0.55	66.0	4.0
Chloroform	245	1.4458	0.57	61.15	4.1
Ethyl acetate	256	1.3724	0.45	77.11	4.4
Acetone	330	1.3587	0.36	56.29	5.1
Methanol	205	1.3284	0.55	64.7	5.1
Acetonitrile	190	1.3441	0.38	81.60	5.8
Water	190	1.3330	1.00	100.0	10.2

Adapted from Snyder et al. (1997). Solvents in the table are arranged in an increasing order of polarity.
cP: Centipoise.

mobile phase should be completely transparent at the required wavelength. Ultra violet transparency is measured in terms of *UV cutoff*, which is defined as the wavelength at which the absorbance of the solvent vs. air in 1-cm matching cells is equal to unity. A practical tip is that if at all possible, the analysis wavelength should be at least 20 nm above the UV cutoff of the solvent.

Solvent volatility (boiling point) is important only for evaporative light-scattering detectors. Solvents with a high vapor pressure at the operating temperature tend to produce vapor bubbles in the detector. If a differential refractometer detector is used, then the solvent refractive index should be considered when selecting solvents. *Refractive index* (RI) is a solvent's property in changing the speed of light. It is computed as the ratio of the speed of light in a vacuum to the speed of light through the solvent. The higher the RI, the slower the speed of light through the solvent. Solvent refractive index (RI) affects the sensitivity of RI detection for a particular sample. For better sensitivity, the RIs of both the solvent and the sample should be as great as possible.

The last, but perhaps the most important solvent property in HPLC method development, is the polarity that is directly related to separation. More polar solvents cause increased retention in reverse phase HPLC and reduced retention in normal phase HPLC. The solvents in Table 10.3 are arranged in an increasing order of *polarity index* (*P*), which is a numerical measure of relative polarity of various solvents. Polarity index (*P*) ranges from $P = 0$ for a nonpolar solvent like pentane to $P = 10.2$ for the most polar solvent, water. Solvent strength increases as solvent polarity decreases in reverse phase HPLC. Hence, in a decreasing order, the solvent strengths are water (weakest) < acetonitrile < methanol < tetrahydrofuran < isopropyl alcohol < methylene chloride (strongest). The example below illustrates how solvent polarity changes the separation in a reverse phase HPLC.

EXAMPLE 10.2.   Two hypothetical chromatograms, obtained by reverse phase HPLC with a C_{18} column and a UV detector, are shown in Figure 10.5. The left one shows the chromatogram obtained with an isocratic 1:1 ratio of acetonitrile to water (50% each). (a) Which of the three analytes is the most polar? (b) What change would you make in order to increase the retention times for better spaced peaks as shown in the right chromatogram?

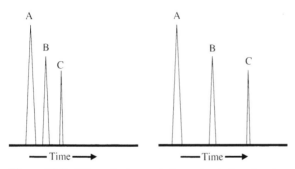

Figure 10.5   Effect of solvent polarity on the elution of analytes in HPLC

SOLUTION: (a) In reverse phase HPLC, the mobile phase is polar and the stationary phase is nonpolar, the polar analytes will be eluted first. Hence, the polarities of three compounds are most likely in the increasing order of C < B < A. (b) Increasing the polarity of the mobile phase will help the column retain the analyte better, and, hence, a longer retention time. Therefore, the percentage of water should be increased, such as 70% or even 90%. A trial run will help decide the optimal mobile phase composition. If this does not work, then try a gradient flow to obtain better-spaced peaks with a higher resolution.

10.2.3 Ion Chromatography

IC is one of the applications of HPLC. A schematic diagram showing the major components of IC is omitted because IC is similar to HPLC, in that it employs a standard HPLC apparatus, including (a) an eluent reservoir holding the mobile phase, (b) a pump, (c) a sample injection system, (d) a column (analytical column and a suppressor column), (e) a detector, and (f) a data-acquisition system (refer to Fig. 10.4). However, since the separation in commonly used IC is based on *ion exchange* of *ionic species* (inorganic and organic) rather than adsorption or partition, as described earlier for normal or reverse phase HPLC, the mobile phase, column stationary phase, and detector are fundamentally different from those described for HPLC.

IC is a combination of ion exchange chromatography, eluent suppression, and conductivity detection. Rather than using organic solvents of various polarities, IC uses acid/base or salt buffer solutions as the *eluent* (mobile phase) to affect the ion exchange process. The type and strength of such eluent is important so that various ionic analytes will have different affinities (retention times).

There are two essential columns arranged in sequence in an IC instrument. The first *analytical column* is used to separate various anions or cations. If anions (negatively charged) are separated, this column is packed with a positively charged cationic exchange resin, and is termed *anion exchange column*. If cations (positively charged) are to be separated, then the analytical column is packed with a negatively charged anionic exchange resin, and is termed *cationic exchange column*. For both types of ion exchange resins, there is an organic polymer to which functional groups are attached. It is either a negatively charged stationary phase ($-SO_3H$ or $-COOH$) for the cation exchange resin, or a positively charged stationary phase ($-NH_3$ or $-NR^{3+}$) for the anion exchange resin.

The second column, called a *suppressor column*, is essential and used to reduce the background conductivity of the eluent to a low or negligible level. This suppression column is placed downstream from the separation column. For anion analysis (A^-), the suppressor column is a large capacity anionic exchange resin used to retain cations and at the same time convert anions into their corresponding acids (HA). For cation analysis (M^+), the suppressor column is a large capacity cationic exchange resin used to retain anions and at the same time convert cations into metal hydroxides (MOH). A schematic diagram showing the principles of ion chromatograph for both anion analysis (left) and cation analysis (right) is given in Figure 10.6.

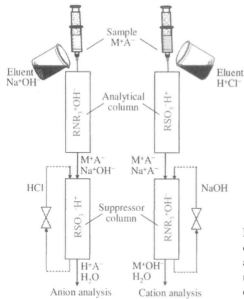

Figure 10.6   Principles of ion chromatography for anion analysis (left) and for cation analysis (right). Only monovalent anion (A^-) and monovalent cation (M^+) are illustrated

During anion analysis, anions are impeded by the attraction of $-NR^{3+}$ groups bound to the resin in the analytical column. These anions are separated into discrete bands on the basis of their affinity for the exchange sites of the resin. Sodium carbonate (Na_2CO_3) and sodium bicarbonate ($NaHCO_3$) are commonly used as the eluents, and when they are used, the following reactions take place in the suppressor column:

$$2R{-}SO_3H + Na_2CO_3 \rightleftarrows 2R{-}SO_3Na + H_2CO_3 \tag{10.9}$$

$$R{-}SO_3H + NaHCO_3 \rightleftarrows R{-}SO_3Na + H_2CO_3 \tag{10.10}$$

The cations (Na^+ and others) are retained in the suppressor column and the separated anions in their acid form are measured using an electrical-conductivity cell. The output of an IC system is a plot of conductance vs. time. It should be noted that the suppressor column will eventually become depleted and will have to be regenerated. *Regeneration* is accomplished using HCl for the cation exchange (anion analysis) or NaOH for the anion exchange (cation analysis) (Fig. 10.6).

10.2.4   Supercritical Fluid Chromatography

SFC is a hybrid of gas chromatography and liquid chromatography that combines some of the best features of each. In Chapter 7 (Sec. 7.3.7), we discussed how supercritical fluid of CO_2 is formed and its use to better extract certain organics. In SFC, the supercritical fluid form of CO_2 is used as a carrier fluid in the separation column just as other mobile phases used in GC or HPLC. Supercritical fluid has some preferred physicochemical properties; some are analogous to gas and some are analogous to conventional solvent. For example, the commonly used supercritical CO_2 fluid can readily dissolve analyte, and the used solvent (i.e., the innocuous CO_2) can be easily removed by reducing the pressure. Equally important is its superior diffusivities that are orders of magnitude higher and viscosities that are orders of magnitude lower than liquid solvents, which are critical for chromatography.

The hardware components of a SFC instrument include (a) a fluid source (a tank of CO_2), (b) a syringe pump that can sustain at least 400 atm pressure, (c) a valve to control the flow into a heated extraction cell, (d) an exit valve leading to the flow restrictor that depressurizes the fluid and transfers it to a collection device, (e) a detector, and (f) a data acquisition system. SFC can use either packed columns or capillary columns that are essentially the same as in HPLC or GC. Both UV-VIS and FID detectors (Section 10.3) are common for SFC. If FID is used, the restrictor must be placed before the detector to remove CO_2. A brief comparison of SFC with GC and HPLC is given in Table 10.4. It should be noted that although SFC has some ideal features, this relatively new technique still has not gained popularity and its wide analytical applications remain to be seen.

Table 10.4 Comparison of SFC with GC and HPLC

	GC	HPLC	SFC
Common mobile phase	Gas (N_2, Helium)	Liquid solvent (e.g., water: methanol)	Supercritical fluid (CO_2)
Common column	Capillary (~50 μm i.d.)	Packed (~mm i.d.)	Capillary/Packed
Separation principle	Partition, Adsorption, and Others	Partition, Adsorption Ion Exchange Size Exclusion	Dissolution – precipitation
Major operating parameter	Temperature	Mobile phase	Pressure, temperature
Common detector	Vary but not UV-VIS	Vary, UV-VIS most common	FID and / or UV-VIS

10.3 COMMON DETECTORS FOR CHROMATOGRAPHY

A chromatographic *detector* is a device that is able to recover chemical information (e.g., concentration, mass, structure) from the column effluent and convert it to a measurable form of signal. Coupled with a data-acquisition system, such a signal may need further amplification before a chromatogram can be generated. The process of converting chemical signals to electric or electronic signals and then generating a chromatogram is much more complicated than the separation process discussed previously. Fortunately, it is usually not necessary for an analyst to have a detailed knowledge of the internal mechanical and electronic devices. The following discussions give the reader an overview of common chromatographic detectors with a focus on their analytical principles and applications.

10.3.1   Detectors for Gas Chromatography

It is estimated that more than 60 types of GC detectors have been developed. For most environmental applications, only a few are commonly used, including thermal conductivity detector (TCD), flame ionization detector (FID), and electron capture detector (ECD). We will describe these three most common GC detectors. A brief comparison will then be made by the inclusion of four additional GC detectors that are of some environmental applications. The mass selective detector or mass spectrometer (MS), because of its uniqueness in both qualitative and quantitative analysis, will be introduced in Chapter 12.

Thermal Conductivity Detector

The thermal conductivity detector was developed prior to the existence of GC. In TCD, two gas streams (carrier gas only or carrier gas with sample) pass over two separate heated resistors or filaments (Fig. 10.7). The resistors are made of materials whose resistance is temperature sensitive. At constant electrical power, the temperature of electrically heated filaments depends upon the thermal conductivity of the surrounding gas. As carrier gas containing analytes passes through the cell, a change in the filament current occurs. The current change is compared against the current in a reference cell with carrier gas only. The difference is measured and a signal is generated. The resistances of the twin-detector pairs are usually compared by incorporating them into two arms of a simple *Wheatstone Bridge* circuit (the classical method for measurement of resistance).

The thermal conductivities of helium and hydrogen are roughly 6–10 times greater than those of most organic compounds. Thus, in the presence of even small amounts of organic materials, a relatively large decrease in the thermal conductivity

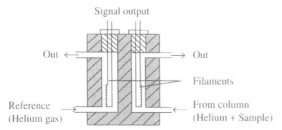

Figure 10.7   Side view of a typical thermal conductivity detector (TCD). From a top view (omitted), an arrangement of two sample detector cells and two reference detector cells can be seen. The filaments in four cells are incorporated into a Wheatstone bridge circuit used to measure electrical resistance

of the column effluent takes place. Consequently, the detector undergoes a marked rise in temperature. Because of the nature of the detector response, TCD is relatively a universal detector but with low sensitivity. Hence, its use in environmental analysis is limited except for the measurement of major constituents of air. They have remained as a popular GC detector, particularly for packed columns and inorganic gas analysis such as H_2O, CO, CO_2, and H_2.

Flame Ionization Detector

As the name implies, the FID uses H_2-air flame to burn organic compounds that undergo a series of ionization reactions. Reactions involved in the H_2-air flame include thermal fragmentation, chemi-ionization, ion molecule, and free radical reactions. However, the exact reaction mechanisms are not well understood. The amount of ions produced is roughly proportional to the number of reduced carbon atoms present in the flame, and hence the number of molecules. Because the flame ionization detector responds to the number of carbon atoms entering the detector per unit of time, it is a mass-sensitive, rather than a concentration-sensitive device. As a consequence, this detector has the advantage that changes in flow rate of the mobile phase have little effects on detector response.

As shown in Figure 10.8, the sample effluent from the column is mixed with H_2 and air and then ignited electrically at a small metal jet. There is an electrode

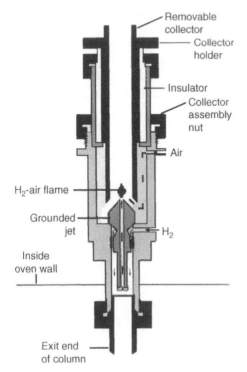

Figure 10.8 Flame Ionization Detector (FID) (© Agilent Technologies, Inc. 2006, Reproduced with permission, courtesy of Agilent Technologies, Inc.)

(cathode) above the flame to collect the ions formed at the H_2-air flame. The number of ions hitting the collector is measured and a signal is generated. In FID, three gases are therefore needed; H_2 and air as the auxiliary gases for the flame, and He used as a carrier gas.

The FID is a universal detector and can be used for all organics with good sensitivity and linearity. Exceptions include some non-hydrogen containing organics (e.g., hexachlorobenzene) and compounds with carbonyl, alcohol, halogen, and amine functionality that yield fewer ions or none at all in a flame. FID does not respond to H_2O and gives little or no response to most inorganic gases, such as CO_2, SO_2, NO_x, and other non-combustible gases. Since FID responds to organics but not water, aqueous solutions can be directly injected. However, gases used should be free of any organic impurities.

Electron Capture Detector

The ECD uses a radioactive chemical that can emit *beta* (β) *particles* (high-energy electrons), which will in turn ionize some of the carrier gas (often N_2) and produce a burst of electrons. The production of electrons will generate current between a pair of electrodes (Fig. 10.9). The β particles (electrons) are normally emitted by radioactive ^{63}Ni. When organic molecules (M) that contain *electronegative* functional groups are introduced from the column, they *capture* some of the electrons and reduce the current measured between the electrodes. Reactions are as follows:

$$Ni^{63} \rightarrow \beta^-$$ (10.11)

$$\beta^- + N_2 \rightarrow 2e^- + N_2^+$$ (10.12)

$$M + e^- \rightarrow M^-$$ (10.13)

where M is an electronegative molecule such as those molecules with halogens, phosphorous, and nitro groups. For the ECD to be detected, the molecule must have these functional groups such that they are electron affinitive (electron-capturing).

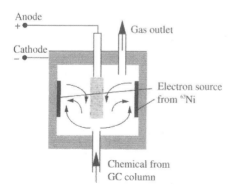

Anode
+
Gas outlet

Cathode
−

Electron source from ^{63}Ni

Chemical from GC column

Figure 10.9   Electron capture detector (ECD)

The above analysis indicates that the ECD is potentially very sensitive but is not a universal one. It is 10–100 times more sensitive than the FID, and approximately 1 million times more sensitive than the TCD. Many chlorinated pesticides such as DDT are very sensitive with the ECD detector. The invention of the ECD in 1961 by Lovelock has in fact revolutionized our understanding toward the global fate and transport of this once widely used pesticide in the world. The ECD is also used to detect purgeable hydrocarbons, phthalate esters, nitro aromatics, halo ethers, and chlorinated hydrocarbons as well as halogenated gases, such as PCE, TCE, DCE, and so forth (refer to Fig. 2.1). The response to chlorinated hydrocarbons increases by about 10 with each additional Cl atom. But the ECD is insensitive toward functional groups such as amines, alcohols, and hydrocarbons. Linearity varies with conditions and analytes, and quite often the ECD has a limited dynamic range because the response is nonlinear unless the potential across the detector is pulsed.

Comparison of Major GC Detectors

The above-described GC detectors are summarized in Table 10.5. Several other GC detectors with environmental uses are also included in this table for comparison purposes. These additional GC detectors are briefly described below:

- *Photo Ionization Detectors* (PID): PID is similar to FID, as both result in the ionization of the analyte. However, PID uses UV light as the energy source to ionize. It does not require support gases, hence it is ideal for portable instruments. For detection of purgeable aromatics, PID is about 35 times more sensitive than FID. It is also non-destructive, so it can be used in series with other detectors.

- *Nitrogen-phosphorous Detector* (NPD): Also called thermion detector (TID) or nitrogen flame-ionization detector (N-FID). NPD is similar to FID but uses a ceramic bead containing alkaline metal that can emit positive ions when heated in a gas stream containing certain analytes. It is sensitive to N-and P-containing compounds such as organophosphorous pesticides or organonitrogen compounds.

- *Flame Photometric Detector* (FPD): The GC effluent is mixed with H_2–O_2 that gives a hot flame. The chemiluminescene emitted from the combustion of S- or P-compounds is then measured at 394 nm (S) or 526 nm (P). FPD is sensitive to S- and P-containing compounds. It is similar to FID but more sensitive.

- *Hall Detector* or *Electrolytic Conductivity* (ELCD): S-, N-, and halogen (X) compounds are converted to ions under a nickel catalyst heated to 850–1000 °C. The conductivity of the dissolved ion in liquid is measured for quantitation. It has been adopted by the U.S. EPA for the detection of halogenated compounds, but generally with limited popular acceptance.

Table 10.5 Comparison of major GC detectors

Detector	Principles	Applications	Sensitivity range	Linearity
TCD	Conduct away heat	Universal	Fair	Good
FID	Burn in H_2–O_2 flame and ionize	Universal for hydrocarbons	Very good	Excellent
ECD	Electronegative compounds capture electron (e^-) from ^{63}Ni	Selective for halogens; No response to hydrocarbons	Excellent	Poor
PID	Analyte ionized by UV light	Selective for aromatics	Excellent	Excellent
NPD	Ionization at the alkali metal anode	Selective for N, P containing compounds	Excellent	Excellent
FPD	Flame and then chemiluminescene measurement	Selective for S, P containing compounds	Very good	Excellent
Hall	Catalytic conversion to ions, then measured by conductivity	Selective for halogens, S, N containing compounds	Excellent	Excellent
MS	Electron impact or chemical ionization	Universal; Provide structural information	Excellent	Excellent

EXAMPLE 10.3. Five U.S. EPA priority pollutants have been identified from a Superfund site and are being considered for a larger scale investigation using GC methods. Available GC detectors proposed for initial screening are TCD, FID, ECD, PID, NPD, Hall, and MS. The five priority pollutants include benzene, DDT, α-BHC (hexachlorocyclohexane), chlordane, and trichloethylene (TCE). In a tabulated form, summarize the applicability of various GC detectors to each of these five compounds. Briefly explain why or why not they are applicable.

SOLUTION: The structure of these compounds can be found in Appendix B. Note that benzene is the only hydrocarbon, whereas all others are chlorinated (halogenated). TCE is a chlorinated aliphatic compound with 2 C and 3 Cl (C_2Cl_3H), and DDT is a chlorinated aromatic pesticide with two benzene rings. α-BHC has a 6-carbon ring structure, but it is not aromatic. Chlordane is a highly chlorinated pesticide with 6 Cl in its molecule. The applicability of various GC detectors is summarized in Table 10.6 on the basis of what we have discussed in this Chapter.

Since TCD and MS are universal detectors, in theory, both detectors should be admissible for these five compounds. NPD will only respond to N and P containing compounds, so it does not respond to any compound listed. The sensitivity of FID

Table 10.6   Applicability of various GC detectors for the analysis of five analytes [§]

	TCD	FID	ECD	PID	NPD	Hall	MS
Benzene	X	X	N	X	N	N	X
DDT	X	X	X	X	N	X	X
α-BHC	X	Poor	X	N	N	X	X
Chlordane	X	Poor	X	X	N	X	X
TCE	X	Poor	X	N	N	X	X

X: Applicable; N: Not applicable

depends on the number of carbons, and its sensitivity is low for non-hydrogen containing organics. Consequently, FID will be sensitive enough for benzene and DDT, but unlikely for α-BHC, chlordane, and TCE. Hall detector, particularly ECD, is very sensitive to halogen-containing compounds. Hence, the non-halogen benzene cannot be measured through these two detectors. Since PID uses UV to ionize the compounds, two saturated halogens (α-BHC and TCE) will be very unlikely to be ionized by UV irradiation.

10.3.2 Detectors for High Performance Liquid Chromatography

Compared to various GC detectors, the numbers of available HPLC detectors are limited. An earlier survey in 1982 indicated three HPLC detectors most commonly used in workplace, 71% of which are UV detectors, 15% are fluorescence detectors, and 5.3% refractive index detectors. Apparently, mass selective detector coupled with HPLC has recently been shown to be a powerful tool that is also complimentary to GC–MS. When LC–MS becomes increasingly affordable, routine use in the future is likely in environmental labs as well. We introduce below these three common HPLC detectors; the details on LC–MS will be given in Chapter 12.

UV Detectors

For a UV detector to be responsive, the analyte should have strong UV absorbing functionalities as introduced in Chapter 8. In principle, *UV detectors* used in HPLC are the same as the UV detectors in the normal UV-VIS spectrometers. The marked difference between these two is the size of the sample cells. In UV-VIS spectrometers, a typical size of sample cells is 1 mL with a light path length of 1 cm. The typical UV cell volumes in HPLC, however, are limited to 1–10 μL and cell lengths to 2–10 mm. The much smaller cell volume is essential to avoid peak broadening for an improved resolution. The most powerful UV spectrophotometric detectors currently used are the photodiode array (PDA) detectors, which permit simultaneous detection of spectral data in a range of wavelengths. Thus, spectral data for each peak can be rapidly collected and stored as a sample passes through to a flow cell from the column.

Fluorescence Detectors

The second most popular HPLC detector, *fluorescence detector*, is similar in design to that of the UV-VIS units. As described in Chapter 8 (Fig. 8.4), however, fluorescence is very different from absorption. For fluorescence detectors, two wavelengths are concerned rather than the single wavelength used in absorption-based UV detectors (Fig. 10.10). The detector employs an excitation source (typically a UV lamp), which emits UV radiation at a range of wavelengths. One or more filters or a grating monochromator in more sophisticated instruments are used to acquire the needed *exciting beam* ($\lambda_{\text{excitation}}$). The emitted light (fluorescence) is most conveniently measured at a 90° angle to the exciting light beam at a wavelength of $\lambda_{\text{emission}}$. This right angle is important because at other angles, increase in light scattering for the solution and the cell walls may result in large errors during fluorescence measurement.

Like UV detectors, fluorescence detectors are not universal. They are, in fact, limited to certain chemicals, but, typically, they are at least an order of magnitude more sensitive than UV detectors. To fully appreciate fluorescence detectors, one should have some knowledge of fundamental chemistry regarding what chemicals fluoresce. In Chapter 8, we have introduced various forms of electronic transition. For the majority of fluorescent compounds, the radiation is produced by either an $n \rightarrow \pi^*$ or a $\pi \rightarrow \pi^*$ transition. The most intense and the most useful fluorescence is found in compounds containing aromatic functional groups with low-energy $\pi \rightarrow \pi^*$ transition levels. This explains why HPLC-fluorescent detector finds its wide applications for the analysis of polycyclic aromatic hydrocarbons (PAHs) of environmental concern. The intensity of fluorescence increases as the number of fused benzene rings increases. Halogen substitution, generally, results in a substantial decrease in fluorescence, and the substitution of a carboxylic acid (COOH) or carbonyl group (C=O) on an aromatic structure may inhibit fluorescence.

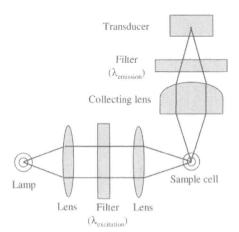

Figure 10.10 Schematic diagram of a fluorescence detector showing the light path (Rubinson, Kenneth A.; Rubinson, Judith F., Contemporary Instrumental Analysis, © 2000, p. 657. Reprinted by permission of Pearson Education, Inc., Upper Saddle River, NJ.)

Refractive Index Detectors

The concept of the refractive index has been introduced in Section 10.2.2. The refractive indexes of various HPLC solvents are also given in Table 10.3. In the presence of solutes, the refractive index of the solution will be changed accordingly. It is, therefore, important to note that the *refractive index detector*, unlike UV and fluorescence detectors, is a universal detector, which responds to nearly all solutes. To some extent, the refractive index detector of HPLC is equivalent to the TCD in GC. Like TCD, the measurement of refractive index is not as sensitive as other properties. Additionally, temperature must be maintained accurately to a few thousandths of a degree centigrade since temperature has a significant impact on refractive index.

The major design components in refractive index detectors are presented in Figure 10.11. As shown in this schematic, a light passes through cell windows, through the sample cell then the window and back to a detector. The same light also passes through a reference cell (behind the page), then reaches to a separate detector. The different light-bending properties from the sample and the reference cells are registered in two photodiodes. The amplified difference in light falling on two photocells provides the detector responses.

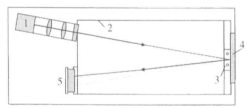

Figure 10.11   Diagram of a differential refractive index detector. 1 = Light emitting diode; 2 = Thermostatted compartment; 3 = Sample and reference cells; 4 = Mirror; 5 = Photodiode (Rubinson, Kenneth A.; Rubinson, Judith F., Contemporary Instrumental Analysis, © 2000, p. 658. Reprinted by permission of Pearson Education, Inc., Upper Saddle River, NJ.)

10.3.3   Detectors for Ion Chromatography

Since IC is essentially a particular application of liquid chromatography for the analysis of ionic species, detectors used in HPLC can also be used in IC if they can respond to ionic species. These may include UV-VIS detectors and refractive index detectors. However, the conductivity detector to date is probably the most commonly employed in environmental analysis and is described below.

Conductivity is the ability of a solution containing a salt to conduct electricity across two electrodes. Three related and sometimes confused terms are used to quantify the "conductivity", namely conductance (G), specific conductance or conductivity (k), and equivalent conductance $(\Lambda$, pronounced as lambda). The *conductance* (G) measured between two electrodes of area A (cm^2) and spacing L (cm) inserted into a conducting solution is the reciprocal of resistance R (unit in Ω, ohms):

$$G = 1/R \qquad (10.14)$$

Since R has a unit of Ω (ohm), G has a unit of Ω^{-1} (mho). In the International System of Unit (SI), G has a unit of siemens (S). Note that $1\ S = 1\Omega^{-1}$ (mho) $= 1\text{ohm}^{-1}$. In practice, microsiemens (μS) is used for relatively dilute solutions.

The second term *conductivity* (k) is commonly referred to as "specific conductance." While conductance (G) is dependent on the cell dimension, the conductivity (k) allows the measure to be independent of the detection cell:

$$k = GK \tag{10.15}$$

where the upper case K (unit cm^{-1}) is termed the cell constant, which is equal to L/A. The K value cannot be obtained by direct measurement, but it is determined using a standard solution for which the conductivity k is known. From Eq. 10.15, the conductivity has a unit of $\Omega^{-1}\ \text{cm}^{-1}$. The third term called *equivalent conductance* (Λ) is introduced to take into account the concentration of the chemical. It refers to the conductivity of an ion with a valence z in a dilute solution at 25 °C, and is defined as:

$$\Lambda = \frac{1000\,k}{Cz} \tag{10.16}$$

where C is the molar concentration of an ion in mol per liter (1000 cm^3), and Λ is the equivalent conductance in a unit of $S\ \text{cm}^2\ \text{mol}^{-1}$. By combining above two equations to eliminate conductivity (k), we obtain:

$$G = \frac{\Lambda Cz}{1000K} \tag{10.17}$$

Eq. 10.17 underlines the principles of conductivity detector, that is, for a given ion (constant Λ and z), the conductance (G) of the solution will be linearly proportional to the concentration of the electrolyte (C). In a conductivity detector, conductance (G in μS) is measured, and through a readout device, a chromatogram is obtained as a plot of conductance (μS) vs. time.

As a summary of HPLC and IC detectors described above, Table 10.7 illustrates the major characteristics of selected HPLC and IC detectors. Mass spectrometry is also listed for the purpose of comparison. For brevity, other less common LC detectors are omitted, including electrochemical (amperometric) detector, solution light scattering detector, evaporative light scattering detector, and FTIR.

10.4   APPLICATIONS OF CHROMATOGRAPHIC METHODS IN ENVIRONMENTAL ANALYSIS

Since the first gas chromatography instrument became commercially available in 1955, chromatographic techniques have evolved to be the premier technique for the separation and analysis of complicated organic mixtures. Now it is hard to imagine an organic analytical lab without any chromatographic instrument. In many labs performing petroleum and petrochemical analysis, clinical and pharmaceutical analysis, forensic and crime lab testing, and environmental monitoring,

Table 10.7 Comparison of major liquid chromatography detectors

Detector	Applications	Mass LOD (State of the art)[a]	Approximate linear range[b]
UV-VIS absorption	Specific for UV-VIS light-absorbing compounds	1 pg	10^4
Fluorescence	Specific for compounds able to fluoresce	10 fg	10^5
Refractive index	Universal detector; No gradient flow can be used	10 ng	10^3
Conductivity	Specific for ionic species	500 pg	10^5
Mass spectrometry	Universal detector	1 pg	10^5

Adapted from [a]Skoog et al. (1998) and [b]Rubinson and Rubinson (2000). 1 gram = 10^9 nanogram = 10^{12} picogram = 10^{15} femtogram. Mass limit of detection (LOD) is calculated from injected mass that yields a signal equal to five times the σ noise. The linear range for mass spectrometry was based on a sector instrument, and other mass detectors such as quadrupole and ion-trap instruments are narrower.

chromatographic instruments are the key player. In the environmental area, early usages started with the residual analysis of pesticides in the 1960–70s. Chromatographic techniques have since become the routine for many environmental professionals who are involved in research, as well as monitoring for regulatory and compliance purposes.

10.4.1 Gases, Volatile, and Semivolatile Organics with GC

GC has gained its popularity largely because of its speed, high resolution, ease of use, and affordability. The compounds can be analyzed directly by GC, however, they are limited to gases, volatile organic compounds (VOCs), and semivolatile compounds (SVOCs). We have defined what constitutes VOCs and SVOCs in previous chapters on the basis of boiling points and Henry's law constant (Section 2.1.2 and 7.3.1).

GC has been at the center of the US EPA's strategy for monitoring organic compounds in the environment since the early to mid-1970s. There are three major categories of GC-based US EPA methods (Table 10.8; for details of these methods, one should review Chapter 5 for their sources). The first group is the US EPA methods for the determination of toxic organic compounds in air, collectively called the *TO series*. The volatile air toxins are generally sampled by adsorbent trap made of Tenax, molecular sieve, activated charcoal, XAD-2 resin, or graphited carbon. Desorption is then accomplished by thermal or solvent using CS_2. VOCs are then analyzed by GC-FID (general detection), GC-ECD (halogenated species), or GC-MS.

Table 10.8   GC-based US EPA Methods

Matrix	Method series	Method number
Air	TO series	13 TO series methods (TO-1 to TO-4; TO-7; TO-9; TO-10; TO-12 to TO-17).
Water/	EPA 500	12 EPA 500 series methods (502–508; 515; 524; 525).
Wastewater	EPA 600	17 EPA 600 series methods (601-604; 606-615; 619; 624; 625)
Waste	SW-846	27 EPA 8000 series methods (8000 to 8430)

The second group of GC-based methods was developed by the US EPA for water and wastewater, including 500 and 600 series methods. The *500 series methods* were published in 1988 to support the Safe Drinking Water Act of 1974 and focus on determining low concentrations of contaminants in drinking waters. These methods underwent a major revision in 1991, although specific methods have been updated since then. There are 12 GC-based analytical methods in 500 series. Six of the methods are for VOCs and certain disinfection by-products, and the other six are designed for the determination of a variety of synthetic organic compounds and pesticides. Five of the methods utilize the purge-and-trap extraction, six methods employ a liquid–liquid extraction, and one method uses a new liquid-solid phase extraction.

The *600 series methods* were published as a result of the Clean Water Act and its amendment for use with the National Pollutant Discharge Elimination System (NPDES) permitting process and enforcement. These methods were last updated in 1984 and still reference packed column methods. The two most important methods in the EPA 600 series are methods 624 and 625. Method 624 is a GC/MS method for purgeable organic compounds that includes compounds also determined by GC methods 601, 602, 603, and 612. Method 625 measures semivolatile organics that includes compounds also detected by GC methods 604, 606, 607, 609, 610, 611, and 612. Because of the larger number of target compounds, methods 624 and 625 are usually specified for organic analysis of contaminated water. These are supplemented by method 608 for pesticides and polychlorinated biphenols and method 613 for 2,3,7,8-TCDD (dioxin). If certain compounds are detected by method 625, they should then be confirmed with a GC method specific to that compound.

The third group is approximately 27 GC-based methods, known as the *8000 series methods* published in SW-846. As noted, various GC detectors are used including FID, ECD, NPD, PID, ELCD, FTIR, low and high resolution MS. In reviewing the above-mentioned GC methods, the reader should realize that improvements of these methods have constantly been made over a period of time. But mostly, the process is slow, such as the case for the change from packed to capillary columns. This is partially because of the bureaucracy and fears of adversely affecting the method performance.

10.4.2 Semivolatile and Nonvolatile Organics with HPLC

It is estimated that approximately 85% of known organic compounds are not sufficiently volatile or stable enough to be separated and analyzed by GC. With this regard, HPLC should have a greater potential in analyzing organic compounds of semivolatile or nonvolatile nature. HPLC has found wide applications in many fields such as environmental, chemical, pharmaceutical, and food industries for research and regulatory compliance with the EPA, FDA, and OSHA. However, the number of HPLC-based methods developed by the US EPA or other agencies is significantly fewer than the number of GC-based methods. Partially, it is because GC has many versatile detectors, and in many cases, derivatization enhanced this technique. HPLC, on the contrary, is still less affordable and involves more troubleshooting. Table 10.9 below is a summary of HPLC-based EPA methods for the analysis of chemicals in air, water, and waste.

Table 10.9 HPLC-based U.S. EPA methods

Matrix	Method series	Method number
Air	TO series	5 TO series methods (TO-5; TO-6; TO-8; TO-11; TO-13).
Water/	EPA 500	1 EPA 500 series method (531).
Wastewater	EPA 600	2 EPA 600 series methods (605; 610).
Waste	SW-846	8 EPA 8000 series methods (8315 to 8332)

Five TO series methods are based on HPLC. These methods are developed to analyze air pollutants including aldehyde, ketone, phosgene, phenol, cresols, formaldehyde, and PAHs. The only 500 series HPLC method deals with the analysis of N-methylcarbomoyloximes and N-methylcarbamates in water. Two 600 series HPLC methods can analyze benzidines and PAHs. Compounds analyzed by HPLC in the SW-846 methods include carbonyl compounds, acrylamide, acrylonitrile and acrolein, N-methylcarbamates, solvent extractable nonvolatile compounds, nitroaromatics and nitramines, tetrazene, and nitroglycerine. Note that all these listed compounds are either semivolatile or nonvolatile. Some of the SVOCs can also be analyzed by GC.

10.4.3 Ionic Species with IC

The environmental applications of IC deserve some special attention. IC is probably the only technique that can separate mixture of anions and detect them at the trace ppb level. Anionic IC has applications for F^-, Cl^-, Br^-, I^-, NO_2^-, NO_3^-, SO_4^{2-}, HPO_4^{2-}, PO_4^{3-}, SCN^-, IO_3^-, and ClO_4^-, as well as organic acids or their salts. The US EPA has released methods, for example, to measure Cl^-, F^-, NO_3^-, NO_2^-, o-phosphate, and SO_4^{2-} in water (Method 300, O'Dell et al., 1984). Several IC-based

SW-848 methods have also been published for the analysis of Cl^-, F^-, Br^-, NO_3^-, NO_2^-, PO_4^{3-}, SO_4^{2-} (method 9056), Cl^- (method 9057), and $Cr_2O_7^{2-}$ (method 7199).

Whereas IC is the method of choice for anions, its applications to cations (metals) analysis are limited in the environmental field. This is because sensitive and powerful alternative instrumental techniques are readily available, such as those introduced in Chapter 9 (atomic spectroscopic techniques). Cationic IC has achieved only certain importance in the analysis of alkali metals (Li^+, Na^+, K^+), alkaline earth metals (Mg^{2+}, Ca^{2+}, Sr^{2+}, Ba^{2+}), and in the determination of NH_4^+-N in drinking water.

Recent advances in IC include the coupling of IC with ICP or ICP-MS that can analyze various species of metals including Hg, Se, and As. The speciations of these three elements are particularly important because of the vast difference in their toxicity and bioavailability in the ecosystem (refer to Sec. 7.2.3). Although less environmentally important, there are many other species that can be measured by IC, including low molecular weight amines, quaternary ammonium compounds, oxyhalides, weak organic acids, silicates, aliphatic and aromatic sulfonic acids, carbohydrates, and amino acids.

10.5 PRACTICAL TIPS TO CHROMATOGRAPHIC METHODS

10.5.1 What Can and Cannot be Done with GC and HPLC

At this point, a beginning chromatographer perhaps wants to know what types of chemicals can or cannot be analyzed by a specific chromatographic technique described above. To some extent, the two major chromatographic techniques *GC and HPLC are complementary.* All gases and volatile compounds can be analyzed by GC, but not by HPLC. Nonvolatile organic compounds and thermally (50–300°C) unstable compounds cannot be analyzed with GC unless their structures are changed through *derivatization* (see Section 7.5). These nonvolatile and thermally unstable compounds, however, can be measured by HPLC. For some semivolatile compounds of environmental interest (e.g., PAHs, nitroaromatic compounds, and explosives), both GC and HPLC can be used. HPLC is preferred in cases when direct analysis of aqueous sample is needed to avoid time-consuming extraction procedure. In other cases, GC is the instrument of choice because a variety of more sensitive detectors are available.

Because volatility is related to boiling point and molecular size, we can extend the above generalizations to some specific compounds that are of common concern. Smaller nonionic organics tend to be more volatile, hence organic compounds (containing up to 25 carbons) can likely be analyzed by GC. For the same reason, HPLC has the advantage of handling compounds with larger molecular weight (MW > 1000), such as those of biomolecules. Fortunately, most chemicals of environmental concerns are the small molecules having MW of less than 500. Besides common environmental pollutants (pesticides, explosives, PAHs, etc), HPLC can analyze amino acids, proteins, nuclei acids, hydrocarbons, carbohydrates, drugs, antibiotics, steroids, metal-organic species, and a variety of inorganic substances.

In addition to the types of compounds being analyzed by GC and HPLC, a general comparison of advantages and disadvantages can be made between these two. One may consider these factors when choosing a specific instrument for a defined analytical task. *Advantages of GC* are the following: (a) low cost, fast analysis, and ease of operation; (b) more sensitive and higher resolution when capillary columns are used; (c) a variety of columns (stationary phases) and variety of general and selective detectors to choose from, providing analytical flexibility; and (d) readily interfaced to mass spectrometer for structural confirmation. *Advantages of HPLC* are the following: (a) direct analysis of aqueous sample without tedious sample preparation and (b) mobile phases of various polarities provide versatility. Compared to GC, HPLCs are generally more expensive, less sensitive, and slower. Besides, HPLC instruments are not readily interfaced to mass spectrometer, and operationally, they are more problematic with hardware.

10.5.2 Development for GC and HPLC Methods

Developing a chromatographic method is to find a combination of instrumental parameters in order to reach a set of performance goals, such as resolution, detection limit and linearity, and certainly *ruggedness* (a rugged method is one that tolerates minor variations in experimental condition and can be easily run by an average analyst). The detailed description of chromatographic method development is beyond the scope of this text. However, a general strategy is suggested for beginners in this field.

- There are many resources available before the analyst actually shoots a sample. To get started, one should always ask help from their colleague or senior graduate students who have experience in a particular instrument method. With the advent of Internet, locating a standard method and the application notes may be just a few clicks away from your office desk.
- Once you have default parameters from a standard method, a similar method developed by your colleague, or a method reported in the literature, make a trial injection using a relatively high concentration of mixed standard solution. The goal at this stage is to make sure the instrument is in working condition and you get signal at least for some compounds of interest. Acquiring a "not-to-do list" for a particular piece of instrument from a skilled person, if at all possible, will help you overcome fears of damaging the instrument. The "not-to-do" list will encourage you to explore what is beyond the textbook and help you quickly get hold of the instrument.
- The next step is to modify the parameters of the default method till you see desired separation and signals of all the test compounds. Knowing which parameter to change is the key and knowing the reason for this change

will make you work more efficient. This is the time to pay off what you have learned regarding the fundamentals of separation and detection theories.

- Optimization could be a trial-and-error process sometimes depending on what the special circumstance is. Sometimes a method can be developed easily by just using a column of appropriate stationary phase with a few adjustments (e.g., mobile phase compositions in HPLC, or temperature gradient in GC). When this is completed, proceed to inject a series of standard solutions to obtain a calibration curve and continue to run the samples for possibly more needed adjustments.

10.5.3 Overview on Maintenance and Troubleshooting

Many books and monographs are available regarding the practical aspects of various chromatographic instruments focusing on maintenance and troubleshooting. Several of these are listed as the suggested readings suitable for readers of various levels and skills (e.g., Agilent, 2002b; Kromidas, 2004; Sadek, 2000; Suyder, p. 997). Provided below are some important but obviously not exhaustive lists of maintenance and troubleshooting guidelines. Generalizations are kept at minimum in an effort to give readers, in particular the first-time users, some specific and useful tips.

General Guidelines on Maintenance

HPLC Maintenance Generally HPLC instruments need more regular maintenance than GC for its proper function. The use of clean mobile phase and samples, frequent flushing and on-time replacement of instrumental parts of injector, autosampler, pump, column, and detector (refer to Fig. 10.4) are among the most important.

Samples and mobile phase frequently contain particles that are not visible with naked eyes but are detrimental to HPLC (cause clogging) and must be removed by membrane filtration (0.25 μm) or centrifugation. When salt buffer is a part of the mobile phase, extensive flushing must be performed after the completion of sample run to avoid accumulation of any salt residual. Several parts in the flow path (as described in Section 10.2.2) are subjective to periodical replacement because of possible clogging, including inlet filter, in-line filter, and guard column. Other replaceable parts include pump seals, injector rotor seals, syringe, and UV lamp. Recommended schedules for average usages are listed as following: replace pump seal after 1–2 years; replace autosampler rotor seal not later than 7000 injections or 3000 injections if salts are used; replace syringe annually; replace UV lamp not later than 4000 h, clean detector cell after every 2 years (Meyer, 2004).

A guard column should always be installed to extend the life of more expensive analytical column. To extend the useful life of UV detector, turn off the UV light when sample run is finished because a regular UV lamp can only last certain hours without substantial decrease in output energy. If HPLC instrument should be idle for an extended period, add some methanol in water. This will avoid potential bacterial growth in tubing and reservoirs.

GC Maintenance The key to GC maintenance is probably to follow manufacterer's recommended maintenance schedule for gas purifier, sample introduction syringe, inlet septum, inlet liner, column, and detector (refer to Fig. 10.3). High purity carrier gases are always required. For instance, oxygen causes rapid column degradation at elevated temperatures, hydrocarbons impurities give high noise to FID detectors, and water and oxygen are detrimental to ECD detectors. Sometimes, do not overlook the obvious–make sure the carrier gas is turned on at all times when the column oven is on.

Replace non-indicating gas trap every 6–12 month or when indicating traps start to change color. Change inlet septum every 50 or so samples or when signs of deterioration are visible (gaping holes, fragments in inlet liner, low column pressure). Replace inlet liner regularly or when dirt is visible in the liner or if chromatography is degraded. For the GC column, the best practice is the weekly–monthly trimming off 1/2–1 meter column front if a large load of sample is routinely analyzed. Columns should not exceed their maximum allowable temperature. Samples with very high or low pH levels, salts and buffers, strong acids, and excess derivatizing reagents will greatly shorten column life, and should be avoided. GC detectors need to be checked on an average of 6–12 month or as needed when a problem occurs.

FID detectors require little maintenance other than occasionally calibrating gas flow rates. Clean FID jet as needed when dirty deposits are present, which cause chromatographic noise and spikes. The primary maintenance for a TCD involves the thermal "bake-out" or replacement of the filaments when wandering baseline, increased noise, or a change in response is present. Thermal cleaning should also be performed for ECD when similar chromatographic problems exist. Cleaning procedures other than thermal, however, should be performed only by licensed personnel who can handle radioactive ^{63}Ni in the ECD detector.

General Guidelines on Troubleshooting

A general troubleshooting strategy is presented here first. This includes the following: (a) take your system apart and think logically. Do not just immediately reach for your toolbox; (b) look at the symptoms as to when and how the patterns change; (c) consider the possibilities and determine the potential cause(s); and (d) do one thing at a time, and eliminate one possibility at a time. Changing too many parameters at one time will make it hard to diagnose the problem.

As introduced in Section 10.1.5, the chromatogram is the basis for all the qualitative and particularly quantitative analysis. Likewise, chromatograms are also the key to diagnose the problem. Since similarities exist between GC and HPLC, we discuss them together for convenience. When chromatographic interpretation is

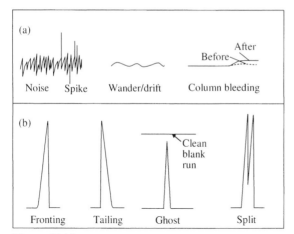

Figure 10.12 Example chromatographic problems used in troubleshooting; (a) Baseline problems and (b) Peak problems

different, it will be noted whether it is for GC or HPLC. Some of the chromatographic problems illustrated below are shown in Figure 10.12.

Baseline Problems In GC, the common causes for a *noisy baseline* could be contaminated gases, inlet, column and/or detector, septum degradation, or leaks when using an MS, ECD, or TCD detectors. When erratic baseline (*wander and drift*) appears, it is a sign of contaminated gas and column, incompletely conditioned column, change in gas flow rate, or large leaks at septum during injection and for a short time thereafter. When a rise in baseline occurs as a column approaches high temperature, this is typical of *column bleeding* when the maximum allowable column temperature is approached or exceeded. Here "bleeding" refers to the breakdown of GC column stationary phase at a high temperature. The column needs to be re-conditioned, trimmed, or replaced. In HPLC, a noisy baseline commonly occurs with a dirty sample, dirty detector cell, trapped air bubbles, or not fully equilibrated mobile phase. Drifting baseline is a sign of not-fully equilibrated system, change in mobile phase because of continuous sparging, or change in temperature.

Peak Problems These include *fronting peaks* (likely caused by sample overloading), *tailing peaks* (such as caused by severe column contamination), *ghost peaks* (caused by contaminated sample or mobile phase, carry-over from a previous injection, inlet, septum bleed), *split peaks* (bad injection technique), and *changes in peak size* (caused by change in detector response, split ratio, or injection volume, etc.).

Retention Time Problems Retention time is not unique to an analyte but should not be changed at a given analytical condition. In GC, changes in column temperature and carrier gas velocity or leaks (in septum, injector or column connection) can result in the retention time shift. In HPLC, changes in mobile phase flow rate and composition, small leaks, small air bubble, column aging and deactivation, or not fully equilibrated system are common causes for the change in retention time. In some cases, an increase in peak width, particularly in early eluting peaks, is because of inappropriate tubing i.d. or mis-set nut and ferrule union. It is almost essential for every HPLC user to have the basic skills in replacing and connecting tubings and unions.

In addition to using chromatogram to guide troubleshooting, pressure in HPLC system is another effective diagnostic tool. No or *low pressure* could be air bubbles, mis-set backpressure regulator, leaks in unions and fittings, broken piston, or major leaks around piston seal. *High pressure* could be because of clogged in-line filter, clogged guard column/analytical column, or clogged connected tubing. High pressure can also be because of the inadequate large flowrate, immiscible solvents, or use of large percentage of high viscosity mobile phase such as isopropanol. *Fluctuated pressure* may be because of damaged pump valves and/or worn seal, air bubble in the pump head, or partially clogged pump inlet filter.

Practical tips

Skilled readers can further consult GC and HPLC troubleshooting guides such as published by Agilent Technologies. As a beginner, you may attend the year-round workshop held by manufactures in many large cities in the United States. Some seminars are free, and others may have high cost. Many instrument manufacturers provide online technical support, training/education and events, application notes, and literature library. The web addresses of a handful of these are provided below.

- Agilent Technologies, Inc.: http://www.chem.agilent.com
- Alltech Association, Inc.: http://www.alltechweb.com
- Dionex Corporation: http://www.dionex.com
- Perkin Elmer, Inc.: http://las.perkinelmer.com
- Shimadzu Corporation: http://shimadzu.com
- Waters Corporation: http://www.waters.com

REFERENCES

Agilent Technologies, Inc. (2001), Maintaining Your Agilent 1100 Series HPLC System.
Agilent Technologies, Inc. (2001), Maintaining Your GC/MS System.
Agilent Technologies, Inc. (2002a), Maintaining Your Agilent GC System.
Agilent Technologies, Inc. (2002b), GC Troubleshooting Guide.
CHRISTIAN GD (2003), *Analytical Chemistry*, 6th Edition, John Wiley & Sons Hoboken, NJ.
FRITZ JS, GJERDE DT (2000), *Ion Chromatography*, 3rd Edition, Wiley-VCH Weinheim.
GROB RL, BARRY EF (2004), *Modern Practice of Gas Chromatography*, 4th Edition, Wiley-Interscience.
KROMIDAS S (2004), *Practical Problem Solving in HPLC*. Wiley-VCH weinheim.
MANAHAN SE (2004), *Environmental Chemistry*, 8th Edition, Lewis Publishers, Boca Raton, FL.

Martin (AJP) and Synge (RLM) (1941), A new form of chromotogram employing two liquid phases. I. A theory of chromotography. 2. Application the micro–determination of the higher Monoamino-acids in proteins. *Biochem. J.* 35: 1358–1368.

*McNair HM, Miller JM (1998), *Basic Chromatography*, John Wiley & Sons, Inc., New York, NY.

Meyer VR (1997), *Pitfalls and Errors of HPLC in Pictures*, Hüthig Verlag Heidelberg.

Meyer VR (2004), *Practical High-Performance Liquid Chromatography*, 4th Edition, John Wiley & Sons, NY.

O'Dell JW et al. (1984), *The Determination of Inorganic Anions in Water by Ion Chromatography–Method 300.0.* EPA-600/4-84-017, Environmental Monitoring and Support Lab, Cincinnati, OH.

Rubinson KA, Rubinson JF (2000), *Contemporary Instrumental Analysis*, Prentice Hall, Upper Saddle River, NJ.

*Sadek PC (2000), *Troubleshooting HPLC Systems: A Bench Manual.* John Wiley & Sons, Inc., NY.

Skoog DA, Holler FJ, Nieman TA (1997), *Principles of Instrumental Analysis*, 5th Edition, Saunders College Publishing, NY.

Snyder LR, Kirkland JJ, Glajch JL (1997), *Practical HPLC Method Development*, 2nd Edition, John Wiley & Sons, Inc., NY.

QUESTIONS AND PROBLEMS

1. Define the following forms of chromatography: GSC, GLC, LSC, and LLC. Which one of these four chromatographic techniques is commonly referred to as GC? Which one is commonly referred to as LC?

2. Describe (a) the difference between adsorption-based chromatography and partition-based chromatography, and (b) the common stationary phases for GLC and LLC.

3. What differs between three capillary columns, WCOT, SCOT, and PLOT?

4. What differs between packed columns and capillary columns regarding their performance in separation? Why?

5. Describe the following (a) Why temperature is important for a good separation in GC, but not in mobile phase? (b) Why solvent mobile phase is important for a good separation in HPLC but not temperature?

6. Describe the effect of the following variables on the resolution and analytical speed (length of retention time) in GC: (a) column diameter, (b) film thickness, and (c) column length.

7. List three methods to improve the resolution in HPLC.

8. In selecting GC column temperature, what factors need to be considered?

9. To separate very polar components by HPLC, would you choose a very polar or non-polar material for the stationary phase? Which component (most polar to least polar) will be eluted first?

10. Define the following terms: separation factor, retention factor, resolution, plate number, and plate height. Which one best defines the column efficiency?

11. Explain why resolution and speed of analysis is always a compromise during a chromatographic analysis.

12. Why a capillary column is more efficient than a packed column in separation?

*Suggested Readings

13. Explain the following (a) How column length affects resolution? (b) Why HPLC column cannot be made very long for a better resolution?

14. The following data apply to a column liquid chromatography: length of packing = 24.7 cm, flow rate = 0.313 mL/min. A chromatogram of a mixture of chemical A and B provided the following data:

	Retention time, min	Width of peak base (w), min
Nonretained solvent	3.1	–
Chemical A	13.3	1.07
Chemical B	14.1	1.16

Calculate the following and make comments regarding the separation:

(a) The number of plates (N) from each peak (chemical A and B)

(b) The plate height (H) for the column on the basis of the average number of plates

(c) The retention factor (k) for each peak

(d) The resolution between chemical A and B

(e) The separation factor (α) between chemical A and B

(f) The length of column necessary to give a resolution of 1.5

15. Refer to Figure 10.2, use a ruler to measure the retention times, and calculate N, H, k, α for chemical A and B. Assume the time unit is in minutes.

16. Explain the difference between the following: (a) reverse phase HPLC and normal phase HPLC, (b) isocratic flow and gradient flow.

17. What will be the effect on retention time (i.e., increased or decreased) if the polarity of mobile phase is (a) decreased in normal phase HPLC, or (b) increased in reverse phase HPLC?

18. You are to separate three chemicals (A, B, C) on a reverse phase HPLC column with a UV detector. You first tried isocratic flow with a mobile phase of 10% acetonitrile and 90% of water at a flow rate of 1 mL/min. The retention times at this condition are 2.5, 5.5, and 5.9 minutes for chemical A, B, and C, respectively. Which chemical is most polar? Since the retention times for chemical B and C are so close and peaks are a little overlapped, what would you do to better separate these two compounds? Why?

19. Using the properties of solvents given in Table 10.3, answer the following: (a) Why acetone is not used for HPLC with UV detector? (b) Why isopropanol often produces high pressure and cannot be used at higher percentage in mobile phase? (c) Which one is most preferred in RP-HPLC with UV detector: methanol, hexane, isopropanol, or acetone?

20. For a reverse phase chromatography, (a) Which of the following, 1:1 water:methanol; 1:1 water:acetone; 1:1 water:THF, has the highest strength in eluting a hydrophobic contaminant from a reverse phase column (use solvent properties in Table 10.3)? (b) Which of the following stationary phase, unbonded silica, cyano, C_3, C_8, C_{18}, has the strongest retention for non-polar, non-ionic compounds?

21. Define UV-cutoff, refractive index, and polarity index. Why are they important in HPLC?

22. Draw a schematic diagram showing the essential components of (a) HPLC, and (b) GC-FID.

23. Describe the general strategy in keeping different temperatures in different zone of GC (i.e., injector, column oven, and detector)? Why is this important?

24. In ion chromatography commonly employed in ion analysis, explain the following: (a) the separation principle, (b) Why eluent is important? and (b) Why suppressor column is essential?

25. For anion analysis, what type (anionic or cationic) of exchange resin is used in analytical column? What type of exchange resin is used in suppressor column? *What if cationic analysis is performed?*

26. Describe the principles of measuring F^-, Cl^-, NO_3^-, and SO_4^{2-} in water by anionic IC with $Na_2CO_3-NaHCO_3$ as the eluting solvents.

27. Name three most commonly used detectors in GC, and one most commonly used detector in (a) HPLC and (b) IC.

28. Briefly explain the following components in GC: (a) septum, (b) inlet, (c) split/splitless injector, (d) liner, (e) on-column injection.

29. Briefly explain the purpose of the following components in HPLC: (a) in-line filter, (b) guard column, (c) degassing unit.

30. Explain the following: (a) Why ECD is highly sensitive to chlorinated pesticides, such as DDT? (b) Why PCE is more sensitive than TCE in ECD? (c) Why C_{12} hydrocarbon is more sensitive than C_6 hydrocarbon in FID?

31. Describe the following: (a) What chemical functionalities will contribute fluorescence and can be measured by fluorescence detectors? (b) How halogen substitution affects a compound's ability to fluoresce?

32. Compare HPLC UV detectors with HPLC fluorescence detectors in terms of sensitivity. Why for the latter, a $90°$ right angle is generally used to measure the fluorescence intensity?

33. Give examples of universal and selective detectors for both GC and HPLC.

34. Given the following: Five GC detectors and five VOCs or SVOCs, find the best one-to-one matching between these two. Available detectors are: (1) TCD, (2) FID, (3) ECD, (4) NPD, (5) FPD; Chemicals to be measured are: (a) Merphos (CAS 150-50-5; $(n-CH_3CH_2CH_2CH_2)_3P$; a pesticide); (b) Methyl methanesulfonate (CAS 66-27-3; $CH_3SO_2OCH_3$); (c) Hexachlorobutadiene (CAS 87-68-3; $Cl_2C=CClCCl=CCl_2$); (d) Ethylbenzene (CAS 100-41-4; $(C_6H_5)C_2H_5$); (e) Mercury vapor.

35. For the following compound and the sample matrix, which chromatographic instrument (GC, HPLC, IC) and detector would you recommend? Suggest an EPA standard method for each chemical as well. (a) trichloroethene (volatile) in air, (b) PAHs in solid wastes, (c) BETX in soil, (d) acid rain composition (SO_4^{2-}, NO_3^-, and Cl^-), (e) species of chromium (Cr^{6+}) in water.

36. Why are HPLC and GC mostly complementary with regard to the chemicals each can analyze? If a chemical can be analyzed by both methods, in what case one is preferred over the other? How and why?

37. List the desirable characteristics of (a) GC, and (b) HPLC.

38. Explain the common causes in HPLC having the following: (a) low pressure, (b) high pressure, and (c) pressure fluctuation.

39. Explain the common causes for both GC and HPLC, regarding: (a) fronting peak, (b) tailing peak, (c) ghost peak, and (d) split peak.

40. Explain the common causes that result in the following baseline problems: (a) noisy and spikes, (b) drifting, (c) a gradual rise at the end of high temperature range (GC only).

41. What causes GC column bleeding and how could it be prevented?

42. Suppose you are involved in a GC-MS lab, you injected a mixed calibration standard that included six benzene series compounds (benzene, toluene, o-xylene, m-xylene, p-xylene, ethylbenzene) and four chlorobenzene series compounds (chlorobenzene; 1,2-dichlorobenzene; 1,3-dichlorobenzene; 1,4-dichlorobenzene). However, somehow you could not find all 10 peaks but observed a total of only seven peaks in the total ion chromatogram (TIC). The GC-MS library search also indicated that benzene is apparently not shown in the chromatogram. In addition to seven major peaks, you also noticed a small peak that does not match any of the chemicals listed above. For each of these three observations, explain what could be the reason(s), and if there are any possible ways, how to solve the problem and/or improve the method.

Chapter 11

Electrochemical Methods for Environmental Analysis

Electroanalytical methods are based on the measurements of *current*, *potential*, and *resistance*, and relate such measurements to analyte concentration. A variety of electroanalytical techniques are available to measure parameters, such as gases, metals, inorganic nonmetallic constituents (pH, anions), physical and aggregate properties (alkalinity, conductivity). Many of these parameters can be determined by other analytical methods introduced in the previous chapters, such as wet chemical methods, atomic spectroscopy, or ion chromatography. In many cases, however, electroanalytical methods offer the competitive advantages of low cost, fast response, and the ability to monitor in situ.

The purpose of this chapter is to introduce the general principles and environmental applications of the common electroanalytical techniques. *Potentiometry* and *voltammetry* are the primary focus, whereas *coulometry* and *conductometry* are only briefly examined. We start with a brief review on *redox chemistry* and important terms essential to the comprehension of electrochemical analysis. For each major variation of potentiometric and voltammetric methods, specific environmental applications are provided to illustrate the underlying chemistry and instrumental components.

Fundamentals of Environmental Sampling and Analysis, by Chunlong Zhang
Copyright © 2007 John Wiley & Sons, Inc.

11.1 INTRODUCTION TO ELECTROCHEMICAL THEORIES

To fully appreciate the electroanalytical techniques that follow, one requires a good understanding of the basic theories and nomenclatures of electrochemistry. To this end, the following section is devoted to an introduction to these subjects followed by the general principles of electrochemical methods. Readers are advised of further readings listed in this chapter for more details (e.g., Christian, 2003; Rubinson and Rubinson, 2000).

11.1.1 Review of Redox Chemistry and Electrochemical Cells

All oxidation-reduction (*redox*) reactions involve electron transfers. As a result of electron transfer, electric current and potential can be generated by the use of a device called an *electrochemical cell*. A familiar example is perhaps the lead-acid battery used in automobiles in which both lead metal (Pb) and lead dioxide (PbO_2) are immersed in a solution of sulfuric acid. The reaction occurs as follows:

$$Pb\ (s) + PbO_2\ (s) + 2H_2SO_4 \leftrightarrows 2PbSO_4 + 2H_2O \qquad (11.1)$$

The above reaction occurs spontaneously, from which chemical energy is released and converted to electrical energy. Since each cell produces about 2 V, a system consisting of six cells in series constitutes a 12-V battery found in most cars today. In this example, Pb^0 in its metallic form is oxidized, whereas the ionic form Pb^{4+} in PbO_2 is reduced.

Perhaps a better example is an electrochemical cell shown in Figure 11.1. This cell consists of two electrical conductors called *electrodes*. The zinc electrode is immersed in $ZnSO_4$ solution and the copper electrode is immersed in $CuSO_4$

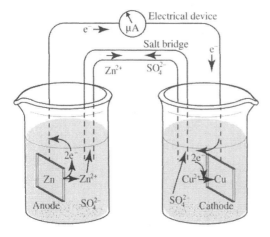

Figure 11.1 An electrochemical cell with a salt bridge

solution. These two separate compartments are called *half-cells*. To complete the cell circuit, an external metal wire is connected between two electrodes and a salt bridge connects two separating electrolyte solutions. The metal wire is needed to allow *electrons* to flow from one electrode to the other and the *salt bridge* allows *charge (ions)* transfer through the solutions but prevents from mixing the solution.

The oxidation and reduction reactions are always coupled. That is to say that no reaction in a single cell can occur by itself. Electrons are also conservative, in a sense that, electrons donated from the *oxidation* of a reducing agent in an *anode* will be equal to the electrons accepted by an oxidizing agent through a *reduction* reaction in a *cathode*. Two half-cell reactions described in Figure 11.1 are

$$Zn\ (s) \leftrightarrows Zn^{2+}\ (aq) + 2e^- \qquad \text{(anode)} \tag{11.2}$$

$$Cu^{2+}\ (aq) + 2e^- \leftrightarrows Cu\ (s) \qquad \text{(cathode)} \tag{11.3}$$

Note that as the oxidation-reduction reaction occurs, the anode would become more positive as Zn^{2+} are produced and the cathode would become more negative as Cu^{2+} are removed from the solution. The salt bridge allows the migration of ions in both directions to maintain electrical neutrality. That is, cations (Zn^{2+}) from the anode migrate via the salt bridge to the cathode, whereas the anion, SO_4^{2-}, migrates in the opposite direction. Combining the above two half-cell reactions to cancel out the electrons will result in a balanced overall reaction

$$Cu^{2+}\ (aq) + Zn\ (s) \leftrightarrows Cu(s) + Zn^{2+}\ (aq) \tag{11.4}$$

The fact that the reaction occurs *spontaneously* indicates that once these half-cells are connected, an electrical energy is produced. This kind of electrochemical cell is a *galvanic (voltaic) cell* in contrast to an *electrolytic cell* where electrical energy is used to force a nonspontaneous redox reaction to occur. The galvanic cells are used in potentiometric analysis described in Section 11.2, whereas electrolytic cells are used in voltammetric analysis described in Section 11.3.

An important characteristic of all electrochemical cells is the difference in potential energy between two electrodes. The difference in potential energy is called an *electrode potential* or *electromotive force (emf)* and is measured in terms of volts. This potential is dependent on the concentrations of the redox species and varies from the standard potential as described by the *Nernst equation*. For the redox reaction in Eq.11.4, the Nernst equation is

$$E = E^0 - \frac{RT}{nF} \ln \frac{[Zn^{2+}]}{[Cu^{2+}]} = E^0 - \frac{0.059}{n} \log \frac{[Zn^{2+}]}{[Cu^{2+}]} \tag{11.5}$$

where E is the electrode potential in volts, E^0 is the *standard potential* for the reaction under standard state conditions (298 K, 1 atm, and unity molar concentrations for all redox species), R is the gas constant (8.3145 J/K/mol), T is the absolute temperature in K, n is the number of moles of charges transferred in the reaction per mole of reactant, and F is the Faraday constant, which represents the number of coulombs in one mole of electrons (96,485 coulombs $\times$ mol^{-1} = 94, 485 J/V/mol).

The standard potential (E^0) in the Nernst equation can be calculated from the tabulated E^0 values of the two half-cell reactions. For example, since the E^0 values for Eqs 11.2 and 11.3 are 0.763 and 0.337 V, respectively, the E^0 value of the standard potential of the overall reaction (Eq. 11.4) is 0.763 + 0.337 = 1.10 V under standard conditions. The *positive* E value indicates the redox reaction in Eq. 11.4 is *spontaneous* as written from left to right. In Eq. 11.5, note the negative sign and by convention the concentrations of all products are in the numerator and concentrations of the reactants are in the denominator. The concentrations of metal Cu (s) and Zn (s) are omitted because, by convention, solids are assumed to be pure with a unity concentration. Note also that, in essence, the concentrations in the Nernst equation should be *activities*. This unique feature of electrochemistry may be an advantage or a disadvantage depending on the applications. In this text, we assume that they are equal in dilute solutions.

11.1.2   General Principles of Electroanalytical Methods

Potentiometry

There are two types of potentiometric techniques: indirect potentiometry (potentiometric titration) and direct potentiometry. With *indirect potentiometry*, an abrupt potential change (ΔE) in a chemical reaction is used as the end point during titration. The major advantages of potentiometric titration over the manual titration (Chapter 6) include the automation and its ability to titrate in colored and turbid solutions. For *direct potentiometry*, the underlying principle for the measurement of chemical concentrations is the relationship between potential (E) and analyte's concentration as described in the Nernst equation (Eq. 11.5). Note that the potential developed between two electrochemical cells is measured in the *absence* of appreciable currents. The potential of the *indicator electrode* with respect to the *reference electrode* responds to changes in the concentration of the chemical species of interest (Again, strictly speaking it responds to changes in activity). The most common devices employed for the measurement of potential include voltmeters, potentiometers, or pH meters. Electrodes of various types will be detailed in the following section.

Coulometry

Instead of measuring potential (E), *coulometry* measures the quantity of electricity or the *charge* (the root word "coul" denotes charge) generated by the complete reaction of an analyte at an electrode. In contrast to potentiometry, an electrical voltage is applied in coulometric measurement to drive the *nonspontaneous* redox reaction (hence electrolytic cell). The charge (Q, unit in coulombs;

1 coulomb $= 1$ A$\times$s) relates to the *current* (I) according to the following equations depending on whether the cell has a constant current

$$Q = It \quad \text{(constant current)} \tag{11.6}$$

$$Q = \int_0^t I \, dt \quad \text{(controlled potential)} \tag{11.7}$$

where the current (I) has a unit of ampere (A) or "amp" for short and t is the time in seconds(s). Two classes of coulometric techniques, therefore, are common: constant current techniques and controlled potential techniques. In *constant current* techniques (also known as coulometric titrations), a reagent is generated at one electrode, this reagent reacts stoichiometrically with the analyte referred to as the titrant. In *controlled potential* techniques, the current resulting from a complete electrochemical reaction of the analyte at the electrode is monitored with respect to time.

The underlying equation relating charge (Q) and the quantity of analtye (m) is the *Faraday's law of electrolysis*, which states that the mass (m) of the analyte reduced or oxidized during electrolysis is proportional to the number of moles of electrons transferred at that electrode. Mathematically, it can be written as

$$m = \frac{Q}{F} \frac{M}{n} \tag{11.8}$$

where F, M, and n are constants representing, respectively, the Faraday's constant (96,485 coulombs per mole electron), molecular weight of the analyte (g per mole analyte), and the number of mole of electrons per mole of analyte. For example, if a constant current of 0.5 A was used to deposit all Cu^{2+} in 10 min in the electrolytic cell described in Figure 11.1, then the amount of copper (Cu) in the solution can be determined as

$$m = \frac{0.5 \, A \times 10 \text{ min} \times 60 \text{ s/min}}{96,485 \text{ coulombs/mol } e^-} \times \frac{63.5 \text{ g Cu/mol Cu}}{2 \text{ mol } e^-/\text{mol Cu}} = 9.87 \times 10^{-2} \text{ g Cu}$$

Voltammetry

Voltammetry is a dynamic electrochemical method, where current (I) is monitored as a function of the changing potential (E) over a period. A voltammetric cell is an electrolytic cell consisting of three electrodes, a micro indicator electrode, a reference electrode, and an auxiliary *counter electrode* (Fig. 11.2). A changing potential is applied on the indicator electrode to drive a redox reaction that can not occur spontaneously, and the counter electrode serves to conduct electricity between two electrodes. The reference electrode has a constant potential throughout the experiment. This unique three-electrode system instead of the two-electrode system allows for an accurate application of potential functions and the measurement of the resultant current. There are various voltammetric techniques. They are distinguished

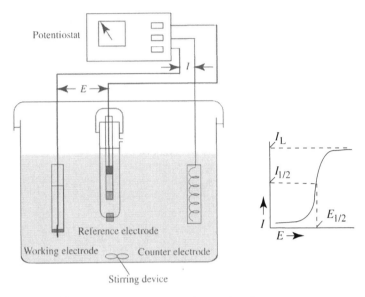

Figure 11.2   Schematic diagram of a voltammetric cell

from each other primarily by the *potential function* that is applied to the indicator electrode to drive the reaction and by the material used as the indicator electrode.

The general principles of voltammetric analysis are as follows. As the potential is applied, the electrolysis of an analyte (redox reaction) begins and the current rises rapidly until it reaches the limiting current. This resulting plot of I vs. E is called *voltammogram*. The *limiting current* is defined as the difference between the background current and the plateau current. The midpoint of the rise (and the point of maximum slope) is termed the *half-wave potential* ($E_{1/2}$), which is a characteristic of the redox reaction of the analyte and can be used for qualitative analysis. For quantitation, the amount of limiting current that is measured is related to the concentration (C) of the analyte according to

$$I_L = k\,C \tag{11.9}$$

where I_L is a limiting current, and k is a constant under specific conditions. When k is constant for a series of standard solutions of various concentrations and an unknown, a calibration plot can be constructed and the unknown concentration can be determined.

Conductometry

Conductometry measures conductance or conductivity, which is related to the ability of a solution to carry an electric current while a constant alternating-current (AC) potential is maintained in a conductivity cell. The concepts of conductance and conductivity have been described in Section 10.3.3, when ion chromatography detector was discussed. The conductivity cell has two plates (or electrodes) placed in a solution, where a potential is applied across the plates and the current is measured. Using the voltage and current values, the resistance is determined by applying *Ohm's law*:

$$R = V/I \qquad (11.10)$$

where R = resistance (ohms), V = voltage (volts), and I = current (amperes). Since electrical conductance (G) is the reciprocal of electrical resistance (Eq. 10.14), electrical conductivity of an electrolytic solution is actually the determination of the electrical resistance of the solution under a known potential gradient using a sensor of known geometry. Conductometry is directly used in water quality measurements, such as conductivity and salinity. It is also the basis for all ion conductivity detectors employed in ion chromatography.

11.1.3 Types of Electrodes and Notations for Electrochemical Cells

Electrodes are the centerpieces of all electrochemical analyses. In the preceding section, we introduced two major types of electrodes (reference electrode and indicator electrode). Various electrodes are further described here along with some shorthand notations for the electrodes.

Reference electrodes provide a constant potential without being affected by the composition of a sample. An ideal reference electrode should also obey Nernst equation. Two common reference electrodes, with their shorthand notations, are given below:

$\parallel$ KCl (saturated), $Hg_2Cl_2(s)|Hg(s)$ (E^0 = +0.214 vs. SHE at 25°C)

$\parallel$ KCl (saturated), $AgCl(s)|Ag(s)$ (E^0 = +0.197 vs. SHE at 25°C)

The first reference electrode, called *saturated calomel electrode* (SCE), is mercury in contact with a solution that is saturated with mercury chloride (calomel), which is in turn saturated with KCl solution. The second one is a *silver/sliver chloride electrode* in which a silver wire coated with a layer of silver chloride is immersed in a saturated KCl solution. In the above notations, $\parallel$ denotes a salt bridge, and $|$ denotes an interface between a solid electrode and a solution. The half-cell standard potential shown above is in reference to a *standard hydrogen electrode* (SHE), which is defined to have zero volts at standard state conditions. If two half-cells are combined to form a complete circuit, by convention, the half reaction involving oxidation (anode) should be written on the left and the reaction involving reduction (cathode) should be on the right.

Indicator electrodes respond directly to the analyte and ideally respond only to it. Two types of indicator electrodes are available, namely, metallic electrodes and membrane electrodes. *Metallic electrodes*, such as metal/metal ion and metal/metal salt/anion, are no longer routinely used as indicator electrodes in environmental analysis. It is the *membrane electrodes* that are found the most in environmental applications today. Membrane electrodes can be classified into glass electrodes, liquid membrane electrodes, and crystalline membrane electrodes, depending on the type of membrane materials. The membrane-based electrode is also termed *ion selective electrode (ISE)* or pIon electrode because its output is usually recorded as p-function (the negative log), such as, pH, pCa, or pNO_3.

11.2 POTENTIOMETRIC APPLICATIONS IN ENVIRONMENTAL ANALYSIS

Potentiometric techniques are probably the most widely used among all electroanalytical methods, particularly, in environmental laboratories. Here illustrate the measurement of pH and ions using direct potentiometry and how potentiometric titration can also be employed for an automated titration.

11.2.1 Measurement of pH

For convenience purpose, the majority of modern pH electrodes are *combination pH electrodes* rather than a set of two separate electrodes. The combination pH electrode, as shown in Figure 11.3, is virtually a tube within a tube. The inner tube housing the pH indicator electrode (pH sensing membrane, Ag/AgCl

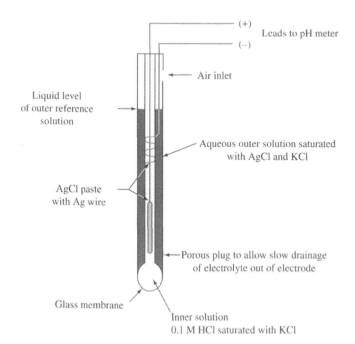

Figure 11.3 Schematic diagram of a combination pH electrode (Courtesy of Prof. David A. Reckhow, University of Massachusetts at Amherst)

reference electrode, and HCl) and the outer one housing the reference electrode (Ag/AgCl) and its salt bridge. The pH sensing component of the indicator electrode is a glass bulb, which is a thin glass membrane of approximately 0.03–0.1 mm thick and made of, for example, 72% SiO_2, 22% Na_2O, and 6% CaO (Corning 015 glass).

When a pH probe is immersed in a sample solution, hydrogen ions (H^+) in the solution will enter the hydrated Si-O lattice structure of the glass membrane in exchange for singly charged cation Na^+. This creates an electrical potential across the membrane interface (referred to as the *boundary potential*) with respect to the internal Ag/AgCl reference electrode. The resulting overall cell potential with regard to the external reference cell can be derived as

$$E_{cell} = \text{constant} + \frac{RT}{F} \ln a_{H^+} = \text{constant} - \frac{2.303RT}{F} pH_{unknown} \qquad (11.11)$$

where a_{H^+} is the activity of H^+. The constant value is unknown and is not of practical interest for pH measurement because standard pH buffers are usually used to calibrate the electrode. The slope factor ($2.303RT/F$) is temperature dependent, therefore, temperature compensation should be made during pH measurement. Note also that Eq. 11.11 is strikingly similar to the Nernst equation (Eq. 11.5) that is used to describe the metallic electrodes. However, the source of potential in pH glass electrodes is totally different from metallic electrodes—one arises from redox reaction whereas the other is due to a boundary potential as a result of ion-exchange reactions.

All modern digital pH meters have an electronic device that can accurately record potential (in millivolts) and transform the voltage caused by H^+ into a direct pH reading. In calibrating the pH meter, *standard buffers* (commonly at pH 4.0, 7.0, 10.0) are readily available from commercial sources. Many pH meters have microprocessors that can automatically recognize the specific pH of the standard buffers and adjust the temperature effects.

Practical tips

The measurement of pH is probably the most widely used potentiometric method in many environmental labs. Perhaps, it is also one of the most poorly understood and maintained apparatus. Do not underestimate the obvious. The hints listed below may help one to optimize the performance of a pH meter.

- The pH probe should be fully hydrated prior to use by soaking the membrane in water for 24–48 h. Never let the glass electrode dry out.

- The pH meter should be calibrated frequently, because many factors can cause errors, including strains within the membrane and mechanical or chemical attack of the external surface. Handle glass electrode with extreme care.

- The pH reading may be sluggish when the pH of a dilute or an unbuffered solution of near neutrality is measured. The solution should be stirred and a sufficient time is then required to obtain a stable pH reading.

- Like other sensor device, general-purpose pH electrodes perform well in certain ranges. Solutions of extreme pH values (pH < 0.5 and pH > 10) will result in *non-Nernstian behavior*, which are termed *alkaline error* and *acid error*, respectively. Use a lithium glass electrode or a full-range electrode (0–14) instead, if the pH of very acidic or basic solutions is measured.

- Make sure to store a pH electrode in its wetting cap containing electrode fill solution (3 M KCl, purchased or prepared by dissolving 22.37 g KCl into 100 mL DI water). Do not store electrode in DI water.

11.2.2 Measurement of Ions by Ion Selective Electrodes (ISEs)

Since the development of H^+ selective pH electrode, many other ion selective electrodes (ISEs) have become available for various applications. Three types of membrane electrodes and the ions each can analyze are summarized in Table 11.1. Ions with standard methods adopted by the U.S. EPA and APHA are also noted.

The general principles and constructions of solid and liquid phase ion selective electrodes are similar to those of glass electrodes used for pH analysis. They are all membrane-based electrodes that respond selectively to ions in the presence of others. The measurement is based on the quantitative relationship between boundary potential (E) generated across a membrane and the concentration (in essence, activity) of the specific ion. All ISEs consist of a cylindrical tube of diameter between 5 and 15 mm and 5–10 cm long. An ion-selective membrane is fixed at one end so that the external solution can only come into contact with the outer surface. The other end is fitted with a low noise cable or gold-plated pin for connection to the millivolt-measuring device.

The specific compositions of the membrane materials vary among different ISEs. In the glass electrodes, for example, variations of membrane composition can be made to increase the affinity of hydrated glass to a specific monovalent cation, such as K^+ and Na^+ (Table 11.1). The boundary potential then depends on these cations, probably through an ion exchange mechanism similar to that described for a glass pH electrode.

As shown in Table 11.1, there are two other major types of membrane material, one based on a solid crystal matrix (either a single crystal or a polycrystalline compressed pellet) and another based on a plastic or rubber film impregnated with a

Table 11.1 Types of ion selective electrodes and their applications[a]

Membrane	Analyte ions
Glass	$H^{+(1,2)}$, Ag^+, K^+, Li^+, Na^+, NH_4^+
Solid phase	$Br^{-(1)}$, $Cl^{-(1,2)}$, $CN^{-(1,2)}$, $F^{-(1,2)}$, I^-, SCN^-, $S^{2-(1,2)}$, Ag^+, Cd^{2+}, Cu^{2+}, Pb^{2+}
Liquid phase	BF_4^-, ClO_4^-, $NO_3^{-(1,2)}$, Ca^{2+}, $K^{+(2)}$, hardness ($Ca^{2+} + Mg^{2+}$)

[a]Standard methods are available from (1) U.S. EPA and (2) APHA (1998).

complex organic molecule that acts as an ion-carrier. Two ISEs with successful environmental applications are the fluoride electrode and nitrate electrode (Fifield and Haines, 2000). The ISE methods for these two ions are particularly important because of their apparent advantages over other methods, including low detection limits and the elimination of tedious sample preparation procedures. These are further described below.

The *fluoride electrode* is of the first type, that is, the solid-state membrane. Direct measurements of the free ion are reliable down to 1 µg/L. The membrane consists of a single lanthanum trifluoride (LaF_3) crystal; within this hexagonal lattice the fluoride ions are relatively mobile. This crystal is doped with europium fluoride (EuF_2) to reduce the bulk resistivity of the crystal. The potential across the membrane is analogous to that of pH electrode (Eq. 11.1). That is,

$$E_{cell} = constant + \frac{RT}{F} \ln a_{F^-} = constant - \frac{2.303RT}{F} pF_{unknown} \qquad (11.12)$$

where a_{F^-} is the activity of F^-. This ISE is extremely selective for F^- ions and is only interfered by OH^- that reacts with the lanthanum to form lanthanum hydroxide, with the consequent release of extra F^- ions.

Interference is a common problem with ISEs. Other limitations include the effects of ionic strengths and the potential formation of complex compounds that are not responsive to ISEs. Many of these problems, however, can be solved by adding a *total ionic-strength adjustment buffer (TISAB)*. The commercially available TISAB for fluoride electrode is a mixture of an acetate buffer (pH 5.0–5.5), 1 M NaCl, and cyclohexylenedinitrilo tetraacetic acid (CDTA). The pH in this range will ensure a low OH^- concentration in the solutions while also avoiding the formation of appreciable HF, which is not responsive to the membrane. The CDTA is a chelating agent, which can release F^- from possible complex compounds such as Al^{3+}, Fe^{3+}, and Si^{4+}.

The *nitrate electrode* is a good example of the second type, that is, the liquid membrane incorporating an ion exchanger. The nitrate electrode method is useful in environmental analysis because it compares favorably with colorimetric methods for the determination of nitrates in soils and water. A method detection limit (MDL) of 2 mg/L and a linear calibration range 5–200 mg/L can be obtained using the U.S. EPA method 9210A. The membrane of commercial nitrate electrode is usually polyvinyl chloride (PVC). As the nitrate ion is a strongly hydrophilic anion, the anion exchanger should be a strongly hydrophobic cation such as tetraalkylammonium salt. For the same reason, when divalent cations, such as Ca^{2+} and Mg^{2+} are analyzed, a cation exchange is used (e.g., calcium organiophosphorous compound for Ca^{2+}).

11.2.3 Potentiometric Titration (Indirect Potentiometry)

Potentiometric titration is a volumetric method in which the potential (E) between two electrodes is measured as a function of the added volume of titrant (V). The indicator electrode can be selected to respond to either a reactant or a product.

A typical plot of E vs. V has a characteristic sigmoid (S-shaped) curve. The part of curve that has the maximum change marks the *equivalent point of titration*. This point can also be determined by the slope of the curve (i.e., the first derivative, $\Delta E/\Delta V$) vs. V plot (Fig. 11.4).

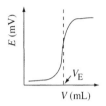 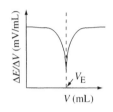

Figure 11.4   A potentiometric titration curve: (a) E vs. V plot and (2) The first derivative plot of $\Delta E/\Delta V$ vs. V

Potentiometric titrations can be used for not only redox reaction but also acid-base, precipitation, and complexation. Examples of acid-base titration include acidity and alkalinity in water quality measurement (Section 6.3.3). The classical volumetric titration methods are based on the chemical color indicators for the determination of end point. The advantage of potentiometric titration for the measurement of acidity and alkalinity is the automated nature of the procedure. Since the end point is based on the millivolts (mV) of the solution, another added advantage is that the titration is not subjected to interference from colored samples.

11.3 VOLTAMMETRIC APPLICATIONS IN ENVIRONMENTAL ANALYSIS

The use of voltammetric techniques, although important within analytical chemistry, is not as widespread as it could be. In environmental labs, the most important applications include dissolved oxygen (DO), chlorine, and metal species analysis. DO can be measured by either galvanic cells or polarographic (electrolytic) cells. Free and residual chlorines are important in monitoring of drinking water quality. Their analyses by volumetric titration have been described in Section 6.3.6. We introduce here the automated amperometric titration technique. Finally, we briefly introduce metal analysis by anodic stripping techniques.

11.3.1 Measurement of Dissolved Oxygen

There are two types of DO measuring probes. If the electrode materials are selected so that the difference in potential is $-0.5\,V$ or greater between the cathode and anode, an external potential is not required and the system is called galvanic. The DO probe of this type is usually Ag or Pb with KOH as the electrolyte. If an external voltage is applied, the system is called *polarographic*. The polarographic probe has a

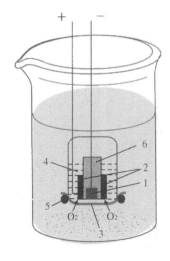

Figure 11.5 A membrane-based dissolved oxygen probe using a polarographic cell [1 = Gold cathode where O_2 is reduced, 2 = Built-in ring-shaped reference electrode (Ag/AgCl), 3 = O_2 permeable membrane (Teflon), 4 = Electrolyte solution (KCl), 5 = O-ring to hold replaceable membrane, 6 = Insulating rod]

Ag anode surrounded by a gold (Au) cathode. The commonly known *Clark oxygen sensor* patented by L. C. Clark in 1956, as shown in Figure 11.5, is the polarographic type. Galvanic probes are more stable and more accurate at low DO levels than polarographic probes. Galvanic probes often operate for several months without electrolyte or membrane replacement, whereas polarographic probes need to be recharged more frequently.

Despite the structural difference between the two, there are similarities in the operational principles. In both types, dissolved oxygen diffuses across a permeable membrane and is reduced in a reaction with a fill solution generating a current flow from cathode to anode. This current flow is proportional to the amount of oxygen that crosses the membrane. Using the Clark-type cell as an example, upon oxygen permeation through the membrane, oxygen is reduced at the gold cathode and the coupled oxidation reaction occurs at the silver reference electrode (anode):

$$O_2 + 2H_2O + 4e^- \leftrightarrows 4OH^- \qquad \text{(cathode reaction)} \qquad (11.13)$$
$$4Ag(s) + 4Cl^- \leftrightarrows 4AgCl(s) + 4e^- \qquad \text{(anode reaction)} \qquad (11.14)$$

In practical uses, one should handle dissolved oxygen probes with extreme care. Inaccurate or even erroneous readings may be caused by a loose, wrinkled, or fouled membrane. The probe should be calibrated on a frequent basis. The calibration can be done by one of the three means, that is, the use of air with 100% relative humidity, air-saturated water prepared by aeration, or by the wet chemical method (Winkler method) introduced in Chapter 6.

EXAMPLE 11.1. The coupled redox reactions shown in Eqs 11.13 and 11.14 have standard potentials of +0.40 and −0.22 V, respectively. (a) Develop an overall balanced reaction and indicate the oxidizing agent and reducing agent, (b) Calculate the cell potential under standard state condition, (c) Write the shorthand notation for the cell, (d) Write the Nernst equation.

SOLUTION:

(a) Combining Eqs 11.13 and 11.14 to cancel out the electrons, we have $O_2 + 2H_2O + 4Ag\,(s) + 4Cl^- \leftrightarrows 4OH^- + 4AgCl\,(s)$. On the basis of the balanced reaction, O_2 is reduced by losing four electrons per mole of O_2. These electrons are moved from anode to the cathode where Ag gained these electrons and Ag is oxidized to Ag^+ in AgCl.

(b) $E_{cell} = E_{cathode} + E_{anode} = +0.40 + (-0.22) = +0.18\,\text{V}$. The positive E_{cell} indicates that the above complete reaction thermodynamically can occur spontaneously.

(c) The shorthand notation for this cell is Ag $(s)|AgCl(s)$, KCl (saturated)$\|H_2O$, Au, where $\|$ is the oxygen permeable membrane used to separate the two electrodes.

(d) $E = E^0 - \dfrac{RT}{nF}\ln\dfrac{[\text{Products}]}{[\text{Reactants}]} = 0.18 - \dfrac{2.303RT}{4F}\log\dfrac{[OH^-]^4}{(pO_2)^2[Cl^-]^4}$. This equation clearly shows the dependence of cell potential on the pressure of oxygen (the analyte). The potential also depends on other factors, such as the temperature, the pH (i.e., $[OH^-]$), and the concentration of Cl^- in the filling solution, which should be kept at constant for an accurate DO reading. The overall reaction in (a) also shows that the DO meter will consume oxygen and generate insoluble AgCl. This also implies that after extended use, silver anode should be cleaned to remove the buildups of AgCl for an improved sensitivity. This example demonstrates that understanding the chemistry will help one obtain the optimal performance of the DO instrument.

11.3.2   Measurement of Anions by Amperometric Titration

Free and residual chlorine are oxidizing agents that can be titrated with a reducing compound such as sodium thiosulphate ($Na_2S_2O_3$) or phenylarsine oxide (PhAsO) (Hach, 1998). If we use Cl_2 to represent both free and residual chlorine, then the redox reaction is

$$Cl_2 + PhAsO + 2H_2O \leftrightarrows 2Cl^- + PhAsO(OH)_2 + 2H^+ \qquad (11.15)$$

When a small voltage of about 0.10 V is applied across the two platinum electrodes, a small electrical current is generated inside the solution. A *reversible* reaction ($Cl_2 + 2e^- \leftrightarrows 2Cl^-$) occurs at both electrodes. That is, the reduced form (Cl^-) is oxidized at the anode and the oxidized form (Cl_2) is reduced at the cathode. The gradual addition of the reducing agent (PhAsO), however, *irreversibly* reduces the oxidized form of the chlorine present. The complete removal of all oxidized chlorine (Cl_2) terminates the reversible reaction and the probe current drops to zero.

In the *amperometric titration* described above, two identical platinum electrodes are used. This is termed as *biamperometric*. Two dissimilar electrodes, such as Ag and Pt, can also be used to conduct current. Other needed instrument components include an *ampere meter* to measure current and a dispenser to

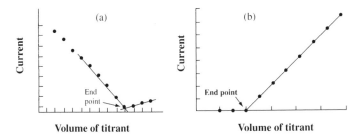

Figure 11.6 Measurement of chlorine by amperometric titration: (a) Forward amperometric titration, (b) Back amperometric titration (Courtesy of Hach Company)

accurately deliver the titrant solution. The output of the amperometric titration, unlike potential titration, is a plot of current vs. titrant volume (Fig. 11.6). As shown in the figure, the abrupt change in current is used as the end point. The concentration of chlorine present in the sample can be calculated by determining the exact amount of titrant (PhAsO) added to extinguish the probe, current.

The titration described above is referred to as *forward titration* (Fig. 11.6a). For water samples containing potential interference, a *back titration* is preferred, in which a known excess amount of the reducing agent ($Na_2S_2O_3$ or PhAsO) is added to the sample. After the reducing agent has reacted with the free and residual chorine in the sample, the amount of remaining titrant is determined through titration with the standard iodine (I_2) solution. The total chlorine is then calculated on the basis of $Na_2S_2O_3$ or PhAsO remaining. The back amperometric end point is signaled when iodide (I^-) is present, which is indicated by a current flow between the electrodes (Fig. 11.6b). The titration reaction is

$$I_2 + 2S_2O_3^{2-} \text{ (excess)} \leftrightharpoons 2I^- + S_4O_6^{2-} \qquad (11.16)$$

By adjusting the reaction condition (pH), the above amperometric titration can be used to further differentiate free chlorine and various residual chlorine compounds (chloramines). The amperometric titration requires more skills and care than the volumetric titration using color indicators.

11.3.3 Measurement of Metals by Anodic Stripping Voltammetry (ASV)

Stripping voltammetry is an elegant electroanalytical technique with the promise of extremely low detection limits for the analyses of both anions and cations. Cations are stripped from anodes and anions are stripped from cathodes, hence the name *anodic stripping voltammetry* and *cathodic stripping voltammetry*. Table 11.2 is a list of common elements that can be determined by anodic and cathodic stripping voltammetry. In environmental analysis, the former is of particular importance because of its ability to analyze multiple metals in a single run. Anodic stripping voltammetry (ASV), therefore, will be the focus of the following discussions.

Table 11.2   List of common elements that can be analyzed by stripping voltammetry

---

By anodic stripping voltammetry:
  Ag[a], As, Au[a], Ba, Bi, Cd, Cu, Ga, Ge, Hg[a], In, K, Mn, Ni, Pb, Pt, Sb, Sn, Tl, Zn
By cathodic stripping voltammetry:[b]
  Br[-], Cl[-], I[-], S[2-], thio compounds

---

[a]Determined on solid electrodes, such as carbon or gold.
[b]These form mercury precipitates on the electrode, which are subsequently stripped off in a negative scan [Rubinson and Rubinson (2000)].

Let us illustrate the analytical principles by assuming that Cd and Cu are the two elements of interest in the solution. The first step in ASV is to *preconcentrate* both metals from the solution into or onto a microelectrode (with a large surface area) by an *electrodeposition* process. The remarkably low detection limit of stripping voltammetry is attributable to the preconcentration step. The second step, called *stripping*, is to *strip* the metals by applying changing potential (Fig. 11.7a). The current vs. potential plot (I vs. E) during the stripping process (Fig. 11.7b) is obtained for quantitative analysis.

In anodic stripping voltammetry, the microelectrode behaves as a cathode where metal ions Cd^{2+} and Cu^{2+} are reduced into metal in their metallic forms (Cd and Cu).

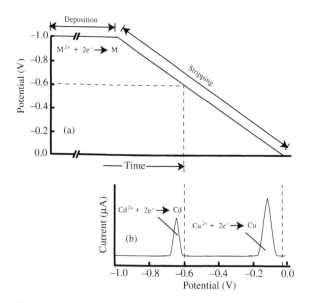

Figure 11.7   Anodic stripping voltammetry (ASV) in a solution containing Cd^{2+} and Cu^{2+}. (a) Potential applied as a function of time. (b) Current vs. potential plot (voltammogram) showing two peaks containing two metals of interest (Cd and Cu). From Principles of Instrumental Analysis 5th edition by Skoog et. al., 1998. Reprinted with permission of Books/Cole. A division of Thomson Learning: www.thomsonrights.com.

The same microelectrode serves as an anode, where Cd and Cu are oxidized into their ionic forms (Cd^{2+} and Cu^{2+}). The reduction reactions in the deposition step are

$$Cd^{2+} + 2e^- \leftrightharpoons Cd(s) \qquad (E^0 = -0.403 \text{ V}) \qquad (11.17)$$

$$Cu^{2+} + 2e^- \leftrightharpoons Cu(s) \qquad (E^0 = +0.337 \text{ V}) \qquad (11.18)$$

For both Cd^{2+} and Cu^{2+} to be reduced, the potential should be low enough. The minimum potential to begin reducing Cd^{2+} can be calculated from the Nernst equation. This is called the *deposition potential*.

After a period of deposition, as shown in Figure 11.7a, the potential is increased linearly (linear potential scan). As a result, Cd starts to be oxidized ($Cd \leftrightharpoons Cd^{2+} + 2e^-$) at an approximate potential of -0.6 V, causing a sharp increase in current. As the deposited Cd is consumed, this current drops to a low level (the first peak of Cd). As potential further increases (sufficiently positive), the second peak for the oxidation of Cu appears.

Note that the indicator (working) electrodes used in ASV differ from the electrodes described previously (Section 11.1.3). The *hanging mercury drop electrode* (HMDE) and the *mercury film electrode* (MFE) are the two most commonly used electrodes for ASV. The larger surface area-to-volume ratio of the MFE compared to the spherical HMDE enables more metal to be concentrated into a given amount of mercury during a specified deposition time. Consequently, a given detection limit can be achieved with a shorter deposition time for the MFE than for the HMDE.

Note also that deposition in stripping voltammetry is not exhaustive. This is in significant contrast to coulometric techniques in which all analytes are electrolyzed. Therefore, it is important in ASV to deposit the same fraction of metal for each stripping voltammogram. To achieve this, the parameters of electrode surface area, deposition time, and stirring must be carefully duplicated for all standards and samples. Deposition times vary from 60 s to 30 min, depending on the analyte concentration, the type of electrode, and the stripping technique. The MFE requires less time than does the HMDE; differential pulse voltammetry requires less time than linear sweep voltammetry as the stripping technique.

To conclude, one should bear in mind that only common environmental applications of electroanalytical techniques are introduced in this chapter. Many electrochemical sensors have become available and are expected to play an increasing role in environmental monitoring. Such sensor devices should allow one to move the measurement of numerous inorganic and organic pollutants from the central laboratory to the field and to perform them rapidly, inexpensively, and reliably. Sensors of various types and applications can be found in the references (Hanrahan et al., 2004; Kellner et al., 2004).

REFERENCES

APHA (1998), *Standard Methods for the Examination of Water and Wastewater*, Washington, DC, 20th Edition, 1998.

BOCKRIS JO'M, REDDY AKN (1970), *Modern Electrochemistry*, Plenum, New York.

*CHRISTIAN GD (2003), *Analytical Chemistry*, 6th Edition, John Wiley & Sons, Hoboken, NJ Chapters 12–15, pp. 354–456.

DOWN RD, LEHR JH (2005), *Environmental Instrumentation and Analysis Handbook*, Wiley-Interscience, Hoboken, NJ Chapters 22 and 23 pp. 459–510.

FIFIELD FW, HAINES PJ (2000), *Environmental Analytical Chemistry*, 2nd Edition, Blackwell Science, Malden, MA pp. 220–279.

Hach Company (1998), Amperometric Titrator, Model 19300, Loveland, CO, pp. 1–84.

HANRAHAN G, PATIL DG, WANG J (2004), Electrochemical sensors for environmental monitoring: Design, development and applications. *J. Environ. Monit.* 6(8):657–664.

KELLNER R, MERMET J-M, OTTO M, VALCARCEL M, WIDMER HM (2004), *Analytical Chemistry: A Modern Approach to Analytical Science*, 2nd Edition, Wiley-VCH Weinhein Chapter 18 pp. 455–499.

KISSINGER PT, HEINEMAN WR (1996), *Laboratory Techniques in Electroanalytical Chemistry*, 2nd Edition, Dekker, New York.

MONK PMS (2001), *Fundamentals of Electro-analytical Chemistry*, John Wiley & Sons, New York, NY.

RUBINSON KA, RUBINSON JF (2000), *Contemporary Instrumental Analysis*, Prentice Hall, Upper Saddle River, NJ, Chapter 7 pp. 204–273.

SAWYER DT, SOBKOWIAK A, ROBERTS JL, Jr. (1995), *Experimental Electrochemistry for Chemists*, 2nd Edition, Wiley, New York.

SKOOG DA, HOLLER FJ, NIEMAN TA (1997), *Principles of Instrumental Analysis*, 5th Edition, Saunders College Publishing, New York, Chapters 22–25, pp. 563–672.

US EPA (1996), SW-846, Method 9210A, Potentiometric determination of nitrate in aqueous samples with an ion-selective electrode.

YSI Incorporated (1999), YSI Model 5239, Dissolved Oxygen Instruction Manual, Yellow Springs, OH, pp. 1–23.

QUESTIONS AND PROBLEMS

1. Explain why electrochemical methods measure the activity rather than concentration?

2. Given the E^0 values of the following two half-reactions:

$$Zn \rightarrow Zn^{2+} + 2e^- \qquad E^0 = 0.763 \text{ V}$$
$$Fe \rightarrow Fe^{2+} + 2e^- \qquad E^0 = 0.441 \text{ V}$$

   (a) Write a balanced complete oxidation-reduction reaction.

   (b) Explain whether the corrosion of an iron pipe (i.e., $Fe \rightarrow Fe^{2+}$) in the presence of Zn/Zn^{2+} is thermodynamically possible or not?

   (c) Explain whether or not Zn will protect the corrosion of iron pipe if metallic Zn is in direct contact with the iron pipe?

3. Construct a complete balanced oxidation-reduction reaction from each pair of two half-reactions:

   (a) $H_2 \rightarrow 2H^+ + 2e^-$ $Fe^{2+} \rightarrow Fe^{3+} + e^-$
   (b) $1/4O_2 \text{ (g)} + H^+ + e^- == \frac{1}{2}H_2O$ $H_2 \rightarrow 2H^+ + 2e^-$
   (c) $2IO_3^- + 12H^+ + 10e^- == I_2 + 6H_2O$ $2I^- = I_2 + 2e^-$

4. Given the electrochemical cell as shown in the figure below: (a) Which one of the following is being oxidized: H_2, H^+, I_2, or both H_2 and I_2? (b) Which one of the

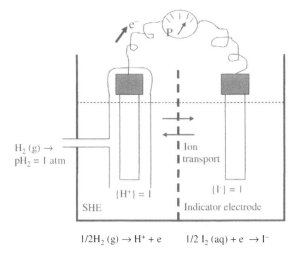

$1/2 H_2\,(g) \rightarrow H^+ + e$ $1/2\,I_2\,(aq) + e \rightarrow I^-$

following is an oxidizing agent: H_2, I_2, or I^-? (c) Which one of the following is incorrect for the reaction: $H_2 + I_2 \rightleftarrows 2H^+ + 2I^-$? (i) Two moles of electron are involved, (ii) The oxidation of H_2 takes place at the anode, (iii) Electron moves from cathode to anode.

5. Write the standard cell notation for the cell shown in Problem 4.

6. Write the standard cell notation for the cell shown in Figure 11.5 employed for dissolved oxygen measurement.

7. Why TISAB is added to the sample solution when ion selective electrode (ISE), such as fluoride electrode, is used to measure the ion concentration?

8. Explain the principles of ion selective electrode used for the measurement of (a) pH and (b) nitrate.

9. Is pH membrane electrode based on the redox reaction that can be described by the Nernst equation? Why or why not?

10. Describe the precautions for the use and maintenance of glass pH electrodes.

11. Give a list of common ions of environmental significance that can be analyzed by ion selective electrodes.

12. The potential (E_1) in a 0.0015 mol/L F^- standard was measured to be 0.150 V by a fluoride electrode in reference to a saturated calomel electrode. For the same electrodes, a voltage of 0.250 (E_2) was measured in a sample containing an unknown concentration of F^-. Calculate the concentration of F^- in the unknown sample (*Hint:* Use Eq. 11.12).

13. Three electrochemical titration methods are discussed in this chapter: potentiometirc, coulometric, and amperometric titration. (a) Describe the fundamental differences of the principles and (b) Give examples of environmental measurements.

14. Use a table to compare the similarities and differences between potentiometry, coulometry, and voltammetry. Consider the following comparisons: (a) The electrical measurement (e.g., current, potential, and charge), (b) The types of cells (galvanic and electrolytic), (c) The fundamental equation employed for quantitative measurement (Nernst, Faraday, and Ohm's), and (d) The ability for qualitative determination.

15. Calculate the mass of Cu in grams that can be deposited from a solution of $CuSO_4$ when 0.1 Faraday electricity is used (atomic weight of Cu = 64, S = 32, O = 16).

16. Two major advantages of anodic stripping voltammetry (ASV) for metal analysis are (a) low detection limits (μg/L to ng/l), and (b) ability to analyze several elements in a sample run. Explain why?

17. Define the following: (a) boundary potential, (b) limiting current, (c) equivalent points of titration, (c) alkaline error, and (d) deposition potential.

Chapter 12

Other Instrumental Methods in Environmental Analysis

12.1 HYPHENATED MASS SPECTROMETRIC METHODS AND APPLICATIONS

12.2 NUCLEAR MAGNETIC RESONANCE SPECTROSCOPY (NMR)

12.3 MISCELLANEOUS METHODS

REFERENCES

QUESTIONS AND PROBLEMS

This chapter introduces several very different topics of instrumental methods. These instrumental techniques are included altogether in Chapter 12, partially because they have not been routinely used in many environmental labs at the present time. The high cost (up to $150,000 and even several million dollars) and the operational requirement for skilled analysts are the major obstacle for the widespread applications. However, the use of these instruments in research communities has been increasing. In fact, two of these techniques (ICP-MS and GC-MS) have recently become common instruments in many environmental labs. It is very likely that the popularity of other instrumental techniques for daily environmental uses will be significantly increased in the future.

These advanced techniques include inorganic (ICP-MS), organic (GC-MS and LC-MS) mass spectrometry and nuclear magnetic resonance (NMR) for quantitative measurement as well as for structural identification (NMR is not commonly used for quantification analysis). Several other specialized techniques are scanning electron microscopes for surface analysis, radiochemical analysis, and immunoassays for screening purposes. Because of the diversity and the specialty of these topics, this chapter is by no means the inclusion of essential details, but rather, it is described in a very elementary and condensed form as compared to the preceding chapters. The purpose of this chapter is to introduce awareness to the readers about these advanced or specialized analytical techniques with regard to

Fundamentals of Environmental Sampling and Analysis,   by Chunlong Zhang
Copyright © 2007 John Wiley & Sons, Inc.

their general principles and environmental applications. At the end of this chapter, further readings are provided for readers at both the elementary and the advanced levels.

12.1   HYPHENATED MASS SPECTROMETRIC METHODS AND APPLICATIONS

A *hyphenated method* is a combination of two instrumental techniques that offer superior analytical results than either of the original individual methods. Three major hyphenated mass spectrometric methods that are of importance in environmental analysis are described below. They are inductively coupled plasma-mass spectrometry (ICP-MS), gas chromatography-mass spectrometry (GC-MS), and liquid chromatography-mass spectrometry (LC-MS). Recall from two previous chapters that we have discussed ICP-OES, GC, and HPLC. The hyphenated methods bear some similarities to these methods, but they are also fundamentally different since a different detector (mass spectrometer) is used. In the following discussion, we therefore focus on the general principles of mass spectrometer and the unique features of hyphenated instrumental components. Major environmental applications and the U.S. EPA methods using ICP-MS, GC-MS, and LC-MS will then be described.

12.1.1   Atomic Mass Spectrometry (ICP-MS)

General Principles In ICP-OES (Chapter 9), elements are ionized by ICP and then measured based on the emission of elemental ions by an optical device at a wavelength characteristic of the element in the UV-VIS range. In ICP-MS, elements are ionized by ICP in the same way as in ICP-OES. However, rather than separating emission light according to their wavelengths, the mass spectrometer separates ions according to their *mass-to-charge ratios (m/z)*. The *m/z* of an atomic ion or molecular ion is the mass of an ion (m) divided by the number of charge (z). For example, the *m/z* of $^{24}Mg^+$ (atomic ion) is 24, whereas the *m/z* of $^{12}CH_4^+$ (molecular ion) is 16. Since most of the ions in mass spectrometers are singly charged, the *m/z* ratio is simply equal to the mass of the ion. Because ions are counted in ICP-MS, the numbers (abundance) of ions are used as the basis for quantitative measurement. It is also important to note that the *mass spectrum* is, therefore, a plot of abundance vs. *m/z* rather than the optical UV-VIS spectrum in ICP-OES which is a plot of light emission vs. wavelength. While the atomic mass spectrum in ICP-MS is used for element determination, the molecular mass spectrum in GC-MS and LC-MS is of prime importance for organic structural identification. Molecular mass spectra will be introduced in Section 12.1.2.

   Instrumental Components A schematic diagram of an ICP-MS system is shown in Figure 12.1. A typical ICP-MS instrument consists of the following: (a) sample introduction system, (b) ICP torch (making ions), (c) interface (sampling ions), (d) vacuum system (pumps), (e) lens (focusing ions), (f) quadrupole

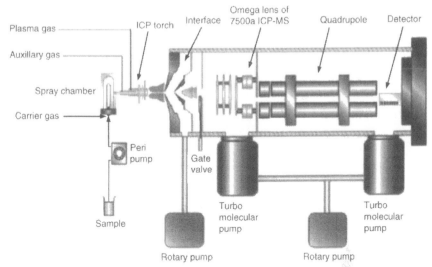

Figure 12.1   Major components of an ICP-MS system (© Agilent Technologies, Inc., 2005, Reproduced with permission, Courtesy of Agilent Technologies Inc.)

(separating ions), (g) dynode detector (counting ions), and (h) data handling and system controller. It should be noted that variations exist among different manufacturers. The quadrupole-based mass spectrometer, as depicted in this figure, and the electron multiplier-based detector are the most commonly used.

To understand the hardware of ICP-MS, one should first recognize the importance of an *interface* between ICP and MS. The ICP torch, used to atomize and ionize materials (Chapter 9), requires a high operational temperature ($\sim$6000 K), whereas the mass spectrometer is operated at room temperature and requires the vacuum condition to avoid collisions with any gas molecules before ions can reach the detector. This task is accomplished through an interface between ICP and MS, and the use of vacuum pumps to remove nearly all gas molecules in the space between the interface and the detector. Second, one should note that in ICP-MS, ions are directly counted rather than measured on the basis of light emission. Hence, there will be no optical device (e.g., optical lens, prisms) in ICP-MS and devices of differing mechanisms are needed.

Ions from the ICP torch are first focused with an electrical field. This is analogous to the optical lens used in ICP-OES to bend the light beams of various wavelengths. These ions of various *m/z* values are then separated in a *quadrupole* system (Fig. 12.2), which consists of four metal rods approximately 20 cm in length and 1 cm in diameter. The quadrupole separates ions by setting up the correct combination of voltages and radio frequencies to guide ions of selected *m/z* between the four rods. The ions that do not match the selected *m/z* pass through spaces between the rods and are ejected from the quadrupole. With this regard, the quadrupole is also referred to as a *mass filter*. The mass filter is equivalent to a monochromator in ICP-OES used to separate out the wavelengths of the light emitted by elemental ions.

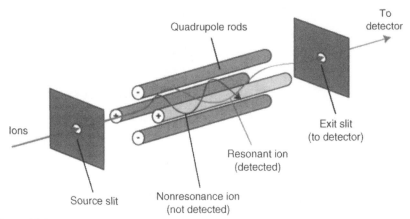

Figure 12.2 Schematic diagram of a quadrupole mass analyzer (Figure used with permission of Dr. P.J. Gates, School of Chemistry, University of Bristo, United Kingdom)

All mass filters can be operated in two different modes. In the *selected ion monitoring (SIM)* mode, selected ion(s) of the target analyte at a specific m/z value(s) is collected and detected subsequently. In the *scan mode*, all ions from the analyte (in a range of m/z values) are collected and analyzed. SIM is often used for quantitative analysis to achieve a high sensitivity because more ions can be collected in a given time period. The scan mode detects all isotopes and fragment ions and contains all structural information needed for identification purposes.

After ions have exited from the mass filter, they are detected by a continuous or discrete dynode-type *electron multiplier*. For example, in a discrete system consisting of a series of *dynodes*, the ions strike the surface of the first dynode that releases electrons. These electrons then strike the second dynode, releasing more electrons. This process continues until a measurable electrical pulse is obtained. By counting the pulses generated by the detector, the system counts the initial ions that hit the first dynode.

In Section 9.3 (Chapter 9), we have compared several atomic spectroscopic techniques, including ICP-MS (now we know that ICP-MS is actually not a *spectroscopic* method but rather a mass *spectrometric* method). From that comparison, we learned that although ICP-MS has a much higher price range, it presents several analytical benefits. These include: (a) better sensitivity than graphite furnace AA, (b) rapid multielement quantitative analysis, (c) wider dynamic range, (d) lower detection limit, and (e) much simpler spectrum as compared to ICP-OES (mass spectrum vs. optical emission spectrum). The optical emission spectrum may have hundreds of emission lines for most heavy elements, demanding a high resolution optical device to minimize potential spectral interference. In ICP-MS, however, there are only 1–10 natural isotopes for all elements in the mass spectrum.

The isotopic information shown in the mass spectrum of ICP-MS actually provides additional analytical advantages that enhance its applications in certain areas. The two primary techniques associated with this capacity are isotope ratio and isotope dilution method. With the *isotope ratio* method, ICP-MS measures specific isotopes of an element, and then the ratio of two or more isotopes can readily be determined. The isotope ratio information can be used in geological dating of rocks, nuclear applications, determining the sources of a contaminant, and biological tracer studies. In *isotope dilution*, an unknown sample is spiked with an enriched isotope (usually of minor abundance) of the element of interest. Isotope dilution mass spectrometry provides high accuracy and precision because the spiked isotope serves as both a calibration standard and an internal standard.

12.1.2 Molecular Mass Spectrometry (GC-MS and LC-MS)

Unlike the *atomic* mass spectrometric ICP-MS methods, GC-MS and LC-MS are *molecular* mass spectrometric methods used for the determination of molecular structure. As we learned in Chapter 10, the GC or HPLC technique alone (without mass spectrometer) provides retention time as the only information to help examine a compound's identity. The structural information provided by GC-MS and LC-MS, on the contrary, are definite. They are also complementary to two other major techniques for structural identification, that is, IR (Chapter 8) and NMR (Section 12.2). For example, GC-MS and LC-MS are able to provide analysts with unique information about the molecular weight. In the following condensed discussion, we describe the general principles and instrumental components of both GC-MS and LC-MS. The readers should, however, be aware that although these two molecular mass spectrometric techniques share some striking similarities in general, the key components, operational principles, and applications areas are fundamentally different.

General Principles and Instrumental Components of GC-MS and LC-MS

Regardless of many variations, the block diagrams shown in Figure 12.3 can be used to illustrate the general principles and instrumental components of GC-MS and LC-MS. Both techniques are based on the chromatographic separation of molecules from the mixture, the generation of gaseous ions from analyte molecules, the separation of ions according to their *m/z* ratios, and the subsequent detection of these ions. Hence the major components include: (a) sample introduction, (b) analyte ionization, (c) interface between chromatographic and mass spectrometric components, (d) vacuum system, (e) mass analysis (mass spectrometer), (f) ion detection, (g) data handling and system controller. Note that in GC-MS and LC-MS, the chromatographic parts are used only for compound separation (sample introduction to the point of ionization). These major components, to a certain extent, are also similar to an ICP-MS system described previously. For example, the quadrupole mass filter is the same as that which can be used in GC-MS and LC-MS, although the *m/z* ranges may be different among various mass spectrometric techniques.

The discussions that follow focus only on the interface and the ionization. The scope of this text prevents us from a detailed discussion on all ionization techniques

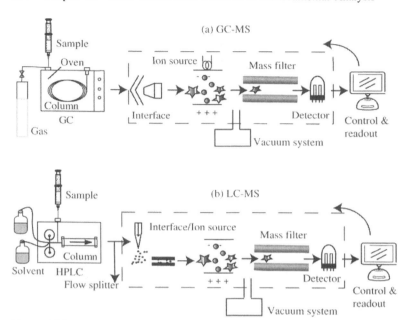

Figure 12.3   Diagram of two molecular mass spectrometers: GC-MS and LC-MS

and mass analyzers commercially available, but we will briefly introduce the most commonly used ion sources in GC-MS and LC-MS. For mass analyzers, since quadrupole (Section 12.1.1) is still the most widely used system in research and commercial labs, we will no longer discuss other mass spectrometers. The readers, however, should be aware of other mass analyzers for certain applications, such as *magnetic sector mass filter, time-of-flight (TOF) mass analyzer,* and *ion trap mass analyzer.* Magnetic sector systems, termed *high resolution mass spectrometry,* have the major advantage of accurate molecular weight determination, but with limitations of cost, size, and weight of permanent magnet. Time-of-flight mass spectrometers have some general applications for volatile compounds but show growing uses in LC/MS for large molecules such as proteins and DNA restriction fragments. Another type of mass spectrometry that has become an extremely powerful tool in many research labs is the *tandem mass spectrometry (MS-MS),* for example, a triple quadrupole system consisting of three sets of quadrupole rods in series. The triple quadrupole systems are the ideal research tools for determining structures of primary molecular fragments.

Interface of GC-MS and LC-MS

Each hyphenated mass spectrometer is unique in the interface, coupling two instruments and the method of ionization. The interface between ICP and quadruple-based MS has been discussed. The interface between GC and MS is relatively simple. GC equipment can be directly interfaced with rapid-scan mass spectrometers with a proper flow adjustment. For capillary columns, the flow rate is usually small enough to feed directly into the ionization chamber of the mass spectrometer. If packed columns are used, a jet separator can remove the carrier gas.

The interface between HPLC and MS is much more challenged and is the key to success. HPLC systems use high pressure for needed separation efficiency and a high load of liquid flow, and hence high gas load (For example, a common flow of liquid at 1 mL/min, when converted to the gas phase, is 1 L/min). The MS part, on the contrary, requires high vacuum and elevated temperate, and does not tolerate high flow of introduced sample. Two interface technologies have become available in the 1990s to address such problems in coupling LC with MS. Since both technologies incorporate the interface and ionization, we will describe them as the ion sources.

Commonly Used Ionization Methods in GC-MS and LC-MS

The four most commonly used ion sources are summarized in Table 12.1. Electron impact (EI) ionization and chemical ionization (CI) are the two most common ion sources for GC-MS, whereas electrospray ionization (ESI) and atmospheric pressure chemical ionization (APCI) are the most common ones for LC-MS. With some similarities in their device, many commercial GC-MS systems can be configured to have both EI and CI, and LC-MS systems to have both ESI and APCI. All these ion sources listed in Table 12.1 can be configured to a quadrupole-based mass analyzer.

There are two types of ionization techniques based on the energy level used to ionize the analyte molecule—hard ionization and soft ionization. *Hard ionization* commonly refers to *electron impact (EI) ionization*. In this EI mode, the "reagent" producing the ionic products is a beam of highly energetic (approximately 70 eV; 1 eV = 23 kcal) electrons. Such electrons are produced by boiling electrons off a narrow strip or coil of wire made of tungsten-rhenium alloy (Fig. 12.4). Note that ionization energies for most organic compounds range from about 5–15 eV and the bond dissociation energies are even smaller. Because of the high energy, the hard ionization usually produces small pieces of *fragment ions* of the sample molecules, thereby offering rich structural information. Since extensive mass spectrum database has been developed, an added advantage of the EI mode GC-MS is the use of

Table 12.1 Most commonly used ionization methods in molecular mass spectrometry

Ionization methods	Ionization agent	Source pressure	Uses
Electron impact (EI)	70 eV electrons	$10^{-4} - 10^{-6}$ torr	GC-MS
Chemical ionization (CI)	Gaseous ions	~1 torr	GC-MS
Electrospray ionization (ESI)	Electric field	Atmospheric or slightly reduced pressure	LC-MS
Atmospheric pressure Chemical ionization (APCI)	Corona Discharge	Atmospheric pressure	LC-MS

1 torr = 1 mm Hg = 1/760 atmospheric pressure at sea level.

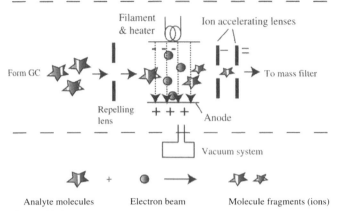

Figure 12.4   Schematic diagram of an electron impact (EI) ion source used in GC-MS

commercially available *mass spectrum library*. The library allows for automated search by comparing the mass spectrum of the unknown to that in the electronic library.

The *soft ionization* used in GC-MS and particularly LC-MS has a variety of techniques. One of the most common soft ionization techniques called *chemical ionization (CI)*, for example, uses a stream of gas such as CH_4, NH_3, isobutene in both GC-MS and LC-MS. Since lower energy is used to bombard the analyte molecule than the electron impact ionization, fragmentation is minimized in soft ionization mode. Mass spectrum is usually simpler with few peaks. This frequently allows intact molecule to be present and detected on the mass spectrum with a soft ionization mode. It is, therefore, particularly useful for characterizing mixtures, although structural information is not as detailed as that acquired in the EI mode.

The *electrospray ionization (ESI)*, first described in 1984, has now become one of the most important techniques for analyzing large molecules (100,000 Da or more). The eluent from HPLC is pumped through a stainless steel capillary needle at a rate of a few μL/min. The needle is maintained at several kilovolts with respect to a cylindrical electrode that surrounds the needle. As shown in Figure 12.5, analyte solution is sprayed (nebulized) into a chamber at atmospheric pressure in the presence of a strong electrostatic field and heated drying gas. The resulting charged spray of fine droplets then passes through a desolvating capillary, where evaporation of the solvent and attachment of charge to the analyte solvent take place. As the droplets become smaller as a consequence of evaporation of the solvent, their charge density becomes greater and desorption of ions into the ambient gas occurs. These ions are then passed through a capillary sampling orifice into the mass analyzer.

The ESI enables mass spectra to be obtained from highly polar and ionic compounds. Another complementary ionization technique using *atmospheric-pressure chemical ionization (APCI)* allows spectra to be obtained from nonpolar

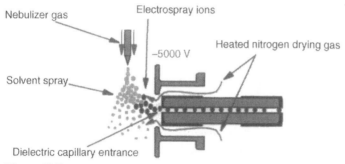

Nebulizer gas

Electrospray ions

−5000 V

Heated nitrogen drying gas

Solvent spray

Dielectric capillary entrance

Figure 12.5 Schematic diagram of an atmospheric pressure electrospray ion (APEI) source used in LC-MS (© Agilent Technologies, Inc., 2005, Reproduced with permission, Courtesy of Agilent Technologies Inc.)

and slightly polar compounds. The APCI method is analogous to CI (commonly used in GC-MS) where primary ions are produced by corona discharge on a solvent spray. In APCI (Fig. 12.6), the LC eluent is sprayed through a heated (typically 250–400°C) vaporizer at atmospheric pressure. The heat vaporizes the

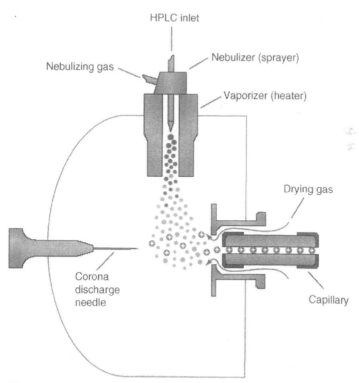

HPLC inlet

Nebulizing gas

Nebulizer (sprayer)

Vaporizer (heater)

Drying gas

Corona discharge needle

Capillary

Figure 12.6 Schematic diagram of an atmospheric-pressure chemical ionization (APCI) used in LC-MS (© Agilent Technologies, Inc., 2005, Reproduced with permission, Courtesy of Agilent Technologies Inc.)

liquid and the resulting gas-phase solvent molecules are ionized by electrons discharged from a corona needle. The solvent ions then transfer charge to analyte molecules through chemical reactions (chemical ionization). The analyte ions pass through a capillary sampling orifice into the mass analyzer. Because the analytes for APCI are usually nonploar, APCI is used in normal phase HPLC more than the electrospray ionization (ESI) is.

Molecule Fragmentation and Basic Terminologies of Mass Spectrum

The above mentioned GC-MS and LC-MS methods, regardless of their variations, all generate two types of plots: a mass spectrum in the scan mode and a total ion chromatogram (TIC) in the SIM mode. Mass spectrum is a plot of ion abundance vs. mass to charge (m/z), and the TIC provides a plot of abundance of ions vs. retention time. TIC resembles any other chromatograms, which is the detector signal vs. time. By displaying data as a TIC, the mass spectrometer does not provide mass spectrum and it is used as a general detector.

We have briefly mentioned that the mass spectrum obtained from soft ionization typically provides a few molecular fragments. These types of mass spectra are relatively easy to interpret, but they do not provide detailed information about the structure of the molecule. It is the hard ionization (i.e., EI mode) that is able to provide rich structural information. Because of the many possible pieces of molecular fragments, however, interpretation for complex organic molecules requires a good understanding of fragmentation chemistry and a good deal of practice.

As a quick overview, we are only allowed to give two simple mass spectrum examples to illustrate the basic terminologies about mass spectra. In most cases, interpretation of mass spectra for unknown compound identifications is not that straightforward. For many of the environmental analysis, a shortcut is to use MS spectrum library because common contaminant analytes are available in the library. For compounds without standard and library source, the interpretation may be a little tricky. A good reference to start with is the book written by McLafferty and Tureček (1993).

EXAMPLE 12.1. The mass spectrum of methanol (CH_3OH) from GC-MS of the EI mode is provided below (Fig. 12.7). Explain the mass spectrum and deduce the fragmentation mechanism of methanol.

SOLUTION: As shown in the above mass spectrum, we can observe several major ions at the m/z values of 15, 29, 31, and 32. Ions at a m/z value of 32 are termed *molecular ions*, since m/z corresponds to the molecular weight of methanol. The ions with a m/z value of 31 are a *base peak*, which is defined as the largest peak in the mass spectrum. Its relative abundance is arbitrarily defined as 100%. A molecular ion peak is the unfragmented (intact) molecule, which is often the largest (heaviest) peak among the highest m/z group (except for any isotopic peaks). The molecular ion represents the parent molecule minus one electron. In addition to

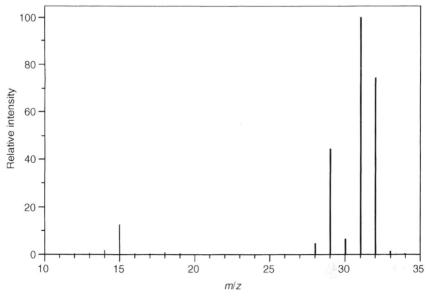

Figure 12.7   Mass spectrum of methanol (SDBSWeb: http://www.aist.go.jp/RIODB/SDBS/ (National Institute of Advanced Industrial Science and Technology, accessed Aug. 2006)

ions at m/z of 31 and 32, the ions of 15 and 29 can be assigned to CH_3^+ and CHO^+, respectively, according to the following fragmentation patterns:

$$CH_3OH + 1 \text{ electron} \rightarrow CH_3OH^+ \cdot \ (m/z\ 32) + 2 \text{ electrons}$$
$$CH_3OH^+ \cdot \ (\text{molecular ion}) \rightarrow CH_2OH^+ \ (m/z\ 31) + H\cdot$$
$$CH_3OH^+ \cdot \ (\text{molecular ion}) \rightarrow CH_3^+ \ (m/z\ 15) + \cdot OH$$
$$CH_2OH^+ \rightarrow CHO^+ (m/z\ 29) + H_2$$

Referring to the above mass spectrum of methanol, there are some small peaks (e.g., m/z at 30 and 33) which are less obvious with respect to their origins. Many of such ions are the results of isotopes of the same element (C and H for methanol). This is better illustrated in the following example.

EXAMPLE 12.2.   Interpret the mass spectrum of 1,4-dichlobenzene (Fig. 12.8).

SOLUTION: The chemical formula of 1,4-dichlobenze is $C_6H_4Cl_2$. Its exact molecular weight depends on the isotopic composition of chlorine atoms (i.e., ^{35}Cl and ^{37}Cl). That is, 146 for $C_6H_4^{35}Cl_2$, 148 for $C_6H_4^{35}Cl^{37}Cl$, and 150 for $C_6H_4^{37}Cl_2$. Note that the natural abundance of ^{35}Cl and ^{37}Cl are 75.53% and 24.47%, respectively. In other words, these two chlorine atoms have an approximate isotopic ratio of $^{35}Cl/^{37}Cl = 3:1$. Such information can be critical to the interpretation of mass spectra, particularly, when atoms such as chlorine and bromine (^{79}Br: $^{81}Br = 50.69\%$: 49.31% $\cong$ 1:1) are present in the organic compounds.

The mass spectrum shows a base peak at m/z of 146, which corresponds to two ^{35}Cl atoms in this molecule (the most abundant isotopic form of 1,4-dichlorobenzene). Two other prominent molecular ion peaks ($m/z = 148$ and 150), although

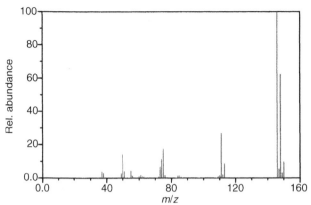

Figure 12.8   Mass spectrum of 1,4-dichlorobenzene. Source: NIST Mass Spec Data Center, S.E. Stein, director, "Mass Spectra" in NIST Chemistry WebBook, NIST Standard Reference Database Number 69, Eds. P. J. Linstrom and W.G. Mallard, June 2005, National Institute of Standards and Technology, Gaithersburg MD, 20899 (http://webbook.nist.gov)

less abundant, are also apparent in the mass spectrum. The ions at 111 and 113 correspond to the dissociation of one of the ^{35}Cl (more abundant) from the molecule ($146 - 111 = 35$; $148 - 113 = 35$). The ions at around m/z of 76 represent the dissociations of the second Cl atom from the parent molecule. The mass spectrum of 1,4-dichlorobenzene shows a clear feature of *isotopic cluster* as a result of the isotopic chlorines in the molecule. Note, however, that isotopic clusters will appear in the mass spectra of all organic compounds. This is because both C and H, the two most common elements of organic compounds, have their naturally occurring isotopes (^{12}C: ^{13}C = 98.89%: 1.11%; ^{1}H: ^{2}H = 99.99%: 0.01%).

12.1.3   Mass Spectrometric Applications in Environmental Analysis

Both atomic (ICP-MS) and molecular (GC-MS and LC-MS) mass spectrometric methods have played a key role in environmental analysis. From the early discovery of disinfection byproducts in drinking water using GC-MS in 1974, mass spectrometry has become the prime tool in many environmental research and regulatory laboratories in helping analysts to identify pollutants in an unequivocal way, and advancing our understanding of the complex environmental processes. In the early 1980s, mass spectrometry was considered to be expensive, complicated, time-consuming, and personnel-intensive, but technology advancement has made such instrument more affordable, user-friendly, and productive. Today there is hardly a GC lab which is not equipped with a GC-MS system for organic analysis. In a similar trend, ICP-MS has grown to become the technique of choice for rapid ultratrace multielement analysis since its first commercialization in 1983.

ICP-MS Applications: ICP-MS can detect almost all elements in the periodic table (few exceptions include C, H, O, N, F, Cl). ICP-MS further offers the analytical benefit of simultaneous multielement analysis. The detection limits are much lower than ICP-OES at or below the ppb range. Additionally, ICP-MS can provide isotopic information as described earlier.

Numerous ICP-MS methods have become available from research communities, the U.S. EPA, and other organizations. An example of development from recent studies includes the use of ICP-MS or IC-ICP-MS for the speciation analysis of several environmentally important elements such as Cr, As, and Se. Several regulatory methods based on ICP-MS have been promulgated by the U.S. EPA. These include Method 200.8 for the analysis of water, wastewater, and wastes. Other U.S. EPA water methods using ICP-MS include Methods 1683 and 1640. ICP-MS based method in SW-846 is Method 6020 for solid wastes, which has been approved for RCRA program since 1995. The latest CLP SOW incorporates ICP-MS instead of GFAA. Besides these U.S. EPA methods, other consensus methods have become available from several agencies such as ASTM, APHA, and DOE.

GC-MS and LC-MS Applications: General applications of GC-MS and LC-MS techniques can be illustrated in Figure 12.9 with regard to molecular weight and compound's polarity. As shown, GC-MS can be used to analyze a small percentage of total compounds (approximately 9 million registered), because GC imparts little or no heat to analyte molecules. A fortunate aspect of GC-MS is that most pollutants of environmental importance have molecular weight less than approximately 500, and some of the low-volatility compounds can be made amenable to GC-MS by derivatization (Chapter 7). Complementary to GC-MS, LC-MS can virtually analyze all semivolatile, nonvolatile, thermally labile, or charged molecules. An added advantage of LC-MS is that it allows direct injection of "aqueous" samples and eliminates the need for time-consuming sample preparation and sometimes chemical modifications.

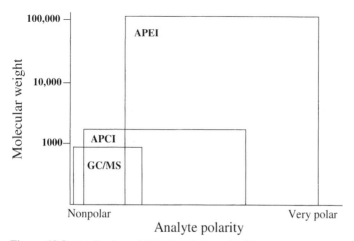

Figure 12.9 Applications of GC-MS and two major LC-MS techniques (© Agilent Technologies, Inc., 2001, Reproduced with permission, Courtesy of Agilent Technologies Inc.)

In Chapter 10, when environmental applications of GC and HPLC were discussed, we have provided a list of EPA methods using GC and HPLC for the analysis of chemicals in air (TO-series), water (500 and 600 series methods), and waste (SW-846 8000 series methods). Of these methods approved by the U.S. EPA, the number of GC-MS methods are too many to be listed here. In a significant contrast is the fact that there are only a few LC-MS based standard methods listed in the EPA method database. For a detailed list, interested reader can find such information at the U.S. EPA's Web site: www.epa.gov/epahome/Standards.html. From this website, there is a link to another important Web site: www.epa.gov/nerlcwww/methmans. html, where a compendium called "The Manual of Manuals" can be found.

Readers can refer to Chapter 5 for specific methodologies of GC-MS and LC-MS. Simply put, GC-MS methods for the analysis of toxic organic compounds in air can be found in TO-series. These methods employ adsorbent tube, liquid impinger, PUF cartridge, and SUMMA canisters as the sampling media. The standard GC-MS methods for the analysis of VOCs can be found in 524 and 624 for drinking water and wastewater, respectively. The equivalent GC-MS methods for SVOCs can be found in 525 and 625 for drinking water and wastewater, respectively. The GC-MS based waste methods in SW-846 include 8260B (VOCs), 8270C (SVOCs), and a particular high resolution GC-MS method for the analysis of dioxin compounds. At the time of writing, there are only two LC-MS methods (8231A and 8235) in the SW-846 that are developed for solvent extractable nonvolatiles.

The readers should be aware that the number of publications in peer-reviewed journals exploded in the last 10 years that explores the use of various mass spectrometric methods. Many of these hyphenated methods are aimed at contaminants of emerging environmental concerns with a low detection limit. Table 5.5 in Chapter 5 is an example list of these emerging compounds, including disinfection byproducts, pharmaceutical chemicals, endocrine disrupting chemicals, chiral contaminants, algal toxin, natural organic mater, and many more. Richardson (2001; 2002; 2003) gave several excellent reviews on the applications of mass spectrometry in environmental sciences.

12.2   NUCLEAR MAGNETIC RESONANCE SPECTROSCOPY (NMR)

12.2.1   NMR Spectrometers and the Origin of NMR Signals

The term "nuclear magnetic resonance (NMR)" refers to a technique in which *nuclei* in the presence of a *magnetic field* are in resonance with a radiation in the *radio frequency* (rf) range (4–900 MHz). As shown in the schematic diagram of an NMR spectrometer in Figure 12.10, the essential components are not hard to understand. It consists of a sample probe, a superconducting magnet to align the nuclear spins in the sample, a radiofrequency transmitter as a source of energy to excite a nuclei from the low to the high energy states, a receiver to detect the

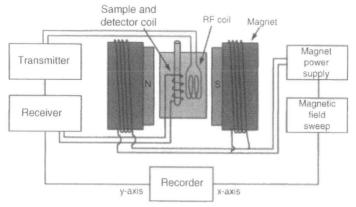

Figure 12.10 Diagram of a nuclear magnetic resonance (NMR) spectrometer (© Rod Nave, HyperPhysics, 2006)

absorption of rf radiation, and a computer console as the terminal for controlling each component and recording NMR spectra. The central superconducting magnet system must be well maintained by regularly filling liquid nitrogen and helium (typically every 10 days for N_2 and every 80–130 days for He). This is also the part in which extreme safety procedures should be carefully followed to keep all ferromagnetic items away from the magnet and avoid suffocation in a confined lab space, where rapid boil-off of cryogenic liquid N_2 and He can occur. To acquire an NMR spectrum, a small amount (mg) of the analyte is dissolved in 0.5 mL of solvent (e.g., $CDCl_3$ or D_2O) and the solution is placed in a long and thin-walled glass tube.

To understand the basic theory of NMR, we need to know the properties of nuclei, the behavior of nuclei in a magnetic field, and the effect of a radio frequency radiation. From elementary chemistry, we learned that a nucleus is made up of protons and neutrons (positive and neutral charge, respectively). Circulating around the nucleus are the negatively charged electrons having the same number as that of protons. For example, ^1H nucleus has one proton and zero neutron in the nucleus and one electron in the 1s orbital (Hence a ^1H nucleus is the same as a proton). Its isotope ^2H (deuterium) has one proton and one neutron in the nucleus and one electron in the 1s orbital. Similarly, ^{12}C has six protons and six neutrons in the nucleus and six electrons around the nucleus ($1s^2\,2s^2\,2p^2$; Refer to Chapter 8 for electronic configuration). Its isotope ^{13}C has six protons and seven neutrons in the nucleus and six electrons.

From *Pauli exclusion principle* in elementary chemistry, we also learned that two identical electrons in the same atom cannot have the same quantum number. These two electrons must possess a property called *spin*, with spin quantum numbers (I) of $+\frac{1}{2}$ and $-\frac{1}{2}$. An electron in the nucleus is analogous to the Earth in relation to the Sun. The Earth rotates once a year around the Sun and it spins along its own axis once a day. Like electrons, nuclei also spin around its own axis at two spin states (quantum numbers of $+\frac{1}{2}$ and $-\frac{1}{2}$). In order for the nuclei to absorb electromagnetic radiation, these two states must have different energies. One way to make them different is to place the nuclei (sample) in a magnetic field. In the presence of a magnetic field (B_0), nuclei behave like tiny bar magnets. Some nuclei will align with the magnetic field (at a lower energy level), whereas others will align against the field (at a higher energy level). Since the nucleus energy is quantized, there will be no other spin states. The energy difference between the two nuclear spin states is as follows:

$$\Delta E = \frac{\gamma h}{2\pi} B_0 \qquad (12.1)$$

where h is the Planck's constant, B_0 is the strength of the magnetic field in the unit of tesla (T), and γ is the *gyromagnetic ratio* in the unit of rad/T/s. One tesla (T) is actually a relatively strong field—the earth's magnetic field is of the order of 0.0001 T.

When an electromagnetic radiation is introduced, the nucleus is resonated if the applied radiation energy ($E = h\nu$) is equal to the energy difference (ΔE) noted in Eq. 12.1. This frequency can be calculated by

$$\nu = \frac{\gamma}{2\pi} B_0 \qquad (12.2)$$

At the resonance frequency, nuclei in the lower energy ($+\frac{1}{2}$) flip to the higher energy state ($-\frac{1}{2}$). Since the value of gyromagnetic ratio (γ) depends on a particular kind of nucleus, different energies are required to bring different kinds of nuclei into resonance. Hence, an NMR signal (rf absorption) can be obtained as a function of the changing frequency. Such a relationship is the basis of an NMR spectrum.

Note that only certain atomic nuclei have the NMR signals. Nuclei with an even mass number and an even charge number (^{12}C) have spin quantum numbers of zero. Nuclei with an even mass number and an odd charge number (e.g., ^{2}H) have integer spin quantum number of one. Neither ^{12}C nor ^{2}H produces NMR signals. Only nuclei with odd mass numbers and half-integer spin quantum numbers ($I = \frac{1}{2}$), such as ^{1}H, ^{13}C, ^{15}N, ^{19}F, and ^{31}P can be used to obtain NMR spectra. Among these, ^{1}H and ^{13}C are the most common, hence the name ^{1}H-NMR (proton NMR) and ^{13}C-NMR (carbon thirteen NMR).

From Table 12.2, it is clear that the most abundant isotope of carbon (^{12}C), unfortunately, cannot be used to obtain an NMR spectrum. The NMR responsive

Table 12.2   Properties of selected nuclei that are important in NMR spectroscopy[a]

Nucleus	Spin (I)	Natural abundance (%)	Gyromagnetic ratio (rad/T/s)	NMR frequency (MHz)[b]	Relative sensitivity[c]
^{1}H	$\frac{1}{2}$	99.99	2.6752×10^{8}	100	1
^{2}H	1	0.01	4.1066×10^{7}	15.35	9.65×10^{-3}
^{12}C	0	98.9	-	-	-
^{13}C	$\frac{1}{2}$	1.1	6.7283×10^{7}	25.15	1.59×10^{-2}
^{14}N	1	99.6	1.9338×10^{7}	7.23	1.01×10^{-3}
^{15}N	$\frac{1}{2}$	0.4	-2.7126×10^{8}	10.14	1.04×10^{-3}
^{19}F	$\frac{1}{2}$	100	2.5182×10^{8}	94.09	8.32×10^{-1}
^{31}P	$\frac{1}{2}$	100	1.0840×10^{8}	40.48	6.65×10^{-2}

[a]Values adapted from Friebolin (2005).
[b]NMR resonance frequency is calculated at $B_0 = 2.3488$ T.
[c]Selectivity is expressed relative to ^{1}H at a constant field for an equal number of nuclei.

isotope ^{13}C has only a 1.1% abundance and its gyromagnetic ratio (γ) is also much smaller than that of ^1H. This implies that the ^{13}C-NMR is much less sensitive than ^1H-NMR. The weak signal of ^{13}C-NMR, however, does not preclude its application in structural elucidation. In fact, the combined information from ^1H-NMR and ^{13}C-NMR spectra helps to determine the exact carbon-hydrogen framework of organic molecules. Since ^1H-NMR and ^{13}C-NMR have the same principle, we focus on ^1H-NMR in the following discussions. For ^{13}C-NMR, only its unique features are briefly described.

12.2.2 Molecular Structures and NMR Spectra

In this section, we use a high-resolution ^1H-NMR spectrum of 2,6-dinitrotoluene (Fig. 12.11) to illustrate how molecular structures are correlated with the spectral characteristics of an *NMR spectrum*. We need to be acquainted with three spectral parameters that are directly related to the structures of molecules: chemical shift, coupling constants, and integral. Chemical shifts are related to the number of signals and the position of signals. In an ^1H-NMR spectrum, chemical shift can be used to tell how many types of H and what types of H are there in the molecule. (Similarly in a ^{13}C-NMR spectrum, chemical shift can be used to tell how many types of C and what types of C are there). The integral is the relative peak area or the magnitudes of signals in an NMR spectrum, so it can be used to tell how many H of each type are there. The third parameter, coupling constant, can be used to tell the connectivity of the chemical bonds.

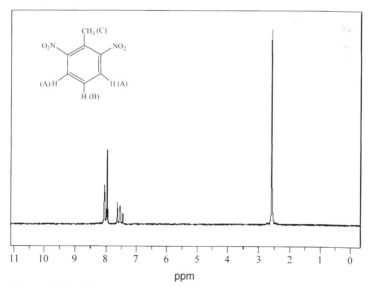

Figure 12.11 ^1H-NMR spectrum of 2,6-dinitrotoluene (NMR conditions: 89.56 MHz; 0.044 g 2,6-dinitrotoluene in 0.5 ml CDCl$_3$) (SDBSWeb: http://www.aist.go.jp/RIODB/SDBS/ (National Institute of Advanced Industrial Science and Technology, accessed Aug. 2006)

Chemical Shift (δ)

As shown in Figure 12.11, an NMR spectrum is a plot of NMR signal intensity (the absorption of rf) vs. chemical shift. *Chemical shift* (δ) measures the difference between the resonance frequency of the nucleus and a reference standard relative to the operating frequency of an NMR spectrometer. This quantity is calculated by

$$\delta = \frac{\nu_{signal} - \nu_{reference}}{\nu_{spectrometer}} \times 10^6 \qquad (12.3)$$

In NMR spectroscopy, this standard is often tetramethylsilane, $Si(CH_3)_4$, abbreviated as TMS. Since the dimensions of ν for both signal and TMS are Hz and the dimension of $\nu_{spectrometer}$ is MHZ, the unit of chemical shift is reported in parts per million (ppm). This unit should not be confused with the ppm used to express the chemical concentration (Chapter 2).

In the preceding section, we know that the signal in the ^1H-NMR originates from the presence of ^1H in a molecule. If all the protons (^1H) in an organic molecule give exactly the same signal, this would tell us nothing about the structure of the compound except that it contains protons. The same would be true for ^{13}C-NMR, and fortunately this is not the case. As we know, the proton of the ^1H nucleus (or carbon) is embedded in a cloud of electrons that circulate around the center of nucleus. In a magnetic field ($B_{applied} = B_0$), the electronic circulation will "induce" a local magnetic field (B_{local}) that is in the opposite direction to the applied magnetic field (Again, recall from elementary physics that magnetism can be induced from moving charges and vice versa). This is equivalent to saying that the ^1H nucleus will "sense" a magnetic field smaller than an isolated nucleus without an electron. Because of the variations of electron densities, various ^1H protons in the molecule will be more or less *shielded* (called *diamagnetic shielding*) and sense different magnetic fields. This will, in turn, requires various frequencies of rf radiation to bring into the resonance of each type of ^1H.

The above analysis indicates that, by looking at how many groups of signals are there in a ^1H-NMR spectrum, one should be able to tell how many types of protons (^1H) are there in an organic molecule. Protons of the same type have the same environment and are called *chemically equivalent protons*. For example, six ^1H protons in CH_3OCH_3 are chemically equivalent, hence only one signal (NMR signal peak) will appear in ^1H-NMR spectrum. 1-Chloropropane ($CH_3CH_2CH_2Br$) has three groups of chemically equivalent protons, therefore, the ^1H-NMR spectrum will have three groups of signals. The three protons in the methyl group (CH_3) are chemically equivalent because of the rotation about the C—C bonds. The two methylene protons and the two protons closest to the bromine atom are chemically equivalent. The NMR spectrum of 2,6-dinitrotoluene (Fig. 12.11) shows three groups of peaks at 8.00, 7.55, and 2.58 ppm. This correlates with three groups of chemically equivalent protons (labeled as A, B, and C) as can be seen from its chemical structure.

The next question is then how these signals are positioned in an NMR spectrum with different values of chemical shift (δ). By convention, we refer a signal in the far left of the x-axis to be the *downfield* and the far right to be the *upfield*. From the upfield to downfield, the resonance frequency is in an increasing order. Whether a nucleus appears in an upfield or a downfield depends on the degree of the shielding, which in turn depends on the electronic environment of the chemical. In general, we can state that protons in *electron-rich* molecular environment are *more shielded*. These shielded protons sense a *smaller magnetic field* and thus come into resonance at a *lower frequency*. These protons (^{1}H) will appear on the right-hand side of the NMR spectrum (upfield). Conversely, protons in *electron-poor* molecular environment are *less shielded*. These less shielded or deshielded protons sense a *larger magnetic field* and, thus, come into resonance at a *higher frequency*. These ^{1}H will appear on the left-hand side of the NMR spectrum (downfield).

Therefore, a chemical shift provides information about the average effective magnetic field due to a nuclear shielding, which is directly related to the variations of the electron density present at various locations within a molecule. For example, the proton of TMS used in an NMR standard appears at the lowest frequency than most signals because silicon (Si) is the least electronegative in organic molecules. Consequently, the methyl protons of TMS are in more electron-dense environmental than most protons in organic molecules. Thus, the protons in TMS are the most shielded. On the contrary, the protons in methyl fluoride (CH_3F) are much less shielded because the strong electronegativity of fluorine atom will draw electrons away from the methyl group. At this point, it should become easy to understand, in the NMR of 2,6-dinitrotoluene, why the signal of two protons (A) closest to the electron-withdrawing functional group (NO_2) will locate in the downfield ($\delta = 8.00\,ppm$), whereas the least affected methyl protons (C) is in the upfield ($\delta = 2.58\,ppm$).

Integral (Signal Intensity)

Integral is the relative area of absorption peaks in the NMR spectrum. Here an *absorption peak* is defined as the family of peaks centered at a particular chemical shift. For example, if there is a triplet (discussed later) of peaks at a specific chemical shift, the number is the sum of the area of the three. Just like the peak areas in chromatographic analysis, integral can be readily obtained by any modern NMR spectrometer. The rule useful to interpret NMR spectrum is that peak area is proportional to the number of a given type of spins (^{1}H proton) in the molecule. We again use Figure 12.11 and a few more examples to illustrate such a relationship.

Consider two chemicals, 1-chloropropane ($CH_3CH_2CH_2Br$) and 1-bromo-2,2-dimethylpropane, ($CH_3)_3CCH_2Br$. As we discussed, the first chemical should give three signals due to three types of chemically equivalent protons. When areas are integrated, the area ratio should give 3:2:2. For ($CH_3)_3CCH_2Br$, we have two types of chemically equivalent protons, namely, nine methyl protons (CH_3) and two methylene proton (CH_2). If these two corresponding peak areas are integrated and normalized, the ratio should be 9:2. Given these examples, now the reader should become confident in deducing the ratio of three types of chemically equivalent protons in 2,6-dinitrotoluene in Figure 12.11. The ratio of protons labeled as A, B, and C should be 2:1:3.

Spin–Spin Coupling

Thus far, we have not explained what caused the spilt peaks as shown in Figure 12.11. Signals in an NMR spectrum can be a single peak (a *singlet*) or can be split into two (a *doublet*), three (a *triplet*), four (a *quartet*), or even more peaks. The number of peaks into which a particular proton is split is called its *multiplicity*. The splitting makes interpretation of an NMR spectrum a little harder, but it provides more structural information about the chemical bonding. We just learned that chemical shift is derived from the density variation of *electrons* that circulate the nucleus. In a significant contrast, signal splitting of a nucleus is caused by a nearby spin-active *nucleus*. These splittings of signals are called *spin–spin coupling*, which are derived from bond interactions between two spin-active nuclei.

The number of splitting peaks (multiplicity of signal) in ^{1}H-NMR spectroscopy can be predicted by *the $n + 1$ rule*, which states that the NMR signal of a nucleus coupled to n equivalent hydrogens will be split into a multiplet with $(n + 1)$ lines. For example, 1,1,2-trichloroethane (Cl_2CHCH_2Cl) has two types of H (hence two groups of signals). The less shielded proton in Cl_2CH (downfield) will have $2 + 1 = 3$ split peaks (triplet) because two equivalent H atoms are present nearby in CH_2Cl group. Conversely, the more shielded protons in CH_2Cl (upfield) will have $1 + 1 = 2$ split peaks because there is only one H atom nearby.

It is, thus, important to keep in mind that it is not the number of protons giving rise to a signal that determines the number of split peaks (multiplicity), instead, it is the number of protons bonded to the *immediately adjacent* carbons that determines the multiplicity. Also keep in mind that equivalent protons do not split each other's signal. Thus, unlike 1,1,2-trichloroethane, 1,2-dichloroethane ($ClCH_2CH_2Cl$) shows only a singlet because four protons in this molecule are chemically equivalent.

In addition to the *number* of split peaks, the *spacing* between the lines of splitting peaks is also important in relating NMR spectrum to chemical structure. The spacing in the frequency unit (Hz) is called the *coupling constant* (denoted by J), which determines the extent of coupling between two nuclei. Since coupling is caused solely by the internal molecular forces, the magnitudes of J are not dependent on the operating frequency of the NMR spectrometers. This characteristic coupling constants (J) can then be used to infer the number and type of bonds that connect the coupled protons as well as the geometric relationship of protons. For example, the *trans*-3-chloropropenoic acid (ClCH=CHCOOH) has a coupling constant of 14 Hz, whereas the *cis*-3-chloropropenoic acid has a coupling constant of 9 Hz.

The above discussions are largely based on ^{1}H-NMR. Since many of these principles are applicable to ^{13}C-NMR, it is essential for us to summarize only the unique features of ^{13}C-NMR. First, the signals in ^{13}C-NMR are not normally split by neighboring carbons and the interpretation of ^{13}C-NMR spectra is easier than ^{1}H-NMR spectra. The reason for the lack of splitting is because the probability of two ^{13}C carbons being next to each other in a molecule is very small ($1.1\% \times 1.1\% = 0.0121\%$). Second, ^{13}C-NMR is less sensitive because of the low abundance of ^{13}C and its gyromagnetic constant (Table 12.1). To alleviate this problem, more sample or a longer run-time is needed. Third, unlike ^{1}H-NMR, the area under a ^{13}C-NMR is *not* proportional to the number of atoms giving rise to the signal, implying that integration cannot be routinely used for ^{13}C-NMR unless special techniques are used. Lastly, the range of the chemical shift in ^{13}C-NMR spectra is much wider than ^{1}H-NMR (220 ppm vs. 12 ppm), so the potential signal overlap is minimized. This is an advantage of ^{13}C-NMR.

12.2.3   Applications of NMR in Environmental Analysis

Recall from Chapter 8 that, UV and IR spectroscopy can be used to identify the presence of conjugation and functionalities in an organic molecule. The underlying chemistry for UV spectroscopy is the electronic excitation upon a molecule's exposure to UV. In IR spectroscopy, the energy is not sufficiently strong to cause the excitation of electrons, but it is strong enough to cause atoms and group of atoms to vibrate. In an NMR spectroscopy, a radiofrequency (rf) of even lower energy radiation is used to cause nuclear spin. As we have described, NMR spectroscopy identifies the carbon-hydrogen framework of an organic compound.

The overall utility in structural identification is in the increasing order: UV < IR < MS < NMR. The combined use of ^{1}H and ^{13}C NMR provides the most useful information. An added advantage of an NMR is that it does not need a standard for structural confirmation. This is advantageous over molecular mass spectroscopic techniques. In many fate and transport studies on the degradation of environmental contaminants, standards of new compounds such as intermediates and products are commonly not available.

Despite its superior power in structural identification, NMR has its major limitation due to its low sensitivity for use in quantitative measurement of contaminants' concentrations. The overall sensitivity is in the order: MS > UV > IR > ^{1}H NMR > ^{13}C NMR. Research is under way to improve the sensitivity of the NMR for environmental uses, such as the detection of pharmaceutical chemicals in the environment using capillary LC-NMR. New NMR sample probes have also been designed for the simultaneous analysis of multiple samples in NMR microcoils which are capable of accommodating very small amount of samples (Borman, 2004; Lens and Hemminga, 1998).

12.3   MISCELLANEOUS METHODS

12.3.1   Radiochemical Analysis

Sources and Properties of Several Important Radionuclides

Radiochemical analysis has become a commonplace in today's environmental labs. An apparent reason is that both naturally occurring and artificial sources of radioactive chemicals can pose significant hazards to human and living organisms.

The hazards due to ionizing radiations and radioactivities are unique from common nonradioactive chemicals. They can be lethal at a high dose, but in most cases the effects are not apparent and are hard to assess. The effects of lower levels of exposure can lead to cell mutation and the development of cancer, which may have consequences for offspring rather than for the parent. Fortunately, the instrumentations for the detection of ionization and radioactivity are readily available.

Table 12.3 is a short list of some common radioactive nuclides (radionuclides). Of the listed radionuclides, carbon-14 ($^{14}_{6}C$) and tritium ($^{3}_{1}H$) are produced in the upper atmosphere as a result of cosmic radiations and their subsequent incorporation into the elemental cycles. Note the common notation $^{Z}_{A}X$, where Z is the atomic number and A is the mass number of the element X. The atomic number is also the number of protons contained in its nucleus and the mass number is the sum of the number of neutrons and protons in a nucleus. Hence, the radioactive $^{14}_{6}C$ has six protons and eight neutrons, whereas its nonradioactive isotope carbon-13 ($^{13}_{6}C$) has the same number of protons but different number of neutrons.

Table 12.3 Characteristics of common radioactive nuclides

Nuclide	Abundance (%)	Half-life	Radiations and energy (MeV)
^{3}H	-	12.26 y	β^- (0.0186)
^{14}C	-	5720 y	β^- (0.155)
^{32}P	-	14.3 d	β^- (1.71)
^{35}S	-	87 d	β^- (0.167)
^{90}Sr	-	29 y	β^- (0.54)
^{137}Cs	-	39 y	β^- (0.51)
^{222}Rn	-	3.8 d	α (N.A.)
^{226}Ra	100	1622 y	α (4.78)
^{234}U	0.0056	2×10^5 y	α (4.77, 4.72)
^{235}U	0.7205	7×10^8 y	α (4.39); γ (0.18, 0.14)
^{238}U	99.27	5×10^9 y	α (4.19)

McV = one million electron volts. Adapted from Pecsok et al. (1968).

The first four radionuclides (^{3}H, ^{14}C, ^{32}P, ^{35}S) in Table 12.3 are of more importance in environmental research on the fate and transport studies using radiolabeled compounds. An important application is to label test contaminant with radionuclides and then study its degradation/transformation pathways. None of these four radionuclides are natural origins. The half-lives for ^{32}P and ^{35}S are short in terms of days but the half-lives of ^{3}H and ^{14}C are very long. For example, ^{14}C has a half-life of 5720 years.

Table 12.3 also includes several other radionuclides with environmental significance. For example, uranium is the basis of nuclear power and nuclear fuel cycle. Uranium present in its two principal isotopes, ^{238}U and ^{235}U, is widely distributed in natural environment and is present in significant levels in many granitic strata. In nuclear industries, uranium is mined, concentrated, processed, and disposed of in the environment in a large quantity. Radium is one of the decay products of uranium (^{238}U). It has four naturally occurring isotopes ^{222}Ra, ^{224}Ra, ^{226}Ra, and ^{228}Ra. Of particular importance is the Radon-222, which is a daughter product of ^{226}Ra. It is of particular environmental importance because radon occurs naturally as a result of the radioactive decays from rocks and soils. Because radon exists in the form of a radioactive gas, its presence in buildings has called for the second leading cause of lung cancer in the United States.

In contrast to stable isotopes (e.g., ^{3}H and ^{14}C as compared to ^{1}H and ^{12}C), one unique property of all radionuclides is their spontaneous decay process (*disintegration*) to their ultimate stable forms. During the radioactive decay process, electromagnetic radiations are emitted. The emissions can be in various forms depending on the nature of the specific radionuclides. These various forms of radiations, that is, α particles (helium nucleus), β particles (electrons), and γ rays (photons) are discussed below.

Beta (β) radiation, as shown in Table 12.3 for several light elements, is the result of the following nuclear reactions:

$$_{Z}^{A}X \rightarrow _{Z+1}^{A}Y + \beta^{-} + \gamma \qquad (12.4)$$

$$_{Z}^{A}X \rightarrow _{Z-1}^{A}Y + \beta^{+} + \gamma \qquad (12.5)$$

where β^{+} and β^{-} (or $_{-1}^{0}\beta$ and $_{+1}^{0}\beta$) denote negatrons and positrons, respectively. *Negatrons* are essentially the electrons in terms of the mass and charge but they have the origin of nuclear reactions. *Positrons* carry the same positive charge equal in magnitude to that of the negatrons. As shown in Eqs 12.4 and 12.5, many β radiations accompany the emission of γ-rays, although ^{3}H and ^{14}C are the exceptions. *Gamma (γ) radiation* is a high-energy electromagnetic radiation (photons) with a wavelength shorter than UV. It is identical to a high-energy X-ray with an exception that their sources are different. Unlike β particles, γ-rays ($_{0}^{0}\gamma$) do not carry charge and mass.

Alpha (α) radiation occurs only for heavy nuclides with an atomic number (A) greater than 90. Thus, radon, radium, and uranium all radiate α particles, which carry $+2$ charge and a mass of 4 atomic mass unit (amu). Therefore, α particles are essentially the helium nuclei that have been stripped of their two electrons, that is, $_{2}^{4}He^{2+}$.

Preservation and Measurement of Radioactive Samples

The general principles of sample preservations for most radioactive nuclides are the same as those described in Chapter 4 for the preservation of metals (Table 12.4). With an exception of tritium, acid preservation to pH < 2 is all that is required. Either plastic or glass containers are acceptable, although, plastic ones are preferred to prevent any breakage during sample transportation. Extra safety precautions that are not usually associated with common chemical analysis should be excised for radioactive materials. For personal protection, it is imperative to wear monitoring devices (personal film badges) at all times. One should also be aware of the different health hazards posted by various particles described in the preceding section. The penetration power is in an increasing order of α, β, and γ

Table 12.4 Sampling handling, preservation, and instrumentation for selected radioactivity analysis (US EPA, 1979)

Parameter	Sample preservation	Container	Instrumentation
Gross α activity	Conc. HNO_3 to pH <2	P, G	Low-background proportional counter
Gross β activity	Conc. HNO_3 to pH <2	P, G	Low-background proportional counter
^{90}Sr	Conc. HNO_3 to pH <2	P, G	Low-background proportional counter
^{226}Ra	Conc. HNO_3 to pH <2	P, G	Scintillation cell system
^{228}Ra	Conc. HNO_3 to pH <2	P, G	Low-background proportional counter
^{134}Cs	Conc. HCl to pH <2	P, G	Low-background proportional counter
^3H	None	G	Liquid scintillation counter
U	Conc. HNO_3 to pH <2	P, G	Fluorometer
Photon emitters	Conc. HNO_3 to pH <2	P, G	Gamma spectrometer

due to their varying degree of mass and energy levels. The protection is warranted with the shielding of a piece of paper only for α particles or a thick layer of aluminum foil for β particles, but it requires several centimeter thick lead brick to fully protect from γ-ray exposure.

The *radioactivity* is measured in numbers of disintegrations (decays) per unit time, which is the dpm, if minute is used for time. By this definition, the radioactivity therefore has a unit of s^{-1}. The Becquerel (Bq) corresponds to one decay per second. The old unit of radioactivity is the Curie (Ci), which is inconveniently a large dose of radioactivity as defined by the activity of 1 g of radium-226. The conversion from Curie is: 1 Curie $= 3.70 \times 10^{10}$ Bq $= 2.22 \times 10^{12}$ dpm. Table 12.4 also lists a number of instrumentations commonly used to measure radioactivity. These are further described below.

Geiger-Muller counter: A G-M counter is based on the principle that the radiations produced by radioactive decay will ionize gases such as argon, helium, or neon. These inert gases are always mixed with approximately 0.1% of a heavier gas such as alcohol, methane, or chlorine which acts as a quenching agent to absorb electrons. The gas is held in a chamber which contains electrodes maintained at a potential difference. The electrodes collect the ion pairs which result from the interaction of the ionizing radiation with the gas. As the ion pairs are collected, electric pulses are produced which are converted to a measure of the radiation.

A G-M counter is a portable device that can be conveniently used for screening purpose, such as the detection of hot spots and surface contaminated with radioactive chemicals. A component of the G-M counter is a window that prevents the passage of α particles and β particles with energies less than 0.25 MeV such as those from ^3H. The G-M counters are, therefore, useful only for monitoring γ radiation and β particles with energies greater than 0.25 MeV such as those from ^{32}P.

The G-M tubes of some G-M counters are covered with a thin mica window which allows less energetic β emitters such as ^{14}C to be detected.

Gas-flow proportional counter: The principle of the gas flow counter is similar to that of the G-M counter. The gas flow counter has a constant supply of gas and can have a thin window or windowless (internal proportional counter). Thin window or windowless gas flow counters will detect α particles and even the weakest β emitters. The gas-flow proportional counter can be used to measure gross α- and β-particles, ^{228}Ra, ^{89}Sr, ^{134}Cs, and ^{131}I.

Liquid scintillation counters: Liquid scintillation counting is the method of choice for measuring β emitters, particularly, weak β emitters such as ^{14}C, ^{3}H, and ^{35}S. The sample is dissolved or suspended in a solution called cocktail. The commercially available cocktails contain an organic chemical which are able to absorb radioactive ionization and reemit (fluoresce) it as UV light. The radiation is thus converted into pulses of light that are detected by a photomultiplier. The amount of light produced is related to the energy of the β emitter.

Gamma scintillation spectrometry: X-rays pass freely through the solutions used for liquid scintillation counting. They are, therefore, detected by using solid fluors containing atoms of high atomic numbers. The most commonly used solid scintillator is a sodium iodide (NaI) crystal containing thallium ions as an intentionally added impurity. Impinging rays excite electrons that produce light when they return to their ground state. The light is detected by a photomultiplier.

The U.S. EPA has prescribed detailed procedures for the detection and measurement of radioactive chemicals in environmental samples such as radio-nuclides in drinking waters and radon gas in atmosphere. Interested readers should refer to these procedures for details, such as EPA 600 and 900 series methods for water and 9000 series methods for hazardous materials.

12.3.2 Surface and Interface Analysis

The characterization of surface and interfaces is of great concern to many areas of environmental studies, such as heterogeneous reactions involving solids (adsorption, diffusion, corrosion), nano-materials, membranes, and aerosol particles, to just name a few. Surface layers can be in the thickness of less than one atomic layer up to several micrometers. Both physical (topology and morphology) and chemical (elemental composition, chemical bonding, and geometric and electronic structure) techniques have been available for the qualitative and quantitative analysis of surface and interfaces. At present, there are more than 30 methods that are actually in use for surface and interface analyses. We only briefly introduce three common microscopic techniques used to characterize the physical nature (image) of the surface and interface, namely, scanning electron microscopy (SEM), scanning tunneling microscopy (STM), and atomic force microscopy (AFM).

The conventional *light microscopes* have a resolution limit of approximately 250 nm (0.25 μ), which is approximately the size of a bacterium. This resolution in nanometer is also approximately the wavelength of the incoming visible light (approximately 360 to 800 nm) used to illuminate the sample (refer to Chapter 8). This implies that features on the sample smaller than this dimension cannot be visualized using the light microscopes and a shorter wavelength is needed. In electron microscopy, this is commonly achieved using a high voltage beam of electrons. When using *scanning electron microscopes (SEM)*, samples are first coated with a metal that readily reflects electrons. This coating is also essential to conducting surface for electrons to avoid charging of the sample. Electrons are emitted from a wire (usually tungsten or lanthanum hexaboride) that has been superheated by an electric current. Since the emitted electrons have a Boltzman energy distribution (see Chapter 9.1, Eq. 9.3), magnetic coils are used to focus and direct a passing electron beam into the sample. This is analogous to the lens and prisms used in optical microscopes to select the desired wavelength. When this incoming condensed electron beam is scanned over the object, an image is formed by the electrons that bounce off the surface of the object under examination. This image is collected onto an imaging screen, so the observer sees a picture of the surface of the sample.

The *scanning tunneling microscopes (STM)*, developed in 1981, are capable of resolving features on an atomic scale on the surface of a conducting solid surface. In STM, the surface of a sample is scanned by a very fine metallic tip. This sharp tip is moved up and down, based on the topographical structure of the surface, by monitoring the tunneling currents between the tip and sample surface in order to maintain the tip at a constant distance from the sample surface. A tunneling current is a current that passes through a medium (vacuum, a nonpolar liquid) that contains no electrons. The mechanism of current flow can be rationalized by quantum mechanisms, but is beyond the scope of this text.

The *atomic force microscopes (AFM)*, developed in 1986, allows for the resolution of individual atoms on both conducting and insulating surfaces. In AFM, a sharp tip mounted on a soft lever is scanned across the sample surface, while the tip is in contact with the surface. The atomic force acting on the tip changes according to the sample topography, resulting in a varying deflection of the lever. The deflection of the lever is measured by means of laser beam deflection of a microfabricated cantilever and subsequent detection with a double-segment photodiode.

12.3.3   Screening Methods Using Immunoassay

An *immunoassay (IA)* is a biochemical test that measures the level of a substance based on the specific reaction of an antibody to its antigen. It is the only simple method introduced in this chapter that does not rely on expensive instrumentations, but is used as an alternative measure of chemical analysis. An *antibody* is a binding protein used by the immune system, which is produced in response to the introduction of an alien substance (an antigen). An *antigen* not only induces an immune response but will also react with antibodies. Many of the environmental contaminants (e.g., pesticides) are not able to induce the production of antibodies but can react with antibodies already present. They are called *haptens*. To generate antibodies specific to a hapten, haptens must be covalently bonded to a protein (e.g., bovine serum albumin, keyhole limpet hemocyanin) for the purpose of stimulating an immune response. This is called *hapten-carrier conjugate*. A common form of immunoassay is the *enzyme immunoassay (EIA)*, in which an enzyme conjugate reagent is used to generate the signal used for interpretation of results. The enzyme mediated response may take the form of a chromogenic, fluorogenic, chemiluminescent, or potentiometric reaction.

The U.S. EPA has adopted several immunoassay methods that use commercially available testing kits for a rapid, simple, and portable test. These products can be used effectively in both laboratory and field settings and require limited training. These test products substantially increase the number of data points that can be generated within a given time period and permit an operator to analyze a number of samples simultaneously within a relatively short period of time and a small budget. The protocols recommended by the U.S. EPA include several specific contaminants, which have been demonstrated to be effective, including pentachlorophenol, 2,4-dichlorophenoxyacetic acid, polychlorinated biphenyls (PCBs), petroleum hydrocarbons, PAHs, toxaphene, chlordane, DDT, TNT, and RDX. Interested readers can find more details in the 4000 method series of the SW-846 methods.

REFERENCES

*Agilent Technologies (2001), Basics of LC/MS: Primer Santa Clara, CA, pp. 1–34.

*Agilent Technology (2005), ICP-MS Inductively Coupled Plasma Mass Spectrometry, A Primer Santa Clara, CA, pp. 1–80.

Ardrey RE (2003), *Liquid Chromatography-Mass Spectrometry: An Introduction*, John Wiley & Sons Ltd., West Sussex, England.

Bakhmutov VI (2004), *Practical NMR Relaxation for Chemists*, John Wiley & Sons, West Sussex, England.

Borman S (2004), Analyzing drugs and the environment, *C&EN*, 80(14):34–35.

*Bruice PY (2004), *Organic Chemistry*, 4th Edition, Prentice Hall, Upper Saddle River, NJ, pp. 480–600.

Carey FA (2006), Spectroscopy, *Organic Chemistry*, 5th Edition, McGraw Hill, NY, Chapter 13, pp. 519–586.

Field LD, Sternhell S, Kalman JR (2003), *Organic Structures From Spectra*, John Wiley & Sons, West Sussex, England.

Fifield FW, Haines PJ (2000), *Environmental Analytical Chemistry*, 2nd Edition, Blackwell Science, Malden, MA.

Friebolin H (2005), *Basic One- and Two-Dimensional NMR Spectroscopy*, 4th Edition, Wiley-VCH, Weinheim.

Hornak JP (1997–1999), The Basics of NMR (http://www.cis.rit.edu/htbooks/nmr/bnmr.htm).

Hübschmann H-J (2001), *Handbook of GC/MS: Fundamentals and Applications*, Wiley-VCH, New York, NY.

de Hoffmann E, Stroobant V (2002), *Mass Spectrometry: Principles and Applications*, 2nd Edition, John Wiley & Sons, West Sussex, England.

Kellner R, Mermet J-M, Otto M, Valcarcel M, Widmer HM (2004), *Organic Mass Spectrometry. I: Analytical Chemistry: A Modern Approach to Analytical Science*, 2nd Edition, Wiley-VCH, Chapter 25 pp. 737–972.

*Suggested readings.

Lens PNL, Hemminga MA (1998), Nuclear magnetic resonance in environmental engineering: Principles and applications, *Biodegradation* 9:393–409.

McLafferty FW, Tureček F (1993), *Interpretation of Mass Spectra*, 4th Edition, University Science Books, Sausalito, CA.

*McMaster M, McMaster C (1998), *GC/MS: A Practical User's Guide*, Wiley-VCH, Weinheim, pp. 167.

Pecsok RL, Sheilds LD, Cairns T, McWilliam IG (1968), *Modern Methods of Chemical Analysis*, 2nd edition, John Wiley & Sons, New York, NY.

*PerkinElmer (2001), The 30-Minute Guide to ICP-MS, Technical Notes D-6355A, Shelton, CT.

Richardson SD (2001), Water analysis, *Anal. Chem.* 73:2719–2734.

Richardson SD (2001), Mass spectrometry in environmental sciences, *Chem. Rev.* 101:211–254.

*Richardson SD (2002), Environmental mass spectrometry: Emerging contaminants and current issues, *Anal. Chem.* 74:2719–2742.

Richardson SD (2003), Water analysis: Emerging contaminants and current issues. *Anal. Chem.* 75:2831–2857.

Rouessac F, Rouessac A (2000), *Chemical Analysis: Modern Instrumental Methods and Techniques*, English Edition, John Wiley & Sons, West Sussex, England.

Skoog DA, Holler FJ, Nieman TA (1997), *Principles of Instrumental Analysis*, 5th Edition, Saunders College Publishing, Orlando, FL.

Smith RM (2004), *Understanding Mass Spectra: A Basic Approach*, 2nd Edition, Wiley-Interscience, Hoboken, NJ.

Snyder JL (2004), Environmental Applications of Gas Chromatography. In: *Modern Practice of Gas Chromatography* (Editor: Grob RL, Barry EF), 4th Edition, Wiley-Interscience, Hoboken, NJ, Chapter 15, pp. 769–882.

Solomons TWG (2000), *Fundamentals of Organic Chemistry*, 7th Edition, John Wiley & Sons, Inc. New York.

US EPA (1980), Prescribed Procedures for Measurement of Radioactivity in Drinking Water. EPA-600/4-80-032.

US EPA (1979), Radiochemistry, *Handbook for Analytical Quality Control in Water and Wastewater Laboratories*, EPA-600/4-79-019, Chapter 11.

QUESTIONS AND PROBLEMS

1. Compare the similarities and the differences between ICP-OES and ICP-MS with regard to the operational principles and the instrumental components.

2. What are the functions of (a) lens, (b) quadrupole, and (c) dynode detector in ICP-MS?

3. What are the uses of isotope ratio method and the isotope dilution method?

4. Explain why interface is important and essential for all hyphenated mass spectrometers. Describe the interface used in ICP-MS, GC-MS (EI), and LC-MS (electrospray).

5. What are the major similarities and differences between the mass spectrometers used in atomic mass spectrometry and those used in molecular mass spectroscopy?

6. Describe the ionization mechanisms for (a) electron impact (EI) ionization, (b) chemical ionization (CI), (c) electrospray ionization (ESI), and (d) atmospheric-pressure chemical ionization (APCI).

7. Why the molecular ions of certain compounds do not show up under EI but show up under CI and ESI?

8. Illustrate the principles of quadrupole-based mass analyzer. Why it is also commonly referred to as a mass filter?

9. List three other major types of mass analyzer besides the most commonly used quadrupole-based mass analyzer.

10. Explain what are the differences between "SCAN" and "SIM" in GC-MS. Why "SIM" is needed to achieve high sensitivity and lower detection limit?

11. Explain molecular ions, fragment ions, isotopic clusters, and base peak in a molecular mass spectrum.

12. What are the abundance ratio of ^{35}Cl to ^{37}Cl and ^{79}Br to ^{81}Br? How do these isotopic ratio affect the mass spectra of compounds containing Cl and Br?

13. The mass spectrum of pentane from an electron ionization is shown below:

    (a) Indicate which peak is the molecular ion peak? Which one is the base peak?

    (b) What causes the line at m/z of 57? at m/z of 43? and at m/z of 29?

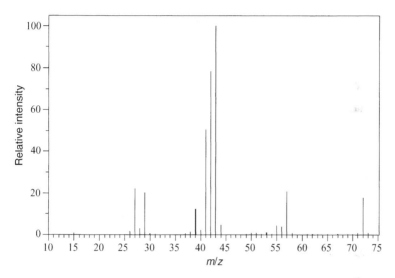

    (c) Can you propose a fragmentation mechanism based on the above mass spectrum?

    (d) Write all the reactions involved in the fragmentation of pentane.

14. Give a list of the major U.S. EPA methods for the GC-MS analysis of (a) VOCs and (B) SVOCs in drinking water, wastewater, and wastes.

15. Explain: Why NMR resolution is increased with the increasing strength of the magnetic field (B_0)? Why ^1H-NMR is more sensitive than ^{13}C-NMR when an equal number of nuclei is compared?

16. If a ^1H-NMR spectrometer is equipped with a magnet with a magnetic field (B_0) of 7.05 T, calculate the required operating frequency of the spectrometer in megahertz (MHz). Use the tabulated value in Table 12.1 for the gyromagnetic ratio (γ).

17. Refer to the above question, what will be the required frequency in MHz if this is an ^{13}C-NMR spectrometer? If the operating frequency is fixed at 200, 300, 500, 600 MHz, what will be the required magnetic field strengths (B_0) at each of these frequencies?

18. Calculate the chemical shift (δ) in ppm if a proton signal is 1150 Hz higher than that of the TMS standard. The ^1H-NMR spectrometer was run at a frequency of 600 MHz.

19. Explain why will a nucleus with more shielding resonate at a lower radio frequency, and why will such a nucleus be located in the upfield?

20. Explain the difference between diamagnetic shielding and spin–spin coupling.

21. Determine how many signals you would expect from the following compounds in the ^{1}H- NMR spectrum: (a) tetrachloroethylene, (b) trichloroethene, (c) 1,1-dichloroethene, (d) 1,2-dichloroethene, (e) vinyl chloride (refer to Appendix B for their chemical structures).

22. Determine how many signals you would expect from the following compounds in the ^{13}C- NMR spectrum: (a) benzene, (b) toluene, (c) ethylbenzene, (d) o-xylene, (e) m-xylene, and (f) p-xylene (refer to Chapter 2 for their chemical structures).

23. For $CH_3CH_2CH_2Cl$, which proton signal would you expect to be at the lowest frequency (upfield) in ^{1}H-NMR? For the same compound, which carbon signal would you expect to be at the lowest frequency (upfield) in ^{13}H-NMR?

24. What would be the order of the chemical shift (δ values) for the compounds: CH_3Br, CH_3Cl, CH_3F, CH_3I in ^{1}H-NMR? Why?

25. Why tetramethylsilane (TMS) is typically used as a reference standard for NMR spectroscopy?

26. Which group of protons in $CH_3CH_2CH_2CH_2NO_2$ will have the highest chemical shift?

27. In a high resolution ^{1}H-NMR of a compound $CH_3CH_2CH_2Cl$, how many split signals would you expect from the methyl group, the protons in the $-CH_2Cl$ group, and the protons of the central $-CH_2-$ group?

28. Describe the advantages and disadvantages of ^{13}C-NMR as compared to ^{1}H-NMR.

29. Describe the difference in atomic structure and properties between ^{13}C (used in NMR) and ^{14}C (used in radio-labeled analysis).

30. Identify the X in the following radioactive decay:

$$\text{(a) } {}^{14}_{6}C \rightarrow {}^{14}_{7}N + X$$
$$\text{(b) } {}^{40}_{19}K + X \rightarrow {}^{40}_{18}Ar + X - ray$$
$$\text{(c) } {}^{238}_{92}U \rightarrow {}^{234}_{90}Th + {}^{4}_{2}\alpha$$

31. Compare ^{1}H, ^{2}H, and ^{3}H in terms of the number of protons, neutrons, mass number, and stability.

32. The current U.S. EPA allowed maximum contaminant level (MCL) for α particles in drinking water is 15 pCi/L. Convert this into Bq/L and dpm/L.

33. What types of shield is needed to protect the exposure of: (a) α-particles, (b) β-particles, and (c) γ-rays.

34. For low-energy β emitters such as ^{14}C, ^{3}H, and ^{35}S, which one of the radioactivity measuring device is desirable?

35. What immunoassay methods have been available and adopted in the SW-846 methods? Give a list of such compounds that can be tested for field screening purpose.

Experiments

The following 15 experiments can be used to fit an one-semester lab course in environmental sampling and analysis or a related course. According to the order of the chapters appearing in the book, the materials cover statistical analysis of environmental data, sampling techniques and sample preservation, in situ measurement using electrochemical probes, wet chemical analysis (gravimetric and volumetric), sample preparation (extraction, digestion, headspace) simple UV-visible colorimetric analysis, and more advanced instrumentations using atomic spectrometry (ICP) and chromatography (IR, HPLC, and GC).

Students are assumed to have some prior knowledge and minimal skills in analytical chemistry, such as pipeting, use of analytical balance, and titration. Chemical reagents and solutions shall be prepared prior to each experiment. These will be provided to the students so that each experiment is doable during a 2–3-h lab session. The experimental procedures are provided in as much details as possible. For instrumental analysis, only the general procedures and the needed operational parameters are provided because the actual procedures will vary depending on the manufacturers and models of the specific instrument. Students are also provided with the assignment for each experiment, which can be included in the lab report. These are designed to help students acquire more insight into each topic.

EXPERIMENT 1

DATA ANALYSIS AND STATISTICAL TREATMENT: A CASE STUDY ON OZONE CONCENTRATIONS IN CITIES OF HOUSTON-GALVESTON AREA

Objectives—The purpose of this experiment is to make students familiarized with several statistical tools that are commonly used for environmental data analysis.

Common statistical analyses available from Microsoft Excel[R] are descriptive statistics, F-test, one-way analysis of variance (ANOVA), two-way ANOVA, and regression analysis. Students will be asked to retrieve environmental monitoring data from the Internet for an exercise in statistical analysis and subsequent data interpretation.

Background—The instructor will introduce monitoring data to the students, such as the ozone data illustrated in this computer laboratory exercise. The instructor may also choose other data sets that are of local or national importance. Many of the environmental monitoring data can be accessed free of charge from the Web sites, such as the U.S. EPA and state environmental protection agencies. Whereas the type of data is not critically important, the utilities of Excel's statistical tools will be the focus of this experiment.

Prior to this experiment, the students should review the basic of statistics (Chapter 2). Students should understand the terms used in descriptive statistics to define the central tendency and dispersion of the population such as mean, standard deviation, median, mode, range, and confidence intervals. From the lectures, students should also have learned how F-test and one-way ANOVA can be used to test the statistical difference between two groups (analysts, analytical methods). Students should also have learned standard calibration curve and how a linear regression can be used to obtain the equation of the standard calibration curve. For an efficient use of the class time, students should also have some prior knowledge in using Excel.

Ozone is listed as one of the six criteria pollutants in the National Ambient Air Quality Standard (NAAQS). In many large cities such as the city of Houston, high levels of ozone have been frequently detected. Ozone is not emitted directly in the atmosphere, but it is produced in the atmosphere by photochemical reactions. The complex reactions require sunlight, volatile organic compounds (VOCs) and nitrogen oxides (NOx). Both VOC and NOx can be related to automobile emission and the petroleum industries in this city. Higher levels of ozone usually occur on sunny days with light winds, primarily from March through October in the Houston region. An ozone exceedance day is counted if the measured 8-h ozone concentration exceeds the National Ambient Air Quality Standard of 85 ppb$_v$.

Resources—This computer exercise is better illustrated in a computer lab so that each student can get an access to a computer with Internet access and Microsoft Excel software installed. If only one computer is accessible to the instructor, then the instructor will give a step-by-step detailed demonstration with a computer projector. Many personal computers do not have a statistical tool package installed initially (to determine whether you have one, select Tools in the menu, you should be able to see Data Analysis if you have it installed). To add this feature into Excel, you need to use the original disk of the Excel software program. This exercise can also be illustrated using Minitab if this program is preferred over Excel.

Time Required—The estimate time of this computer exercise is approximately 2.5 hours.

Procedures

1. Data retrieval and input into Excel spreadsheet. The data shown in Table E-1 are retrieved from Texas Commission on Environmental Quality (TCEQ) Web site (http://www.tceq.state.tx.us/nav/data/ozone.html). Included in this table are the daily maximum ozone concentrations in Houston-Galveston area. Two monitoring sites are selected for a comparison purpose: Houston downtown area at monitoring station C81 (US EPA site 48-201-0070) and Galveston area at monitoring station C34 (US EPA site 48-167-0014). Galveston Island is located on the Gulf of Mexico, about 50 miles south of Houston. The City of Houston is subject to sea breeze from the Gulf coast, but the predominant wind in Houston is northwest.

    The instructor will first present the source data. By specifying the monitoring station, month, year, and the report format, a report will be generated, which include the O_3 concentrations as requested. To input the data from the Web site to Excel, just cut-and-paste the data into Microsoft Excel spreadsheet.

2. In the Excel menu bar, select Tools|Data Analysis. You will see a list of available statistical tools. Select Descriptive Statistics, and then follow the instruction. You will generate four summary statistics tables by separately highlighting four rows of data shown in the table. The descriptive statistics include the mean, standard deviation, median, mode, and range for O_3 concentration for January and August of the two areas.

3. Under Tools | Data Analysis, select ANOVA: Single Factor to perform a one-way analysis of variance. For each one-way ANOVA test, you need to select two groups of data for comparison (January vs. August, or Galveston vs. Houston). Use Houston data, for example, to statistically test whether O_3 concentration in August is significantly higher than that in January, you specify the significance level (α) at 0.05 (equivalent to 95% confidence level). You should be able to generate one-way ANOVA output that contains F-value, F-critical, and p value. The instructor will discuss with the students how to interpret the results.

Table E-1 Daily maximum 8-h ozone averages in January and August 2005 (Ozone concentration unit in ppb_v)[a]

Site	Date	1	2	3	4	5	6	7	8	9	10	11	12	13	14	15	16	17	18	19	20	21	22	23	24	25	26	27	28	29	30	31
Galveston	January	28	30	30	29	30	15	7	26	29	25	28	32	33	32	28	33	27	33	37	39	28	33	35	43	37	32	21	22	20	18	14
	August	61	56	50	51	64	81	63	52	24	15	27	32	22	18	20	21	22	30	31	18	30	63	48	27	49	53	76	55	57	66	49
Houston	January	26	21	25	17	24	6	1	19	25	18	24	25	23	27	25	30	30	17	23	32	29	31	29	31	37	26	12	13	15	13	9
	August	88	89	74	68	58	78	61	52	16	17	23	30	27	18	14	18	28	25	30	30	68	63	59	40	41	76	69	47	42	67	64

[a]Galveston: Data obtained at Galveston Airport (C34), 8715 Cessna Street, Galveston, Texas; Houston: data obtained at Houston TCEQ Regional Office (C81), 5425 Polk Avenue, Houston, Texas.

4. Repeat Step 3 for the one-way ANOVA analysis of other pairs of data (e.g., January vs. August in Galveston, Houston vs. Galveston in January, Houston vs. Galveston in August). Use $\alpha = 0.05$.

5. The third statistical test to be illustrated will be regression analysis. This is better illustrated by using a standard calibration data, which is very important in environmental analysis. Regression analysis can be used to define the relationship between two variables (e.g., the dependent variable instrumental signals vs. the independent variable contaminant concentration) and whether such relationship is significant at a specified α.

   For illustration purpose, we can use the same data shown in the table. If the predominant wind in August 2005 is northeast, then O_3 from Houston downtown (independent variable) can contribute to the O_3 in Galveston (dependent variable). To perform a regression analysis to determine if there is a significant relationship between the concentrations of O_3 in two locations, select Regression under Tools | Data Analysis menu and select "Line Fit Plots." Use Houston's O_3 concentration as the independent variable (x) and the Galveston's O_3 concentration as the dependant variable (y). You should be able to generate a report of the linear regression model ($y = ax + b$) and a linear fit plot. The instructor will show how to extract the slope (a), intercept (b), regression coefficient (R^2), the p-value and F-value from the output.

6. Include a printout of the above statistical output in your lab report. If time permits, the instructor can also demonstrate other useful tools in Excel. Tools relevant to environmental sampling and analysis include t-test, random number generation, and various mathematical functions such as the statistical functions used to generate all the tabulated values in Appendix C.

Post Lab Assignment

1. Write a lab report containing all the Excel outputs (descriptive statistics, one-way ANOVA, and regression).

2. For both locations, are O_3 concentrations in August significantly higher than that in January? In both months, are O_3 concentrations in Houston downtown significantly higher than that in Galveston Island?

3. What is the linear regression model describing the O_3 concentrations of Galveston as a function of O_3 concentrations in Houston? Perform separately in January and August. Are the two models significant at 95% confidence level?

REFERENCES

Chapter 2: Basics of Environmental Sampling and Analysis (this book).

BILLO EJ (2001), Excel [R] for Chemists: A Comprehensive Guide, 2nd Edition, Wiley-VCH Weinheim.

KELLEY WD, RATLIFF TA, Jr, NENADIC C (1992), Basic Statistics for Laboratories: A Primer for Laboratory Workers, John Wiley & Sons, Hoboken, NJ.

SINCICH TL, LEVINE DM, STEPHAN D (2002), Practical Statistics by Example Using Microsoft Excel and Minitab, Prentice Hall, Upper Saddle River, NJ, 2nd Edition.

EXPERIMENT 2

COLLECTION AND PRESERVATION OF SURFACE WATER AND SEDIMENT SAMPLES AND FIELD MEASUREMENT OF SEVERAL WATER QUALITY PARAMETERS

Objectives— The objectives of this experiment are to learn (1) sampling techniques for surface water and sediment; (2) the use of a hand-held Global Positioning System (GPS) to document sampling locations; and (3) field analytical techniques (particularly electrochemical probe methods) for the measurement of temperature, conductivity, salinity, pH, and dissolved oxygen (DO). Sampling is preferred through a field trip to a nearby water body (pond, stream, or lake) that is within the walking distance.

Background—To assess whether a specific water body meets water quality standards, water quality monitoring is always the first step. The QA/QC requires that some water quality parameters be measured in the field. The most common *field tests* for water quality purposes are temperature, pH, DO, conductivity, salinity, and CO_2. Many other water quality parameters can be analyzed *in the lab* after the samples have been transported to the lab with or without the preservation of samples. Most environmental parameters are in this category, including physical and aggregate properties, metals, inorganic nonmetallic constituents, and aggregate/individual organic compounds.

Temperature affects many biological processes such as metabolism and growth of many organisms, as well as physicochemical processes such as sorption/desorption, dissolution/precipitation, and solubility of gases (hence DO). pH or the activity of hydrogen ions has major impact on many aquatic processes. Unpolluted natural water is often in equilibrium with carbon dioxide of the atmosphere, hence the pH of natural water is slightly acidic. However, factors such as algal growth and industrial or municipal wastewater discharge will possibly change the water pH dramatically.

Conductivity is the measure of the ability of an aqueous solution to carry an electric current. It depends on the presence of the ions, their total concentrations, mobility, valence, and the temperature of the measurement system. Pure water does not have many dissolved ions in it and is a poor conductor. On the contrary, seawater contains a large amount of electrolytes, and can have a very high conductivity.

Distilled water produced in a lab, generally has a conductivity in the range of 0.5–3 μmhos/cm. The conductivity of portable waters in the U.S. ranges generally from 50 to 1500 μmhos/cm. Obviously, water with a very high conductivity is not suitable for domestic and industrial uses.

Salinity is a measure of the mass of dissolved salts in a given mass of solution. The experimental determination of a salt content by drying and weighing presents some difficulties because of the loss of some components. Thus, salinity is often indirectly measured from a related physical property such as conductivity, density, sound speed, or refractive index. Salinity has a unit of parts per thousands (ppt).

Dissolved oxygen (DO) levels in natural and wastewater depend on the physical, chemical, and biochemical activities in the water body. Algae produce oxygen during photosynthesis whereas aerobic bacteria consume oxygen. DO is also known to directly influence many chemical processes like the redox reactions and dissolution of toxic metals in water. Many aquatic creatures such as fish need a DO level of greater than 4 mg/L. The concentration of DO is controlled by other parameters, such as water temperature, degree of agitation, and the extent of organic pollution.

Method Principle—Temperature can be precisely measured with a mercury-filled thermometer or an electronic thermometer. The most popular electronic device for temperature measurement is a thermo-resistor, whose resistance to current flow is temperature dependent.

The most accurate pH measurement is by a pH meter. The basic principle is to determine the activity of the hydrogen ions by potentiometric measurement, using a glass electrode and a reference electrode. The most popular, called combination electrode, incorporates the glass and the reference electrode into a single probe. The glass electrode is sensitive to the hydrogen ions, and changes its electric potential with the change of the hydrogen ion concentration. The reference electrode has a constant electric potential. The difference in the potential of these electrodes, measured in millivolts (mV), is a linear function of the pH of the solution. The scale of the pH meters is designed so that the voltage can be read directly in terms of pH.

Conductivity (k) is commonly referred to as "specific conductance." While conductance (G) is defined as the reciprocal of resistance R (i.e., $G = 1/R$), conductivity (k) is the constant of proportionality such that:

$$G = k(A/L) \qquad \text{(E-1)}$$

where A is the electrode surface area (cm^2), and L is the distance between the electrodes (cm). Since R has a unit of ohm, the unit of k is ohm*cm^{-1} or mho/cm. Conductivity is customarily reported in μmhos/cm. In the SI system, conductivity has a unit of mS/m (milli-siements per meter). Note that: 1 S = 1/ohm = 1 mho; 1 mS/m = 10 μmhos/cm; 1 μS/cm = 1 μmhos/cm.

The salinity of sea water is given a practical salinity of 35 part per thousand (ppt) and can be interpreted as 35 g of solutes per kg of sea water. In practice, salinity is most conveniently measured with a conductivity meter by using one of several empirical equations. Advanced digital conductivity meters may have these built-in empirical relationships; as a result, salinity measurement is carried out along with the conductivity measurement.

An accurate measurement of DO can be done with a DO meter which has oxygen-sensitive membrane electrodes. These two solid metal electrodes are in contact with supporting electrolyte separated from the test solution by a selective membrane. The polyethylene and fluorocarbon membranes allow molecular oxygen to diffuse. The "diffusion current" across the membrane is directly proportional to the DO concentration.

Time Required—If traveling time to a sampling location is minimized, the field sampling and measurement of several water samples can be finished in 2.5 h.

Apparatus and Reagents

Equipment

(a) Water sampler (grab sampler and/or Kemmerer sampler); (b) Sediment sampler (Ponar dredge); (c) Handheld Global Position System (GPS, Magellan Model 315 or equivalent); (d) Battery operated DO meter (YSI Model 58 or equivalent); (e) Handheld battery operated metering device for salinity, conductivity and temperature (YSI Model 30 or equivalent); (f) pH meter (Beckman Model 3500 or equivalent); (g) One liter plastic bottle, bucket, beaker, ice-chest, markers, tapes, and notebook.

Reagents

(a) Oxygen saturated water (Instructor prepared for DO calibration), (b) 0.01 M KCl (Instructor prepared for conductivity calibration), (c) pH standard solutions (commercially available pH 4.0 and 7.0 solutions for pH calibration).

Procedures

1. Instructor demonstrates the calibration of DO meter, conductivity meter, and pH meter in the lab.
2. Bring sampling equipment, GPS, and meters (except pH meter) to the field.
3. Instructor demonstrates the operation of sampling tools. Depending on the accessibility of the water body, the instructor may choose a grab sampler (collect water sample through a river bank) or a Kemmerer sampler (collect water sample through a bridge or a boat). Ponar sediment sampler can be used to collect from a boat or a bridge for most sediment types. Take a water sample first and then a sediment sample. Have students practice these sampling tools.
4. Document the sampling location using GPS. Measure the following in the field: temperature, DO, conductivity, and salinity. Record the data, time, and sampling location in your lab notebook.
5. Collect additional samples in a cooler packed with ice. These samples may be preserved for future lab uses depending on the type of analytes. Before

Table E-2   A check list of water quality parameters in three test samples

Parameters	Surface water sample	Lab tap water	Ultrapure water[a]
1. Temperature	√	⊕	⊕
2. Dissolved oxygen (DO)	√	⊕	⊕
3. Conductivity	√	√	√
4. Salinity	√	√	√
5. pH	√	√	√

[a]Such as Millipore, MilliQ or any other distilled and/or deionized water.

returning to the lab, the instructor can go over the general procedures for sample's Chain-of-Custody. For example, have students fill up the Chain-of-Custody form.

6. The pH is measured immediately upon return to the lab. It is measured in the lab because pH meters require electricity to operate. (1) Calibrate with two pH standard solutions (pH 4.0 and pH 7.0). (2) Transfer 200 mL of water sample into a beaker and measure the pH.

7. Depending on the time allowed, the instructor may require students to measure two additional samples in the lab for comparison (Table E-2). The assignments labeled ⊕ in the table are optional. Include these results in the lab report.

Safety and Health Issues—Watch for passing traffic during walking and sampling. When a sampler is collecting water in a deep and flowing water body, life vest must be worn. When water/sediment samples are collected from a boat, a qualified person must accompany who would be in charge of the boat. The instructor will also prepare insect/mosquito repellent if needed.

Post Lab Assignment

1. Give a list of water quality parameters that must be measured in situ.

2. You are recently hired as an entry level Environmental Field Specialist in a new firm dealing with groundwater remediation where a sampling and analysis plan (SAP) has not been developed. You are assigned to work as an assistant to the Project Manager for groundwater sampling, and are asked to do the official preparation for this sampling event. Make a list of major items you may need in the field. Exclude containers and sampling tools in this list.

3. Suppose your firm has developed a monitoring program for a river in Houston area. This program is designed to detect the presence of certain industrial chemicals in the river water or sediments. The sampling sites and frequency were selected on the basis of sound statistical reasoning. The laboratory analytical protocols have all been established. Your QA/QC program is in place. All that remains to be done is to go to the field, obtain samples, and safely transport them back to the lab. You are ready to do this fieldwork, but you must assemble your sampling equipment and preservative

Table E-3 Preparation of sampling equipment, container, and preservatives

Sample Type	Analyte	Sampling equipment	Container material	Preservation technique
River water, taken at a depth of twenty feet	Semivolatile organic chemicals			
River water, taken at the river's edge near an industrial outfall	Heavy metals			
Surface water samples, taken near a gas station	BTEX			
Soil samples from river bank	Pesticides and pharmaceuticals			
An abandoned barrel of granular waste product, found near the river	Material of unknown origin			

reagents. You have a table of sample types, and need to fill in the blanks in Table E-3 by deciding the: (a) Sampling equipment (e.g., dipper, Kemmerer, autosampler, Coliwasa, auger, etc), (b) Container material (e.g., glass, plastic, Teflon), and (c) Method for preserving the samples.

REFERENCES

Chapter 11: Electrochemical Methods for Environmental Analysis (this book).

BOEHNKE DN, DELUMYEA RD (2000), Laboratory Experiments in Environmental Chemistry, Prentice Hall, Upper Saddle River, NJ.

DUNNIVANT FM (2004), Environmental Laboratory Exercises for Instrumental Analysis and Environmental Chemistry, Wiley-Interscience, Hoboken, NJ.

EXPERIMENT 3

GRAVIMETRIC ANALYSIS OF SOLIDS AND TITRIMETRIC MEASUREMENT OF ALKALINITY OF STREAMS AND LAKES

Objectives—The objectives of this experiment are to learn (1) gravimetric analytical techniques for the measurement of total solids and (2) volumetric analytical techniques for the measurement of alkalinity.

Background—Gravimetric and volumetric methods are indispensable methods for some of the environmental parameters (Table 6.3). Solids and alkalinity tested in this study are two such parameters that rely on classical wet chemical methods. *Solids* refer to matter suspended or dissolved in water or wastewater. They may affect water or effluent quality adversely in a number of ways. Waters with high dissolved solids generally are of inferior palatability. A limit of 500 mg dissolved solids/L is imposed for drinking waters. When suspended solids are very high, the water may be esthetically unsatisfactory for such purposes as bathing.

Solids have different forms depending on its operation principles. "Total solids" (TS) is the term applied to the material residue left in the vessel after evaporation of a sample and its subsequent drying in an oven at a defined temperature. Total solids (TS) include "total suspended solids" (TSS) and "total dissolved solids" (TDS). TSS is the portion of total solids retained by a filter, whereas TDS is the portion that passes through the filter. The temperature at which the residue is dried has an important bearing on results, because weight losses as a result of volatilization of organic matter, mechanically occluded water, water of crystallization, and gases from heat-induced chemical decomposition as well as weight gains because of oxidation, depend on temperature and time of heating. Each sample requires close attention to desiccation after drying.

Alkalinity of a water is its capacity to neutralize acids. Alkalinity of many surface waters is primarily a function of carbonate, bicarbonate, and hydroxide content. Waters with a high alkalinity may not fit certain applications because of the high cost of acid neutralization of large amount of water, but the advantage is the resistance to acidic precipitation. A change in alkalinity may be an indication of pollution problems.

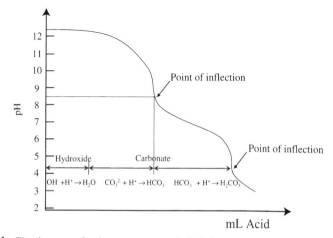

Figure E-1 Titration curve for the measurement of alkalinity

Alkalinity can be divided into two types according to its titration procedure – the phenolphthalein alkalinity and total alkalinity (methyl orange alkalinity). The phenolphthalein alkalinity is the acid-neutralizing power of hydroxide (OH^-) and carbonate ions (CO_3^{2-}) present in the water sample, whereas total alkalinity represents all the bases in it (OH^-, CO_3^{2-}, and HCO_3^-). The procedure utilizes indicators such as phenolphthalein and methyl orange to signal the end points in titration. The amount of acid used to reach the end point of phenolphthalein (pH 8.3) is taken as the phenolphthalein alkalinity, whereas the total amount of acid consumed to reach the end point of methyl orange (pH 4.4) is the total alkalinity (Fig. E-1). The results of alkalinity are usually reported as mg $CaCO_3$ per liter of water.

Time Required—The evaporation of water for SS measurement is a lengthy process. Be certain to heat SS sample as soon as the lab gets started, and measure alkalinity while the evaporation is in process. Given a good time management, the experiment should be completed within 3 h.

Apparatus and Reagents

Apparatus

(a) analytical balance (Mettler AE 260 Delta Range or equivalent), (b) 12" × 12" hot plate (1600 watts Thermolyne Type 2200 or equivalent), (c) dessicator, (d) burettes and stands, (e) 250-mL beaker, 100-mL graduated cylinder, and two 400-mL flasks for each student group.

Reagents

(a) methyl orange indicator solution, (b) phenolphthalein indicator solution, (c) 0.0245 N HCl solution standardized by the instructor.

Procedures

1. Use a water sample from Experiment 1 or any local source. Measure total solids (TS) according to the following steps:

   1) Weigh a 300-mL glass beaker on an analytical balance (A) capable of measuring ± 1 mg (0.0001 g).

   2) Mix the water sample by hand shaking and use a graduated cylinder to transfer 200 mL of water into the pre-weighed 300-mL beaker.

   3) Place the beaker on a hotplate. Gently heat the contents to boiling slightly, allow the evaporation to continue until it is dry. Perform this under the hood. The evaporation may take approximately 2 h. Keep eye on it to avoid overheating or sample splashing.

   4) Immediately after the evaporation, place the beaker into a desiccator. Allow 5–10 min to cool down to room temperature.

   5) Weigh the beaker again on an analytical balance (B)

6) Calculation of TS: mg total solids/L $= (B-A)*10^6$/sample volume (mL), where A = weight of beaker (g), and B = weight of dried residue + beaker (g).

2. Measure alkalinity by titration:

1) Transfer 150-mL (V) well-mixed water sample into a 400-mL flask. Rinse a burette three times with 0.0245 N HCl and then fill it with this same solution. Clear bubbles from the burette, and verify that the titrant level is at 0.00 mL.

2) Add six drops of phenolphthalein indicator solution into the sample.

3) If pink color develops, titrate the sample with a standard HCl solution till the pink color just disappears (the end point pH 8.3). Record the volume of HCl consumed (C). Then add six drops of methyl orange indicator solution. Titrate with standard HCl solution. The end point (pH 4.4) is a color change from orange to red. Record the volume (D) of acid consumed.

4) If no pink color develops after the addition of phenolphthalein (why?), then C = 0 mL. Add two drops of methyl orange indicator solution. Titrate with HCl as described in Step 3).

5) Calculation of alkalinity:

Phenolphthalein alkalinity, mg $CaCO_3$ /L $= (C \times N \times 50000)/V$

Total alkalinity, mg $CaCO_3$/L $= (C + D) \times N \times 50000)/V$

Where C is the volume in mL of HCl solution used to reach pH 8.3, D the volume in mL of HCl solution used to reach pH 4.4, N the concentration of standard HCl solution in normality, and V the volume of sample in mL

Safety and Health Issues—Hot plate can be very hot and can cause skin burn. When preparing dilute HCl solution from concentrated HCl, prepare under a ventilation hood to avoid inhalation of vapor. Add concentrated HCl to DI water rather than adding water into concentrated HCl.

Post Lab Assignment

1. Give a separate list of common environmental parameters that can be measured by gravimetric method or volumetric (titration) method.

2. What are the natural and pollution sources of acidity and alkalinity? Why is water considered to be healthy and beneficial only when the alkalinity is in a certain range?

3. A sample of water collected from the overflow of a recarbonation basin has a pH of 9.0; 200 mL of the water require 1.2 mL of 0.02 N H_2SO_4 to titrate it to the phenolphthalein end point and an additional 20.8 mL of 0.02 N H_2SO_4 to titrate it further to the orange end point. Assuming the sample contains no calcite particles, what are the total and carbonate alkalinities of the sample in meq/L and the total alkalinity in mg/L as $CaCO_3$?

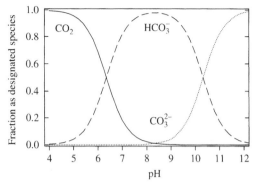

Figure E-2 Species distribution diagram showing the fraction of each species (CO_2, HCO_3^- and CO_3^{2-}) as a function of pH change in water

4. Instead of using a titration procedure, the alkalinity of natural water can also be calculated if the pH and either bicarbonate or carbonate concentration is measured. The formula is:

Carbonate alkalinity (mg $CaCO_3$/L) = 0.820 [HCO_3^-, mg/L] + 1.667 [CO_3^{2-}, mg/L]

where the coefficients 0.082 and 1.667 are the ratios of equivalent weight of $CaCO_3$ (100) to HCO_3^- (61.0) and $CaCO_3$ to CO_3^{2-} (30.0), respectively. The key to such calculation is species distribution diagram shown in Figure E-2, where [CO_3^{2-}] can be calculated from given pH and [HCO_3^-] or vice versa. Calculate the alkalinity if a groundwater sample has a pH of 8.5 and HCO_3^- concentration of 275 mg/L.

REFERENCES

Chapter 6: Common Operations and Wet Chemical Methods in Environmental Laboratories (this book).
APHA-AWWA-WEF (1998), Standard Methods for the Examination of Water and Wastewater, APHA-AWWA-WEF, Washington, DC, 20th Ed.
Weiner ER (2000), Environmental Chemistry: A Practical Guide for Environmental Professionals, Lewis Publishers, Boca Raton, FL.

EXPERIMENT 4

DETERMINATION OF DISSOLVED OXYGEN (DO) BY TITRIMETRIC WINKLER METHOD

Objectives—The objectives of this experiment are (1) to learn the specific sampling techniques for dissolved oxygen (DO) (sampling container and sample preservation),

(2) to understand the chemical principle of Winkler Method for DO measurement, (3) to understand potential errors/interferences and the corresponding mitigation methods.

Background—The importance of DO in water is evident. Without an appreciable level of DO, many kinds of aquatic organisms cannot exist in water. For example, fish will die if the DO is normally below 4 mg/L. Many fish kills are caused not from the direct toxicity of pollutants but from a deficiency of oxygen. Water without an appreciable amount of dissolved oxygen will undergo anaerobic decay, and should be avoided at all costs for uses.

The level of DO in water is dependent on many physical, chemical, and biochemical factors—aeration, wind, velocity of water flow, algae, temperature, atmospheric pressure, organic compounds, salt content, bacteria, and animals. Most of the elemental oxygen comes from the atmosphere, which is 20.95% oxygen of the volume of dry air. Therefore, the ability of a water body to reoxygenate itself by contact with the atmosphere is an important characteristic. The reoxygenation is enhanced by aeration (e.g., swimming, boating) as well as wind and higher flow velocity of water.

Algae produce oxygen during photosynthesis under sunlight. However, this process is really not an efficient means of oxygenating water because some of the oxygen formed by photosynthesis during the daylight hours is lost at night when the algae consume oxygen as part of their metabolic processes. When algae die, the degradation of their biomass also consumes oxygen.

Temperature has a significant impact on the solubility of oxygen in water (Table E-4). Increasing the temperature will normally decrease the DO concentration in water. It is important to distinguish between oxygen *solubility*, which is the maximum DO concentration at equilibrium, and the *actual concentration* of DO, which is generally not the equilibrium concentration and is limited by the rate at which oxygen dissolves. Water saturated with oxygen at 25°C contains 8.4 mg/L. The oxygen in water (hence the dissolved oxygen) in equilibrium with atmospheric oxygen can be predicted by Henry's law constant.

Dissolved oxygen in polluted water is highly dependent on the amount and types of pollutants and the presence of bacteria, and therefore is an important parameter for the assessment of water quality. Most often, oxygen-consuming pollutants are organic compounds (protein, sugar, and fatty acids) which can be easily biodegraded by aerobic bacteria, although in some cases inorganic contaminants in reduced form will also consume a significant amount of oxygen. Many of the synthetic environmental pollutants (e.g., some detergent and pesticides) cannot be

Table E-4   Solubility of oxygen in water of various temperatures and salinities

Temperature (°C)	0	5	10	15	20	25	30
Fresh water ($Cl^- = 8$ mg/L)	14.6	12.8	11.3	10.2	9.2	8.4	7.6
Sea water ($Cl^- = 20,000$ mg/L)	11.3	10.0	9.0	8.1	7.4	6.7	6.1

biologically degraded by bacteria, therefore these compounds will not affect the DO level unless their concentrations are high enough to inhibit the activity of aerobic bacteria. Aerobic bacteria use organic compounds for their food and energy uses.

Method Principle—Two methods for DO analysis commonly used are the Winkler or iodometric method and its modifications and the electrometric method using membrane electrodes (refer to Experiment 2). The iodometric method is a titrimetric procedure based on the oxidizing property of DO. The choice of procedure depends on the interference present, the accuracy desired, and in some cases, convenience or expedience. The iodometric test is the most precise and reliable titrimetric procedure for DO analysis. However, it is not suited for field-testing and cannot be adapted easily for continuous monitoring or for DO determination in situ.

A key step for the Winkler method is sampling technique in the field. Samples should be collected very carefully. Improper sample transport to the laboratory may bring additional errors. For example, agitation of the water may saturate the sample with oxygen from the air, regardless of its original content. Therefore, in the Winkler method, the oxygen is "fixed" immediately after sampling in the field by the reaction with Mn^{2+} followed by the addition of strong alkali (see reaction in Eq. 6.9). After transport to the laboratory, the sample is acidified with sulfuric acid. In the presence of iodide ions in an acidic solution, the oxidized manganese reverts to the divalent state, with the liberation of iodine equivalent to the original DO content (Eq. 6.10). The released iodine can then be titrated with standard solution of sodium thiosulfate using a starch indicator (Eq. 6.12). The titration end point can be easily detected visually with a starch indicator.

It is important to keep in mind that the standard iodometric method (APHA 4500-O) has several modifications depending on possible interference, particularly oxidizing or reducing materials that may be present in the sample. Certain oxidizing agents liberate iodine from iodides (positive interference) and some reducing agents reduce iodine to iodide (negative interference). Most organic matter is oxidized partially when the oxidized manganese precipitate is acidified, thus causing negative errors. A list of the modifications is provided in Table E-5.

Table E-5   The Winkler method and its modifications

Iodometric Methods	Modification	To remove interference of
APHA 4500-O C.	Azide modification	NO_2^-
APHA 4500-O D.	Permanganate modification	Fe^{2+}
APHA 4500-O E.	Alum flocculation modification	Suspended solids (SS)
APHA 4500-O F.	$CuSO_4$-Sulfamic acid flocculation modification	Activated-sludge mixed liquor

Time Required—The DO measurements can be completed in 2 h allowing for two samples to be analyzed per student group. The time may be longer, if the sampling time is significant.

Apparatus and Reagents

Apparatus

(a) 300-mL narrow-mouth glass-stoppererd BOD bottles with tapered and pointed ground-glass stoppers and flared mouths (2 per group), (b) burettes and stands, (c) two 250-mL flasks, (d) 100-mL graduated cylinder.

Reagents

(a) manganase sulfate solution: Dissolve 48.0 g $MnSO_4 \cdot 4H_2O$ or 36.4 g $MnSO_4 \cdot H_2O$ in 100 mL DI water. Filter and dilute to 100 mL, (b) alkali-iodide-azide solution: 50.0 g NaOH (or 70.0 g KOH) and 13.5 g NaI (or 15.0 g KI) in distilled water and dilute to ∼ 90 mL. Add 1.0 g NaN_3 dissolved in 4.0 mL. Combine these two solutions and dilute it to 100 mL, (c) concentrated sulfuric acid (H_2SO_4), (d) starch solution: 2 g in 100 mL hot DI water, (e) standard sodium thiosulfate solution (0.025 M): 6.205 g $Na_2S_2O_3 \cdot 5H_2O$ (equivalent weight = 248.12) in distilled water. Add 0.4 g solid NaOH and dilute to 1 L. Store in a tightly closed glass bottle, (f) DO-deficient water (Instructor prepared using bubbling of helium gas).

The pre-prepared solution of sodium thiosulfate is not stable, and need to be standardized with a known concentration of $K_2Cr_2O_7$ solution prior to use during the day of experiment. Steps are as follows: (1) Dry $K_2Cr_2O_7$ at 120°C and keep it in a desiccator, (2) Dissolve 2.4520 g $K_2Cr_2O_7$ (MW = 294.2; EW = MW/6 = 49.04) in 500 mL DI water to yield 0.1 N solution, (3) To 80 mL DI water, add with constant stirring, 1 mL concentrated H_2SO_4, 5 mL 0.1 N $K_2Cr_2O_7$ solution, and 1 g KI; (4) Let it stand for 6 min in the dark. It will have a yellow color from the liberated iodine, (5) Titrate with 0.1 N $Na_2S_2O_3$ solution until the yellow color is almost discharged, (6) Add 1 mL starch indicator. The color of the solution turns blue, (7) Continue titration until the blue color disappears. (8) Calculate the normality of $Na_2S_2O_3$: N = mL of $K_2Cr_2O_7 \times 0.1$/mL $Na_2S_2O_3$ consumed.

Procedures

1. Sampling and sample preservation in the field:
    (1) Collect surface water samples in BOD bottles of 300-mL capacity (2 bottles per group). Avoid entraining or dissolving atmospheric oxygen.
    (2) Record sample temperature to nearest degree Celsius (why?).
    (3) To the sample, add 1 mL $MnSO_4$ solution, followed by 1 mL alkali-iodide-azide reagent. Hold pipet tips just above liquid surface when adding reagents. Stopper carefully to exclude air bubbles and mix by

inverting bottle a few times. You should see a brown solution with precipitation.

2. Upon return to the lab, add 1 mL concentrated H_2SO_4 into the sample bottles. Restopper and mix by inverting several times until dissolution of the brown precipitate is complete.

3. Transfer 201 mL (which corresponds to 200 mL original water sample, why?) of sample to a 250-mL flask using a graduated cylinder. Titrate with 0.025 M $Na_2S_2O_3$ solution to a pale straw color (manual titration).

4. Add sufficient amount of starch solution ($\sim$3 mL) till a blue color develops and continue titration till first disappearance of blue color. Record the volume (V_1) of titrant.

5. Repeat Step 1 to 4 for an instructor prepared DI water sample which has been sparged with helium gas in the lab. Record the volume of tritrant solution (V_2).

6. Calculation of DO concentration: For titration of 200 mL sample, 1 mL 0.025 M $Na_2S_2O_3 = 1$ mg/L. Express DO in mg/L. You should also report your results in percentage saturation using the solubility data (Table E-4). If $Na_2S_2O_3$ is not exactly equal to 0.025 M, use Eq. 6.13 to calculate DO concentration.

Safety and Health Issues—The DO procedure contains the use of concentrated H_2SO_4. Make sure there are no spills of concentrated H_2SO_4 and always clean-up your work place if you see any suspicious liquid. Wear gloves when handling this chemical, and do not pour the remains down the drain when the test is complete. Follow the instructor's directions for disposing of the waste containing $K_2Cr_2O_7$.

Post Lab Assignment

1. List the potential errors in the measurements of dissolved oxygen. Oxidizing or reducing agents will give either positive or negative interference. Use chemical reactions to explain why?

2. Write the chemical reactions involved in DO measurement using azide modification? How does the classical Winkler method compares to the electrochemical probe method used in Experiment 2?

3. On the basis of stoichiometry, justify the answer why 1 mL of 0.025 M of $Na_2S_2O_3$ is equivalent to 1.0 mg/L of DO? Calculate the DO value if your 250 mL sample consumes 18 mL of 0.014 M $Na_2S_2O_3$?

4. Theoretically estimate the concentration of dissolved oxygen (O_2) in a small campus pond. Assume air above the pond water has a temperature of 20°C, 1 atm of atmospheric pressure and the average O_2 concentration in the air is 21%. Henry's law constant for oxygen is 43.8 mg/(L × atm). Calculate the

saturated DO in water? What is the percentage of saturation of DO in this water?

REFERENCES

Chapter 6: Common Operations and Wet Chemical Methods in Environmental Laboratories (this book).

APHA-AWWA-WEF (1998), Standard Methods for the Examination of Water and Wastewater, APHA-AWWA-WEF, Washington, DC, 20th Edition.

DUNNIVANT FM (2004), Environmental Laboratory Exercises for Instrumental Analysis and Environmental Chemistry, Wiley-Interscience, Hoboken, NJ.

EXPERIMENT 5

DETERMINATION OF CHEMICAL OXYGEN DEMAND (COD) IN WATER AND WASTEWATER

Objectives—The objective of this experiment is to determine the chemical oxygen demand (COD) of a water sample using a standard method—a potassium dichromate ($K_2Cr_2O_7$) digestion in an open-reflux condenser followed by titration with ferrous ammonium sulfate (FAS). Secondary objective of this experiment is to understand the chemical principles, interference, and advantages/disadvantages of the COD as opposed to biochemical oxygen demand (BOD).

Background—In Experiment 4, the dissolved oxygen (DO) in natural waters was measured and the maximum value was found to be about 9 mg/L at 20°C. Not only is the DO level itself important, but the ability of a water body to reduce this parameter, due to organic matter present, is also important. This capacity is quantitated through the concept of BOD or COD. A BOD test provides the closest measure of the processes actually occurring in the natural water system. However, this test is very time-consuming (5 days) and involves many uncertain factors such as the origin, concentration, pollutants, and the number and viability of active microorganisms present to affect the oxidation of all pollutants.

The COD test is used as a measure of the oxygen equivalent of the organic matter content of a sample that is susceptible to oxidation by a strong chemical oxidant. In a COD test, results can be obtained in 2 h or less. Also the method is simple and inexpensive. When wastewater contains only readily oxidizable organic matter and is free from toxins, the results of a COD test provide a good estimate of the BOD. One disadvantage of the method is that dichromate can oxidize materials that would not ordinarily be oxidized in nature. Therefore, COD test is unable to differentiate between biologically oxidizable and biologically inert organic matter. The COD test can also generate a large volume of liquid hazardous waste (acid, Cr, Ag, Hg).

The dichromate oxidation is normally 95–100% complete for most organic substances. However, it will not oxidize a number of refractory molecules including aromatic hydrocarbons, pyridine and related compounds, and straight-chain aliphatics. Volatile organic compounds are oxidized only to the extent that they remain in contact with the oxidant.

The COD test is used extensively in the analysis of industrial wastes. It is particularly valuable in surveys designed to determine and control losses to sewer systems. Results may be obtained within a relatively short time and measures taken to correct errors on the day they occur. In conjunction with the BOD test, the COD test is helpful in indicating toxic conditions and the presence of biologically resistant organic substances. The test is widely used in the operation of treatment facilities because of the speed with which results can be obtained.

Method Principle—The water sample being measured is refluxed with excess of $K_2Cr_2O_7$ in concentrated sulfuric acid for $\sim$2 h. The chemical oxidation (Eq. 6.15) requires strong acidic conditions and elevated temperature. During the COD determination, organic matter is converted to CO_2 and water, and organic nitrogen is oxidized into NH_4^+ or further into nitrates. Silver sulfate (Ag_2SO_4) may be included to catalyze the oxidation process if samples contain alcohols or low molecular weight fatty acids. After dichromate digestion, the excess of dichromate is titrated with ammonium iron (II) sulfate (Eq. 6.16). The indicator for this titration is a chelating agent 1,10-phenanthroline (ferroin, $C_{12}H_8N_2$). When all $Cr_2O_7^{2-}$ is reduced, ferrous ions (Fe^{2+}) react with ferroin to form a red-colored complex $Fe\{C_{12}H_8N_2\}_3$ (Eq. 6.17).

Note that $Cr_2O_7^{2-}$ has a yellow to orange brown color depending on the concentration, and Cr^{3+} has a blue to green color. So the color of the solution during the titration starts with an orange brown, and then there is a sharp change from blue-green to reddish brown, which corresponds to the color of $Cr_2O_7^{2-}$, Cr^{3+}, and $Fe\{C_{12}H_8N_2\}_3$, respectively.

Reduced inorganic species such as Fe^{2+}, S^{2-} and Mn^{2+} will be oxidized under the test conditions, resulting in interference. The most common interference is chloride ions, which give a positive interference through the reaction:

$$Cr_2O_7^{2-} + 6Cl^- + 14H^+ \rightarrow 2Cr^{3+} + 3Cl_2 + 7H_2O \qquad \text{(E-2)}$$

The interference is reduced by addition of mercury sulfate ($HgSO_4$). Mercuric ion combines with the chloride ions to form a poorly ionized mercuric chloride complex:

$$Hg^{2+} + 2Cl^- \rightarrow HgCl_2 \qquad (\beta_2 = 1.7 \times 10^{13}) \qquad \text{(E-3)}$$

In the presence of excess mercuric ions the chloride ion concentration is so small that it is not oxidized to any extent by dichromate.

Time Required—This experiment should be completed within 2.5 h if each student group has two sets of reflux device. The instructor should have one group to perform the COD reflux and titration for a control sample (DI water), which requires an additional reflux device. The data for the controlled group are needed for all groups during subsequent calculation of COD values. Time can be reduced from the standard 2 h if samples are known to contain only readily oxidizable organics.

Apparatus and Reagents

Apparatus

(a) 250 mL Erlenmeyer flask with ground-glass 24/40 neck and 300-mm jacket Liebig, West, or equivalent condenser with 24/40 ground-glass joint, (b) a hot plate (one per sample).

Reagents

(a) 0.04167 M standard potassium dichromate solution: Dissolve 12.259 g $K_2Cr_2O_7$ (MW = 294.2) in 1 L distilled water (what is the concentration in normality? You need this number for later calculation), (b) Standard ferrous ammonium sulfate (FAS) titrant (approximately 0.25 M): 98 g $Fe(NH_4)_2(SO_4)_2 \cdot 6H_2O$ (MW = 392.14) in 1 L DI water. Standardized by $K_2Cr_2O_7$ solution (see Procedures), (c) Potassium hydrogen phthalate (KHP) COD standard: Dissolve 425 mg $HOOCC_6H_4COOK$ (MW = 204) in 1 L distilled water (what is the theoretical COD value of this solution?), (d) Concentrated sulfuric acid with Ag_2SO_4: 5.5 g Ag_2SO_4/kg H_2SO_4, (e) Ferroin indicator solution: 1.485 g 1,10-phenanthroline monohydrate and 695 mg $FeSO_4 \cdot 7H_2O$ in distilled water and dilute to 100 mL.

Procedures

1. Wash three 500-mL flasks thoroughly to exclude extraneous contamination of organic compounds. Each group should do two samples: one blank sample with Millipore water and one water sample from a local water, wastewater, or a calibration standard with a known value of COD (i.e., KHP solution).

2. Transfer proper volume (e.g., 50, 50, and 25 mL for blank, sample, or COD calibration standard, respectively) into three 250-mL flasks. If a volume of less than 50 mL is used (because of higher COD), dilute to 50 mL with water. Label your flasks. Add a few boiling chips.

3. Carefully and very slowly add 5.0 mL sulfuric acid reagent into each flask.

4. Cool while mixing to avoid possible loss of volatile organics.

5. Use a pipet to transfer accurately 25.0 mL 0.0417 M $K_2Cr_2O_7$ solution and mix.

6. Very carefully add additional 70 mL sulfuric acid reagent. Swirl and mix while adding the concentrated sulfuric acid reagent. Wipe the outside of the flasks and make sure there are no spills. *Mix reflux mixture thoroughly before applying heat to prevent local heating of flask bottom and a possible blowout of flask contents.*

7. Attach to a condenser and turn on cool water.

8. Place your flasks on hot plates under the hood. Depending on the specific organics to be measured, as long as 2 h of digestion may be needed. Generally 1 h is sufficient. For the blank, a period of 10–15 min is sufficient.

9. Turn off the hotplate. Wait till cool and then wash down with distilled water. Disconnect the reflux condenser and remove the flasks.

10. Add approximately 150 mL DI water to a total volume of about 300 mL.

11. Add six drops of ferroin indicator. Titrate the excess $K_2Cr_2O_7$ with standard FAS solution. Your sample should have an initial orange brown color, the titration should have a sharp color change from blue-green to reddish brown at the end points. Record the volumes for the blank (A), your water sample (B) and the COD standard sample (C).

12. Calculation:

COD as mg O_2/L = (A–B) × N × 8000/mL water sample, or

COD as mg O_2/L = (A–C) × N × 8000/mL COD standard sample

where A is the mL FAS used for blank, B the mL FAS used for your water sample, C the mL FAS used for your COD standard sample, and N the normality of FAS as determined in Step 13.

13. Calibration of standard ferrous ammonium sulfate (FAS) solution: FAS solution can be slowly oxidized by oxygen, hence standardization is required. *You may do this while your samples are being digested.*

(1) In a 250-mL flask, dilute 10.0 mL standard $K_2Cr_2O_7$ to about 100 mL. Add 30 mL concentrated H_2SO_4 and cool.

(2) Add two to three of drops ferroin indicator; Titrate with FAS titrant. You should expect the color change at the end point from blue–green to reddish brown. Record the volume (V_2) of FAS used.

(3) Calculate the concentration of FAS:

Normality (N_2) of FAS solution = $N_1 V_1 / V_2$, where

where N_1 is the normality of $K_2Cr_2O_7$, V_1 the volume of $K_2Cr_2O_7$ ($V_1 = 10.0$ mL), and V_2 the volume of FAS used in titration (mL).

Safety and Health Issues—(a) Safety glasses (goggle) must be worn at all times in this experiment, especially during digestion. (b) Use the digestion block only in a fume hood and do not exceed its recommended temperature (150°C). Shield the block and keep the door of the hood closed during digestion. (c) The flasks contain concentrated sulfuric acid, potassium chromate (a strong oxidizing agent and possible carcinogen), and a very toxic mercury salt (if your sample contains Cl^-). Use gloves when handling these solutions. (d) Draining disposal of the waste (acid, chromium, silver and mercury) is prohibited. The spent liquid waste must be disposed into waste storage tanks. (e) Flowing cool water and a hot plate is a combination with potential safety concern. Make sure the connections between the tubing and the condenser are firm. Do not leave reflux reaction unattended. (f) If sample of unknown characteristics is analyzed, such as high concentration wastewater, use only a small volume of the sample to start with to avoid potential violent reaction.

Post Lab Assignment

1. Discuss your COD values as fully as possible. Describe the samples you used, and what substances may contribute most to its value.

2. List three possible sources of error in this experiment. How might they be eliminated?

3. Indicate what would affect your COD results if (a) mercuric sulfate was not added for a sample containing Cl^-, (b) silver sulfate was not added, (c) the concentration of ferrous ammonium sulfate was calibrated several days before the experiment.

4. Indicate what modification(s) you would make: (a) to measure samples with a very low COD value for an increased sensitivity, (b) to reduce hazardous waste generation.

5. What is the theoretical COD value of a solution containing 0.340 g/L potassium hydrogenphthalate ($C_8H_5O_4K$; MW = 204)? (Similar calculation can also be performed for other compounds, such as ethanol and glucose). This compound is oxidized according to: $C_8H_5O_4K + 7\frac{1}{2} O_2 \rightarrow 8CO_2 + 2H_2O + K^+ + OH^-$.

6. COD of a wastewater sample (50 mL) is measured by standard dichromate digestion followed by the titration with 0.25 M FAS solution. If 15.0 mL and 10.0 mL of FAS were consumed for the blank (DI water) and the wastewater sample, respectively, what is the COD value in mg O_2/L?

7. Explain why BOD_5 is typically smaller than COD? If sample A has BOD_5 of 220 mg/L and COD value of 280 mg/L, and sample B has a BOD_5 value of 50 and a COD value of 500, what can you infer from the results of these two wastewater samples regarding the biodegradability of each?

8. The oxidation reduction reaction involved in COD measurement is given in Eq. 6.16 (a) What is the equivalent weight of $K_2Cr_2O_7$ (MW = 294.2)? (b) What is the equivalent weight of Fe^{2+} (MW = 55.8)? (c) To prepare 1 L 1.0 M $K_2Cr_2O_7$, how many grams of $K_2Cr_2O_7$ would you need? (d) What is the concentration in normality (N) of 0.1 M $K_2Cr_2O_7$?

REFERENCES

Chapter 6: Common Operations and Wet Chemical Methods in Environmental Laboratories (this book).

APHA-AWWA-WEF (1998), Standard Methods for the Examination of Water and Wastewater, APHA-AWWA-WEF, Washington, DC, 20th Edition.

SAWYER CN, MCCARTY PL, PARKIN GF (1994), Chemistry for Environmental Engineering, McGraw-Hill, NY.

EXPERIMENT 6

DETERMINATION OF NITRATE AND NITRITE IN WATER BY UV-VISIBLE SPECTROMETRY

Objectives—The objectives of this experiment are: (1) to understand the chemical principles for the colorimetric measurement of both nitrite (NO_2^-) and nitrate (NO_3^-), (2) to learn the principles and techniques of using UV-visible spectrophotometry, and (3) to apply Beer-Lambert's Law and determine the unknown sample concentration using a standard calibration curve.

Background—Nitrogen in natural waters and wastewaters has various forms. In order of increasing oxidation state, they are organic nitrogen, ammonium (NH_4^+), nitrite (NO_2^-), and nitrate (NO_3^-). Nitrate (NO_3^-) generally occurs in trace quantities in surface water but may attain high levels in some groundwater. It is found only in small amounts in fresh demonstrated wastewaters, but in the effluent of nitrifying biological treatment plants nitrate may be found in concentrations of up to 30 mg NO_3^--N/L.

Nitrite is an intermediate oxidation state of nitrogen, which can be originated from the oxidation of NH_4^+ or the reduction of nitrate. Such oxidation and reduction may occur in wastewater treatment plants, water distribution systems, and natural waters. Nitrite can also enter a water supply system through its use as a corrosion inhibitor in industrial process water.

In excessive amounts, nitrate contributes to the illness known as methemoglobinemia in infants. However, nitrite is the actual etiologic agent of methemoglobinemia. Nitrous acid, which is formed from nitrite in acidic solution, can react with secondary amines to form nitrosamines, many of which are known to be carcinogenic.

Method Principle—Nitrite (NO_2^-) is determined through formation of a reddish purple azo dye produced at pH 2.0 to 2.5 by coupling diazotized sulfanilamide with N-(1-naphthyl)-ethylenediamine dihydrochloride (NED dihydrochloride). The color product obeys Beer's law to 180 µg N/L with a 1-cm light path at 543 nm. Higher NO_2^- concentrations can be determined by diluting a sample. Common interferences include colored compounds and suspended solids. Color interference should be removed and solids should be filtered out before analysis.

Nitrate (NO_3^-) analysis is often difficult because of the relatively complex procedures required, the high probability that interfering constituents will be present, and the limited concentration ranges of various techniques. The most common standard method used for the nitrate analysis is "Cadmium Column Method." In this method, NO_3^- is reduced almost quantitatively to nitrite (NO_2^-) in the presence of Cd granules packed in a glass column. However, this method is time-consuming and too tedious to be employed as a routine field test.

The method introduced in this experiment for students' practice is a modification of the Cadmium Column Method recommended by Hach Company. HACH has two methods available for nitrate analysis: (1) High Range (0–4.5 mg NO_3^--N/L) and (2) Low Range (0–0.4 mg NO_3^--N /L). For the Low Range nitrate test used in this experiment, cadmium metal is used to reduce nitrate to nitrite (Eq. E-4). The Cd is provided in NitraVer 6 Reagent Powder Pillows. Nitrite ions react with sulfanilic acid to produce an intermediate diazonium salt, which then forms a red-orange colored complex (maximum absorption wavelength $\lambda = 507$ nm) with chromotropic acid in direct proportion to nitrate concentration in the sample (Eqs. E-5 and E-6). Sulfanilic acid and chromotropic acid are contained in NitriVer 3 Reagent Power Pillows.

$$NO_3^- + Cd + 2\,H^+ \rightarrow NO_2^- + Cd^{2+} + H_2O \tag{E-4}$$

$$NO_2^- + \text{Sulfanilic acid} + 2\,H^+ \rightarrow \text{Diazonium salt} + 2H_2O \tag{E-5}$$

$$\text{Diazonium salt} + \text{chromotropic acid} \rightarrow \text{Red-orange color} + H^+ \tag{E-6}$$

Time Required—Three hours. Time can be reduced by the use of fewer data points in the standard calibration curve.

Apparatus and Reagents

Apparatus

(a) Spectronic 20 (Bausch & Lomb or equivalent), (b) Votex mixer, (c) Fifteen 20 mL test tubes per student group, (d) 1 mL and 10 mL pipettes.

Reagents

(a) 1.0 mM $NaNO_2$ (MW = 69.00) stock solution: 0.0690 g $NaNO_2$ dissolved with DI water in 1.0 L volumetric flask, (b) NO_2^- color reagent: To 80 mL water add 10 mL 85% phosphoric acid and 1 g sulfanilamide. After dissolving sulfanilamide completely, add 0.1 g N-(1-naphthyl)-ethylenediamine dihydrochloride. Mix to dissolve, then dilute to 100 mL with water. This solution is stable for 1 month when stored in an amber bottle in refrigerator, (c) 1.0 mM $NaNO_3$ (MW = 84.99) stock solution: 0.0850 g $NaNO_3$ dissolved in 1.0 L volumetric flask, (d) NitraVer 6 Reagent Powder Pillows (cadmium powder, Hach), (e) NitriVer 3 Reagent Power Pillows (sulfanilic acid and chromotropic acid, Hach), (f) Two unknown samples labeled "Unknown 1" and "Unknown 2" prepared by instructor by spiking a known amount of NO_2^- and/or NO_3^-.

Procedures

1. Preparation of nitrite (NO_2^-) calibration curve
   (a) Wash five 7-mL test tubes thoroughly. Label each tube.

(b) Use 1.0-mL pipette to transfer measured volumes of 1.0 mM $NaNO_2$ stock solution into each tube. Use 10-mL pipette to add DI water to each tube until the total volume is 10 mL. An example is shown in Table E-6 below.

(c) Use 1.0-mL pipette to add 0.4 mL NO_2^- color reagent into each tube. Mixed well and let stand for 10 minutes for a full color development. Transfer solution to absorption cells (i.e., test tubes).

(d) Measure absorbance (A) at a wavelength of 543 nm for each sample.

    (1) Switch on the power and allow 15 minutes to warm up the Spectronic 20. Set transparency (T) to zero using the adjustment knob. Set wavelength (λ) at 543 nm.

    (2) Insert an absorption cell with blank (DI water). Adjust to full scale of transparency (i.e., $T_0 = 100\%$ or $A_0 = 0.0$).

    (3) Measure solutions in Test Tube 1 to 5, and record absorbance readings (A_1, A_2, A_3, A_4, and A_5).

(e) Construct a standard calibration curve by plotting absorbance values against μM NO_2^-. Do the regression analysis to obtain the regression coefficients ($x = \mu M$ NO_2^-, $y =$ absorbance).

2. Determination of NO_2^- in two unknown samples

(a) Into two 20-mL test tubes, pipet a measured volume (e.g., $V = 1.0$ mL) of Unknown Sample 1 and Unknown Sample 2. Add DI water to the final volume of 10 mL in each tube.

(b) Proceed to the previous procedures as described in standard calibration curve (Step 1c & 1d). Record the absorbance reading for Unknown Samples (S_1 and S_2). *If the absorbance reading exceeds that of standard solutions, dilute the sample and redo.*

(c) Determine the NO_2^- concentrations in unknown samples using standard calibration curve, and calculate as follows:

μM $NO_2^-/L = \mu M$ NO_2^- calculated from calibration curve $\times (10/V)$

where $V = $ mL original sample, and 10/V is the dilution ratio of the original samples.

3. Preparation of nitrate (NO_3^-) calibration curve

(a) Wash five 20-mL test tubes thoroughly. Label each tube.

(b) Use 1.0-mL pipette to transfer measured volumes of 1.0 mM $NaNO_3$ (not $NaNO_2$) stock solution into each tube. Use 10-mL pipette to add

Table E-6 Preparation of standard solutions for NO_2^-

Test tube No.	1	2	3	4	5
Concentration (μM NO_2^-)	0	5	10	15	20
mL stock solution (1.0 mM $NaNO_2^-$)	0	0.05	0.10	0.15	0.20
mL DI water	10.0	9.95	9.90	9.85	9.80

Millipore water to each tube until the total volume is 10 mL. An example is shown in Table E-7 below.

(c) Add one bag of NitraVer 6 (not NitriVer 3) Reagent Powder Pillow in each tube. Mix for at least one minute in each tube. Let it stand for 5 minutes.

(d) Add one bag of NitriVer 3 (not NitraVer 6) Reagent Power Pillow in each tube. Mix for 30 seconds. Let it stand for 10 minutes.

(e) Measure absorbance (A) at a wavelength of 507 nm (not 543 nm) for each sample on Spectronic 20. Use DI water as a blank.

    (1) Switch on the power and warm up the Spectronic 20 for 15 min. Set transparency (T) to zero using the adjustment knob. Set wavelength (λ) at 507 nm.

    (2) Insert an absorption cell with blank (DI water). Adjust to a full scale of transparency (i.e., $T_0 = 100\%$ or $A_0 = 0.0$).

    (3) Measure solutions in Test Tube 1 to 5 and record absorbance readings (B_1, B_2, B_3, B_4, and B_5).

(f) Construct a standard calibration curve by plotting absorbance values vs. $\mu M\ NO_2^-$. Do the regression analysis to obtain the regression coefficients ($x = \mu M\ NO_3^-$, $y =$ absorbance).

4. Determination of NO_3^- in two unknown samples

(a) Into two 20-mL test tubes, pipet a measured volume (e.g., $V = 1.0$ mL) of Unknown Sample 1 and Unknown Sample 2. Add DI water to the final volume of 10 mL in each tube.

(b) Proceed to the previous procedures as described in standard calibration curve. Record the absorbance reading for Unknown Samples (S_1 and S_2).

(c) Determine the NO_3^- concentrations in unknown samples using standard calibration curve, and calculate as follows:

$\mu M\ NO_3^-/L = \mu M\ NO_3^-$ calculated from calibration curve $\times\ (10/V)$

where $V = $ mL original sample, and $10/V$ is the dilution ratio of the original samples.

Safety and Health Issues—(1) Note that cadmium in the powder is toxic. Wear gloves at all times. (2) Draining disposal of the waste containing cadmium is prohibited. The spent waste must be disposed into a waste storage tank.

Table E-7 Preparation of standard solutions for NO_3^-

Test tube No.	1	2	3	4	5
Concentration ($\mu M\ NO_3^-$)	0	10	20	30	40
mL stock solution ($NaNO_3$)	0	0.10	0.20	0.30	0.40
mL DI water	10.0	9.90	9.80	9.70	9.60

Table E-8   Absorbance vs. concentration data for a series of standard solutions containing NO_2^- or NO_3^-

Concentration (μM NO_2^-)	0	5	10	15	20
Absorbance (A_{543} nm)	0	0.220	0.41	0.59	0.80
Concentration (μM NO_3^-)	0	10	20	30	40
Absorbance (A_{507} nm)	0	0.152	0.33	0.44	0.52

Post Lab Assignment

1. In the lab report, provide calibration curves and the equations calculated from linear regression for both NO_2^- and NO_3^-. For two unknown samples, report the concentrations in the units of μM, mg/L, and mg/L as N. Note that the nitrate method measures both NO_2^- and NO_3^-, so you need to subtract NO_2^- from the total to obtain the NO_3^- concentration.

2. Perform a linear regression analysis using Excel for both nitrate and nitrite data obtained by the instructor (Table E-8). What are the calibration equations and R^2? Indicate whether the linear relationship is significant at $p < 0.05$.

3. Give a list of water quality parameters that are measured by simple colorimetric analysis that are still routinely used in environmental labs (*Hint:* Chapter 8).

4. The Hach Method adopted in this experiment is a simple and rapid method, but it is not a standard method. Search through the National Environmental Methods Index (NEMI) at http://www.nemi.gov to acquire a list of standard methods developed by the U.S. EPA, ASTM, and APHA.

REFERENCES

Chapter 2: Basics of Environmental Sampling and Analysis (this book).
Chapter 8: UV-Visible and Infrared Spectroscopic Methods in Environmental Analysis (this book)
Hach Company (1992), Hach Water Analysis Handbook, 2nd Edition, Loveland, CO.

EXPERIMENT 7

DETERMINATION OF ANIONIC SURFACTANT (DETERGENT) BY LIQUID-LIQUID EXTRACTION FOLLOWED BY COLORIMETRIC METHODS

Objectives—The objectives of this experiment are to learn (1) the classical sample preparation technique using solvent extraction for interference removal and concentration, and (2) the chemical principles for the colorimetric measurement of trace anionic surfactants using a UV-visible spectrometer and the limitations of this method.

Background—The major sources of surfactants in the environment are discharge from household and industrial laundering and other cleansing operations. Surfactants possess unique molecular structure and properties. A surfactant molecule has a hydrophobic "tail" and a hydrophilic "head." As a result, surfactant molecules tend to concentrate at the interfaces between the aqueous medium and the other phases such as air, oily liquids, and particles, which are related to properties such as foaming and emulsification.

There are generally three types of surfactants based on the charge of the hydrophilic group in the surfactant molecule—nonionic, anionic, and cationic surfactants. A nonionic surfactant does not have an ionizing moiety, it commonly contains a polyoxyethylene hydrophilic group ($ROCH_2CH_2OCH_2CH_2....OCH_2CH_2OH$). Anionic surfactant is negatively charged (e.g., $RSO_3^- Na^+$), whereas a cationic surfactant is positively charged (e.g., $RMe_3N^+Cl^-$). In the United States, ionic surfactants account for approximately two thirds of the total surfactants used and nonionic surfactants for about one third. Cationic surfactants amount to less than one tenth of the ionics and are used generally for disinfecting, fabric softening, and various cosmetic purposes rather than for their detersive properties.

The concentration of surfactant in raw domestic wastewater is in the range of about 1 to 20 mg/L. Most domestic wastewater surfactants are dissolved in equilibrium with proportional amounts adsorbed into particulates. Primary sludge concentrations range from 1 to 20 mg (surfactant)/g on a dry basis. In environmental waters, the surfactant concentration is generally below 0.1 mg/L except in the vicinity of an outfall or other point source of entry.

Method Principle—In this method, anionic surfactants are measured and reported as methylene blue active substances (MBAS). Anionic surfactants, or MBAS, can form ion pairs with a cationic dye, that is, methylene blue. This blue color ion pair product can be extracted from an aqueous solution into an immiscible organic solvent using chloroform. The intensity of the resulting blue color in the organic phase is a measure of MBAS. The method comprises three successive extractions from acid aqueous medium containing excess methylene blue into chloroform ($CHCl_3$), followed by an aqueous backwash and measurement of the blue color in the $CHCl_3$ by spectrophotometry at 652 nm.

Anionic surfactants are among the most prominent of many substances showing methylene blue activity. Although soaps are anionic in nature, they do not respond in the MBAS method. This is because soaps are so weakly ionized that an extractable ion pair is not formed under the test conditions. Nonsoap anionic surfactants commonly used in detergent formulations are strongly responsive. These include principally surfactants of the sulfonate type $[RSO_3]^- Na^+$, the sulfate ester type $[ROSO_3]^- Na^+$, and sulfated nonionics $[RE_nOSO_3]^- Na^+$. Note that the MBAS method measures only the total amount of these substances. To differentiate sulfonate- from sulfated-type surfactants, or linear alkylbenzene sulfonate (LAS) from alkylbenzene sulfonate (ABS), other methods must be employed, such as HPLC, GC, and GC–MS.

Time Required—The classical liquid–liquid extraction is a very tedious and time-consuming process. Time required for this experiment can be as long as 3 h. The time can be reduced if students are instructed to work on a fewer samples. Extraction will be progressively faster when students have acquired needed proficiency in extraction procedures.

Apparatus and Reagents

Apparatus

(a) Spectronic 20 (Bausch & Lomb or equivalent), (b) separatory funnels and stands (two sets of funnels are required for one sample, it is preferred to have four sets of funnels for each student group to expedite the experiment), (c) 50 mL volumetric flasks (cleaned and dried by instructor prior to the class), (d) 10 mL pipettes, graduated cylinders, filtration funnel, (e) glass wool: pre-extracted with $CHCl_3$ to remove potential interferences.

Reagents

(a) Stock surfactant solution: Dissolve 0.5000 g of sodium dodecyl sulfonate (SDS) into 500 mL water. Use magnetic bar for stirring instead of vigorous shaking to avoid foam formation (1.00 mL = 1.00 mg), (b) Standard SDS solution: Dilute 10.00 mL stock SDS to 1000 mL water (1.00 mL = 10.0 μg), (c) Phenolphthalein indicator solution, alcoholic: 1.0 g phenolphthalein into 100 mL ethyl alcohol and 100 mL water, (d) 1 N NaOH, (5) H_2SO_4, 1 N and 6 N, (e) Chloroform ($CHCl_3$), (f) Methylene blue reagent: Dissolve 100 mg methylene blue in 100 mL water. Transfer 30 mL to a 1000-mL flask. Add 500 mL water, 41 mL 6 N H_2SO_4, and 50 g $NaH_2PO_4 \cdot H_2O$. Dilute to 1 L, (g) Wash solution: Add 41 mL 6 N H_2SO_4 to 500 mL water in a 1 L flask. Add 50 g $NaH_2PO_4 \cdot H_2O$ and shake until dissolved. Dilute to 1 L, (h) An unknown sample containing spiked anionic surfactant (Instructor prepared).

Procedures

1. Preparation of calibration curve
   (a) Wash and label two separatory funnels. No soap/detergent wash is used.
   (b) Five 50-mL volumetric flasks (per group) have been thoroughly cleaned for you. The flasks should be free of water before use. Label each flask (4 for standards and 1 for unknown sample).
   (c) Use 10-mL pipette to transfer measured volume of standard surfactant solution (10 μg/mL) into each separatory funnel. For example, a series

of 2.5, 5, 10, and 15 mL corresponds to 25, 50, 100, and 150 µg of SDS, respectively. Add Millipore water to each funnel until the total volume is approximately 50 mL. *Each group may do one sample at a time. Note that two funnels are needed for each sample.*

(d) Make alkaline by dropwise addition of 1 N NaOH, using phenolphthalein indicator. Discharge pink color by dropwise addition of 1 N H$_2$SO$_4$ (*Skip this step if pH adjustment is not needed for your samples*).

(e) Under a ventilation hood, add 8 mL chloroform (CHCl$_3$) and 25 mL methylene blue reagent. Shake funnel vigorously for approximately 30 sec. Release the pressure build-up in the funnel periodically during extraction, and avoid excessive agitation, which may cause emulsions and difficult phase separation.

(f) Swirl the funnel gently, and then let it settle for a few minutes for phase separation. Draw off the bottom CHCl$_3$ layer (blue color) into a second funnel.

(g) Repeat the extraction for two additional times, using 8 mL CHCl$_3$ each time.

(h) Discard the contents in the first separatory funnel. In the secondary separatory funnel that has the combined CHCl$_3$ phase, add 50 mL wash solution and shake vigorously for 30 sec. Let it settle, then swirl, and draw off CHCl$_3$ layer through a funnel containing a plug of glass wool into a 50-mL volumetric flask. Make sure not to let water enter the volumetric flask.

(i) Add 8 mL of CHCl$_3$ and extract the wash solution. Draw off CHCl$_3$ to the flask through the glass wool.

(j) Repeat Step 1i for an additional extraction with 8 mL of CHCl$_3$.

(k) The total volume in the flask should be approximately 40 mL (5 × 8). Dilute to 50 mL with CHCl$_3$ and mix. The color is stable for a minimum of 24 hrs.

(l) Repeat Step 1c through 1k for three remaining standards.

(m) Transfer solution into four test tube cuvettes. Using pure CHCl$_3$ as a blank, determine the absorbance of each solution at 652 nm.

(n) Construct a standard calibration curve by plotting absorbance values (*y*) against micrograms surfactant (*x*) in 50 mL final volume. Do the regression analysis to obtain the regression coefficients.

2. Determination of surfactant concentration in an unknown sample

(a) Use sample volume of 50 mL (depending on the spiked concentration of surfactant)

(b) Proceed to the previous procedures as described in standard calibration curve. Record the absorbance reading for the unknown sample.

(c) Determine surfactant concentration in the unknown samples using standard calibration curve, and calculate the concentration according to: mg surfactant/L is the μg surfactant/V, where V is the mL original sample (e.g., 50 mL).

Safety and Health Issues—(a) A large amount of chloroform is used in the extraction during this experiment. Chloroform is very volatile and a suspected carcinogen. Take appropriate precautions against inhalation and skin exposure. You are required to perform the extraction under the hood. Do not leave the container open if it has chloroform. Always remember to minimize the exposure to any volatile chemicals regardless of its toxicity. (b) Chloroform will dissolve plastic containers. Do not use plastic containers or disposable cuvettes to store $CHCl_3$ solution. Do not use parafilm. (c) Draining disposal of the waste containing chloroform is prohibited. The spent liquid waste must be disposed into an organic solvent waste storage tank.

Post Lab Assignment

1. Why the methylene blue active substances (MBAS) method is a nonspecific measurement of surfactant? What is the standard method for the measurement of nonionic surfactants? What methods other than colorimetric methods should be sought in order to measure a specific surfactant?

2. Describe the extraction principles of MBAS method. Is surfactant directly extracted by chloroform?

3. In performing the surfactant extraction by chloroform, why should water be eliminated in the final extract solution?

REFERENCES

APHA-AWWA-WEF (1998), Standard Methods for the Examination of Water and Wastewater, APHA-AWWA-WEF, Washington, DC, 20th Edition.

ZHANG C, VALSARAJ KT, CONSTANT WD, ROY D (1999), Aerobic biodegradation kinetics of four anionic and nonionic surfactants at sub- and supra-critical micelle concentrations (CMCs), *Water Res.*, 33(1):115–124.

EXPERIMENT 8

DETERMINATION OF HEXAVALENT AND TRIVALENT CHROMIUM (Cr^{6+} AND Cr^{3+}) IN WATER BY VISIBLE SPECTROMETRY

Objectives—The objectives of this experiment are (1) to understand sample handling, sample pretreatment, interference removal, and the chemical principles for the colorimetric measurement of both hexavalent (Cr^{6+}) and trivalent (Cr^{3+}) forms of chromium, (2) to learn the principles and techniques in trace metal analysis using UV-visible spectrometry and its pros and cons as compared to other instrumental techniques such as AA and ICP, (3) to further familiarize the Beer-Lambert's Law and determine the unknown sample concentration using a standard calibration curve, and (4) to acquire UV-Visible spectrum and understand the origin of optimal wavelength used in quantitative analysis.

Background—Chromium (Cr) is of crucial importance because of its widespread use and its extent of environmental contamination in the U.S. and worldwide. While the average natural abundance of Cr in the earth's crust is as high as 122 mg/kg, much lower background concentrations are expected in soils ($11 \sim 22$ mg/kg), surface waters (1 µg/L), and ground waters (100 µg/L). Major pollution sources of chromium are from alloys, electroplating, pigment industries. Chromate compounds are also added to cooling water for corrosion control.

The toxicity of Cr and its mobility in aquatic and terrestrial environments depend on its oxidation states. Two most stable oxidation sates of chromium in the environment are hexavalent (Cr^{6+}) and trivalent (Cr^{3+}). In natural waters, trivalent chromium exists as Cr^{3+}, $Cr(OH)^{2+}$, $Cr(OH)_2^+$, and $Cr(OH)_4^-$, whereas hexavalent chromium exists as CrO_4^{2-} and $Cr_2O_7^{2-}$. Cr^{3+} is expected to form strong complexes with amines, and can be adsorbed by clay minerals. In portable water system, chromium exists predominantly in its hexavalent form. Chromium is considered nonessential for plants, but its trivalent form is an essential trace element for animals. Hexavalent Cr compounds are carcinogenic to human and are corrosive to tissue.

Method Principle—The colorimetric method measures only haxavalent chromium (Cr^{6+}). In order to determine total chromium, Cr^{3+} must be first oxidized to the hexavalent state by potassium permanganate ($KMnO_4$). The hexavalent chromium reacts with diphenylcarbazide in acid solution to produce a red-violet colored complex (Eq. 8.16). This reaction is very sensitive, the molar absorptivity (ε) based on chromium being about 40,000 L/g $\times$ cm at 540 nm.

The reaction with diphenylcarbazide is highly specific for chromium, but some minor interference may be originated from hexavalent molybdenum (Mo^{6+}) and mercury salts (Hg^+ and Hg^{2+}). These ions react to form color with the reagent, which can be minimized by pH adjustment to significantly reduce color intensities. If vanadium (V^{5+}) is present, it can interfere strongly because of the formation of a yellow complex, but the color will fade after 15 min and concentrations up to 10 times that of Cr will not cause trouble. Potential interference from the red-violent reagent permanganate can be readily eliminated by prior reduction with sodium azide. Iron (Fe^{3+}) in concentrations greater than 1 mg/L may produce a yellow color but the ferric ion color is not strong and no difficulty is encountered normally, if the absorbance is measured at the specified wavelength.

Time Required—The time needed for this experiment is 2.5 h. Although a little more time may be needed for a manual acquisition of the UV absorption spectrum, it is preferred over the auto spectral scan that can be obtained by most modern spectrometers.

Apparatus and Reagents

Apparatus

(a) Spectronic 20 (Bausch & Lomb or equivalent), (b) a 10' × 10' hotplate, (c) 100 mL volumetric flasks, (d) pipettes of various sizes (1, 2, 5, 10 mL), (e) boiling chips.

Reagents

(a) Standard chromium solution (1.00 mL = 5.00 µg Cr): Dilute 10.0 mL stock chromium solution (141.4 mg $K_2Cr_2O_7$ in 1 L; 1.00 mL = 50.0 µg Cr) to 100 mL, (b) 1 + 1 H_2SO_4, (c) Potassium permanganate solution: Dissolve 4 g $KMnO_4$ in 100 mL water, (d) Sodium azide solution: Dissolve 0.5 g NaN_3 in 100 mL water, (e) Diphenylcarbazide solution: Dissolve 250 mg 1,5-diphenlcarbazide (1,5-diphenylcarbohydrazide) in 50 mL acetone. This solution in unstable, and should be kept in a refrigerator. Discard when solution becomes discolored, (f) 6 N H_2SO_4 (dilute 16.7 mL conc. H_2SO_4 to 100 mL), (g) Two unknown samples containing chromium (Cr^{6+} and/or Cr^{3+}) prepared by instructor.

Procedures

1. Preparation of calibration curve

   (a) Wash five 100-mL volumetric flasks thoroughly and label each flask.

   (b) Use 5-mL and 10-mL pipettes to transfer measured volumes of standard chromium solution (5 µg/mL) into each flask. For example, a series of 0, 5, 10, 15, and 20 mL corresponds to 0, 25, 50, 75, and 100 µg of Cr, respectively. Add Millipore water to each flask until the total volume is approximately 40 mL.

   (c) Add two drops of 6 N H_2SO_4 to acidify the samples. Add additional water to dilute to a final volume of 100 mL.

   (d) Add 2 mL diphenylcarbazide solution to each flask. The final volume is 102 mL.

   (e) Seal the flask with parafilm. Immediately mix the samples and let it stand for 15 minutes for a full color development. Transfer solution to absorption cells (curvet).

   (f) Measure absorbance (A) at a wavelength of 540 nm for each sample specified in 1b).

(g) Construct a standard calibration curve by plotting absorbance values against micrograms chromium in 102 mL final volume. Do the regression analysis to obtain the regression coefficients (x in µg Cr, y in absorbance).

2. Determination of Cr^{6+} in two unknown samples

   (a) Into two 100-mL volumetric flasks, pipet measured volume (V) of Unknown Sample 1 and Unknown Sample 2.

   (b) Proceed according to the procedures as described in standard calibration curve. Record the absorbance reading for Unknown Samples (S_1 and S_2).

   (c) Determine the Cr^{6+} concentrations in unknown samples using standard calibration curve, and calculate the following: mg Cr/L = µg Cr (in 102 mL final volume)/V, where V = mL original sample.

3. Determination of total chromium in two unknown samples: Oxidation of Cr^{3+}

   (a) Into two 100-mL beakers, pipet measured volume (V) of Unknown Sample 1 and Unknown Sample 2.

   (b) Add Millipore water to final volume of approximately 40 mL.

   (c) Add 1 mL 1:1 H_2SO_4 to acidify the samples, and add 2 drops of $KMnO_4$ to give a dark red color. If fading occurs, add $KMnO_4$ dropwise to maintain an excess of about two drops.

   (d) Add few acid-washed boiling chips. Boil for 2 min on a hotplate.

   (e) Add 1 mL NaN_3 solution and continue boiling gently. If red color does not fade completely after boiling for approximately 30 s, add another 1 mL $NaNa_3$ solution. Continue boiling for 1 min after color fades completely, then cool.

   (f) Transfer to two 100-mL volumetric flasks. Rinse the beakers at least twice with water. Add water to bring the final volume to 100 mL.

   (g) Add 2 mL diphenylcarbazide, mix and let it stand for 15 min. Record the absorbance reading (S_3 and S_4).

   (h) Determine total Cr concentrations in unknown samples using the standard calibration curve, and calculate as follows: mg Cr/L = µg Cr (in 102 mL final volume)/V, where V = mL original sample.

   The concentration of Cr^{3+} in the unknown sample is calculated by the difference between total Cr and Cr^{6+}, that is, Cr^{3+} = Total Cr − Cr^{6+}

4. Determination of absorption spectrum: Use any one standard or sample, measure absorbance at an interval 5–10 nm in the range of 400–650 nm. Plot absorbance vs. wavelength.

Safety and Health Issues—(a) Hexavalent chromium is toxic. Wear gloves to avoid skin contact. (b) Draining disposal of the waste containing chromium is prohibited. The spent liquid waste must be disposed into an inorganic waste storage tank.

(c) The color compound is very hard to remove if there are spills on floor or lab bench. Be extremely careful to avoid any spills.

Post Lab Assignment

1. Include absorption spectrum in your report. What is the λ_{max}?
2. Given the cell length of 1.0 cm, calculate the molar absorptivity (ε) from the calibration data.
3. What sample preservation procedures will differ if the water sample is used for dissolved Cr^{6+} vs. total Cr analysis?
4. The standard method calls for filtration if samples are turbid. Explain why.
5. Until the widespread use of atomic spectrometric techniques, visible spectrophotometry was the most commonly used technique for metal ion analysis. Standard methods were developed for all commonly found metals on the basis of the color-forming complexing agents. Measurement of chromium by colorimetric method is one such excellent example of UV-VIS application trace metal analysis (MDL: 5 µg/L). Seventeen other metals can be analyzed colorimetrically by dithizone complexing agent. Interference is a common problem, but selectivity is improved by precise control of pH and the use of masking agents. Describe the pros and cons of colorimetric analysis for metals as compared to other instrumental techniques (AA and ICP).

REFERENCES

Chapter 8: UV-Visible and Infrared Spectroscopic Methods in Environmental Analysis (this book).
APHA-AWWA-WEF (1998), Standard Methods for the Examination of Water and Wastewater, APHA-AWWA-WEF, Washington, DC, 20th Edition.

EXPERIMENT 9

DETERMINATION OF GREENHOUSE GASES BY FOURIER TRANSFORM INFRARED SPECTROMETER

Objectives—The objectives of this experiment are (1) to learn the use of Fourier transform infrared spectrometer, (2) to learn the basic of sample preparation principles for infrared analysis, and (3) to understand the infrared spectrum and the origin of infrared absorption from molecular structure.

Background—Global warming and ozone depletion are the two global environmental problems faced by every nation. There have been significant progresses toward ozone depletion issue since the early scientific discovery in 1970s that anthropogenic Freons were the cause of the huge ozone hole in the South Pole. This also resulted in a successful international cooperation, as witnessed by the signing of the Montreal Protocol on substances that deplete the ozone layer in 1987. The global warming issue, on the contrary, continues to be the topic of many scientific and political debates. There are conflicting views and difficulties in predicting the trend and impact of global warming. The future is uncertain even though the Kyoto Protocol aimed at the reduction of global warming gases which has been ratified by 163 countries as of April 2006.

Nevertheless, the origin of global warming is scientifically well-established. Global warming is primarily caused by the increased concentration of CO_2 in the atmosphere. Simply put, CO_2 absorbs the infrared range of solar radiation, resulting in the rising temperature on earth. The increased CO_2 concentrations are believed to be the result of the increased human activities, such as the combustion of fossil fuels.

One should recognize that CO_2 is the major contributing gas for global warming, but it is not the only gas. Many other gases can absorb infrared light, therefore, they too are the contributing factors in global warming. These common gases, in a decreasing order of concentrations, are H_2O, CO_2, CH_4, N_2O, and Freons. If we compare the global warming potential (GWP), CO_2 has, in fact, the least effect considering the heat absorbing ability and the decay of the gas (Table E-9). The GWP has an increasing order of $CO_2 < CH_4 < N_2O <$ Freon-11 < Freon-12.

In this experiment, four gases (H_2O, CO_2, N_2, and O_2) are tested using a FTIR spectrometer. Of these four gases, only CO_2 is considered the global warming gas. Although water is IR active and has the highest concentration in the average atmosphere, it is not usually listed among the global warming gas because the concentration fluctuation of water is all natural. N_2 and O_2 consists of 78 and 21% of the atmosphere; they are not IR active.

Method Principle—The energy of infrared radiation is not high enough to break covalent bonds or cause electronic vibration, but it is in the range that can cause vibrational or rotational motion of a molecule. For a molecule to absorb IR radiation,

Table E-9 Important infrared-active gases in the atmosphere

	H_2O	CO_2	CH_4	N_2O	Freon-11	Freon-12
Current concentration	0.4%	370 ppm$_v$	1.745 ppm$_v$	0.314 ppm$_v$	0.268 ppb$_v$	$\sim$1 ppb$_v$
Percentage increase/year	-	1.5	7.0	0.8	-1.4	5
GWP§	-	1	21	310	3400	7100
Absorbing band (μm)	2.5–3.4 10–13	4.0–4.5 $\sim$3	3.0–4.0 7.0–8.5	3.0–5.0 7.5–9.0	9.1–10 11.1–12.5	9.5–11.5

§The GWP describes the long-term contribution of any gas by comparison to that of CO_2.

the frequency of IR and the frequency of the motions should match. A molecule must also undergo a net change in dipole moment as a result of its vibrational or rotational motion. Symmetric molecules such as N_2 and O_2 will not have net change in dipole moment, no matter how much the covalent bonds are stretched. On the contrary, polyatomic molecules such as H_2O, CO_2, CH_4, N_2O, and Freons have many vibrations that will result in change in dipole moment, hence absorb infrared radiation.

The detector in FTIR measures the amount of light absorbed by the analyte molecule, and records the amount of light transmitted as a function of wave number (IR spectrum). These are respectively the basis for quantitative and qualitative measurement using FTIR. Because it measures the transmittance rather than the absorbance used in UV-VIS spectrometers, the IR spectrum is a series of upside down peaks. An IR spectrum usually provides more structural information than a UV spectrum of the same compound.

Another unique feature of FTIR is its ability to analyze gas, liquid, or solid samples, although different sample preparation procedures should be followed. For solid sample, 50–200 mg is desirable, but 10 µg ground with transparent matrix (such as KBr) is the minimum. For qualitative determinations, 1–10 µg minimum is required if solid is soluble in suitable solvent. For liquid samples, a 0.5 µL is needed if neat, less if pure. For gas samples, 50 ppb and a longer sample cell is needed.

Time Required—Sample preparation will take most of the time for this experiment since the measurements of FTIR takes only a few minutes per sample. The instructor can reserve some time for retrieving the FTIR spectra of other molecules, and provide students with a better understanding of the interpretation of IR spectra.

Apparatus and Reagents

Apparatus

(a) FTIR spectrometer, (b) Sample cells for gas or plastic sample bags depending on the configuration of FTIR spectrometer.

Reagents

(a) Dry ice (frozen solid carbon dioxide), (b) N_2 gas cylinder, (3) water.

Procedures

1. Prepare four samples in four plastic bags or gas cells as follows:
   (a) To prepare a gas sample containing N_2, fill a plastic bag with nitrogen gas from a nitrogen tank.

(b) To prepare a gas sample containing CO_2, insert a piece of dry ice into the bag. Squeeze out any remaining air. Tie the top of the bag securely.

(c) To prepare a gas sample containing H_2O vapor, pipet approximately one mL of water into a plastic bag. Fill the bag with N_2 and tie the top of the bag securely. Shake and roll the bag until the water vapor saturates the gas.

(d) To prepare a gas sample containing atmospheric gas, open the bag, wave it around until it is $3/4$ full of gas and tie the top of the bag securely.

2. Scan gas samples with the FTIR

(a) Follow instrument instruction manual to turn on and warm up.

(b) Use N_2 as the blank, and scan the other three samples. You should keep a printout of the IR spectrum for each gas in your report.

Safety and Health Issues—The temperature of dry ice is very cold $(-109°C)$, which can cause severe frostbite. If you suspect you have frostbite seek medical help immediately. Wear safety goggles and gloves when doing experiments with dry ice.

Post Lab Assignment

1. In your lab report, label each absorption peak with respect to the vibration that causes the absorption. Indicate why N_2 is used for blank?

2. There are four vibrational modes for CO_2 molecule, that is, asymmetric stretch, symmetric stretch, vertical bend, and horizontal bend. Which one(s) can have the absorption of infrared radiation?

3. Which of the following atmospheric components/pollutants are IR active: Argon (Ar), carbon dioxide (CO), sulfur dioxide (SO_2), nitrous oxide (N_2O), nitric oxide (NO), nitrogen dioxide (NO_2), chlorine (O_2), ozone (O_3), and VOCs (hydrocarbons)?

REFERENCES

Chemistry Department, University of Maine (2001), FTIR Analysis of Greenhouse Gases, http://icn2.umeche.maine.edu/genchemlabs/Greenhouse_Gases/gasesv4.html

Dunnivant FM (2004), Environmental Laboratory Exercises for Instrumental Analysis and Environmental Chemistry, Wiley-Interscience, Hoboken, NJ.

Girard JE (2005), Principles of Environmental Chemistry, Jones and Bartlett Publishers, Sudbury, MA.

Lashof DA, and Ahuja DR (1990), Relative contributions of greenhouse gas emissions to global warming, *Nature*, 344, 529–31.

EXPERIMENT 10

DETERMINATION OF METALS IN SOIL – ACID DIGESTION AND INDUCTIVELY COUPLED PLASMA – OPTICAL EMISSION SPECTROSCOPY (ICP-OES)

Objectives—The objectives of this experiment are to learn: (1) the principle and operation of an inductively coupled plasma–optical emission spectroscopy (ICP-OES) for metals analysis, and its pros and cons over atomic absorption spectroscopy, and (2) the techniques of soil sample digestion for metal analysis and determine the concentrations of metals (Al, Cd, Cu, Zn, Fe, Cr, Pb, and Ni) in a water and a soil sample.

Background—Of all the elements on the periodic table, metals make up about 75% of the total. The concentration of metal species in the environment covers a wide range. In the oceans, the concentration of Na^+ is approximately 0.46 M and Mg^{2+} is present in sufficiently large amounts that extraction from marine water is a viable source of the element. Other metal concentrations range down to ppm, ppb or even ppt levels.

Metals are of environmental interest and importance because of their influence on the environment and biological processes. Metals such as K and Ca are important nutrients required in substantial amounts by plants, animals, and microorganisms. Other metals such as Cu and Zn are also nutrients, but the amount required by organisms is very small. These metals, if present in excessive amounts can be toxic, so there is a range of concentrations, sometimes narrow, that is suitable for supporting life processes. Still other metals such as Cd and Hg are not essential nutrients and even very small concentrations can be toxic to many living organisms.

Method Principle—There are two common instrumental methods for the analysis of metals in water, atmospheric particulate, soils, and biological samples: (1) the atomic *absorption* spectrometry (AA), and (2) the inductively coupled plasma-optical *emission* spectrometry (ICP–OES). The atomic absorption technique is based upon the absorption of a monochromatic light by a cloud of atoms and of the same atoms as those being analyzed. The source produces intense electromagnetic radiation with a wavelength exactly the same as that absorbed by the atoms, resulting in extremely high selectivity.

ICP–OES is a special type of atomic emission technique that is based on the spectral lines emitted when they are heated to a very high temperature. The "flame" in which analyte atoms are excited in plasma emission consists of an incandescent plasma (ionized gas) of argon, heated inductively by radio-frequency energy at 4–50 MHz and 2–5 kW. The energy is transferred to a stream of argon through an induction coil, producing temperatures up to 10,000 K. A sample aerosol is generated in an appropriate nebulizer and a spray chamber and is carried into the plasma through an injector tube located within the torch. The sample atoms are subjected to temperature around 7000 K, twice more than those of the hottest conventional flames (e.g., nitrous oxide-acetylene). As the name implies, ICP–OES relies on the light emission properties of sample atoms excited by the high temperature of argon plasma. Theoretically, the concentration of a sample is proportional to the emitted light intensity of the atoms.

Time Required—The experiment described herein, because of its focus on the use of ICP–OES instrument, can be completed in 2.5 h. If the entire procedures are desired at instructor's discretion (including soil sample digestion and the preparation of mixed standard solution), this experiment can be modified to fit two 2.5-h lab sessions.

Apparatus and Reagents

Apparatus

(a) ICP–OES instrument (Perkin-Elmer 400 ICP–OES or equivalent), (b) 50 mL volumetric flasks, (c) Graduate cylinder, 5-mL pipette, (d) a hot plate.

Reagents

(a) Stock solutions (pH < 2) for each individual metal (1000 mg/L or 1 ml = 1 mg Al, Cd, Cu, Zn, Fe, Cr, Pb, and Ni, (b) Unknown water sample 1 (tap water), (c) Unknown sample 2 (digested soil sample), (d) Argon gas cylinder, (e) De-ionized Millipore water (used as a calibration blank), (f) Concentrated HNO_3 and 5% HNO_3, (g) Five mixed standard solutions (Table E-10).

Procedures

1. Acid digestion of soil samples: A soil sample will be acid-digested for you according to the procedures described below. You do not even need to

Table E-10   Preparation of mixed standard solutions (in 1% HNO_3)

Metals	S1 (mg/L)	S2 (mg/L)	S3 (mg/L)	S4 (mg/L)	S5 (mg/L)
Al	10	25	50	75	100
Cd	0.50	1.25	2.50	3.75	5.00
Cu	2	5	10	15	20
Zu	2	5	10	15	20
Fe	2	5	10	15	20
Cr	2	5	10	15	20
Pb	2	5	10	15	20
Ni	2	5	10	15	20

perform acid digestion in this lab; it is important that you understand the principle of soil sampling, soil sample pretreatment and the operational procedures for acid digestion.

(a) A typical procedure for the pretreatment of a soil sample or any other solid samples (sediments, sludge, hazardous waste) will involve air drying, grounding, sieving and homogenization. The entire procedure may be very time-consuming and tedious, especially when air-drying and sieving are involved. For metal analysis, it is important to keep in mind that no metal equipment should be used during sampling or sample pretreatment.

(b) Weigh to the nearest 0.01 g soil sample (dry weight) and transfer it to a 250-mL Erlenmeyer flask. Add 10 mL 1:1 HNO_3, mix the slurry, and cover with a watch glass. Heat the sample to $95 \pm 5°C$ for 15 minutes without boiling. Allow the sample to cool, add 5 mL of concentrated HNO_3, replace the cover, and reflux for 30 minutes. If brown fumes are generated, indicating oxidation of the sample by HNO_3, repeat this step (addition of 5 mL of conc. HNO_3) again until no brown fumes are given off by the sample.

(c) Add 10 mL concentrated HCl to the sample digest and cover with a watch glass. Place the sample on the heating source and reflux at $95 \pm 5°C$ for 15 minutes or longer till the disappearance of dense brown fume.

(d) Filter the digestate through Whatman No. 41 filter paper (or equivalent) and collect filtrate in a 100-mL volumetric flask. Dilute to 100 mL and the digestate is now ready for ICP–AES analysis.

2. Preparation of a water sample: The relatively clean water sample such as groundwater or the tap water sample as described below does not require acid digestion. If other water samples with more complicated matrix (e.g., wastewater) are being analyzed, a prior acid digestion should be followed.

(a) Use a graduated cylinder, measure 25.0 mL of tap water, transfer to a 50-mL volumetric flask.

(b) With a 5-mL pipette, measure 2.5 mL of concentrated HNO_3, carefully add the acid to the flask. Shake the contents gently.

(c) Add Millipore water to the flask to make the final volume of 50-mL (*Note that your sample is now diluted by a factor of 2*). Cover the flask with a piece of parafilm, shake the flask to mix the contents. The sample is now ready for ICP–AES analysis.

3. Analysis of metals by ICP–OES

(a) Instrument start-up procedure

(i) Turn on computer, printer, main power of the ICP.

(ii) Open valve on argon tank and confirm tank volume is ≥ 500 lbs; if argon tank volume is < 500 lbs, change out tank before starting analyses. Verify that gas regulator on argon tank is set at 80 psi, if not adjust to 80 psi.

(iii) Set up inlet and outlet tubing on peristaltic pump and turn the pump switch on to the low setting. Insert inlet tubing into water/blank; verify counterclockwise movement of fluid through tubing (inlet tube goes from sample to nebulizer and outlet tube goes from nebulizer to the collection vessel on the floor under the instrument); make sure tubing is not crimped or impeded anywhere along the pathway. Make sure the pump is continually drawing up liquid.

(iv) Verify the nebulizer is set at the proper flow rate.

(v) Ignite the plasma; allow the instrument to warm up to about 20 minutes to stabilize the wavelengths, then it will be ready to begin analyses.

(b) Running samples

(i) Retrieve prior developed method file from the operating software. Verify that the method file has all the elements of interest, the right wavelength for each element, and the concentrations of the calibration standards (Table E-10).

(ii) Follow instructions to initiate running sample analysis in the order of blanks (Millipore water), standards, Unknown Sample 1 (tap water), and Unknown Sample 2 (soil digestate).

(iii) Monitor the pressure in the argon tank; if it drops below 500 lbs, get to a stopping place in the analysis, shut down the ICP (see Shut Down Procedure) and change out the argon tank; when you restart, you must repeat the start-up procedures from the beginning and rerun your blanks and standards.

(iv) Keep a printed copy of the analytical data including the calibration curves and the instrumental signals for two unknown samples.

(c) Shut-down procedure

(i) Following completion of your sample analyses, run 5% HNO_3 through the ICP for 5 minutes, followed by DI water for 10 minutes.

(ii) Follow instruction manual to extinguish the plasma.

(iii) Turn off peristaltic pump; loosen the two screws and flip the lever back into the up (locked) position to release tension on the inlet and outlet tubes.

(iv) Turn off main power. Close valve on argon tank.

(v) On the computer, escape out of the ICP analysis program; turn off computer and printer.

(d) Data analysis: Data generated by the ICP will either be the instrumental signals or the concentration or both. The concentrations reported are the concentration of the liquid digest sample in, for example, milligrams per liter (mg/L). This will need to be converted to micrograms per gram dry weight for solid samples using the formula: $mg/kg = mg/L \times V/W$, where V is the volume of digested solution (mL) and W the dry soil sample weight (g).

Safety and Health Issues—(a) Acid digestion should be conducted in a hood with a good ventilation. On the completion of digestion, the hood is preferred to be rinsed with water to protect it from corrosion. (b) Special safety precautions should be exercised for the use of strong acid, particularly $HClO_4$, which can be explosive when samples of high organic materials are digested. (c) It is always a good practice to run through dilute HNO_3 and water before turning off ICP.

Post Lab Assignment

1. List the wavelength used for the measurement of each element in this experiment. Are there any other characteristic wavelengths that can be used? What is the typical range of wavelength (i.e., UV, VIS, IR)? What are the similarities of the optical device in ICP as compared to UV–VIS spectrometer?

2. ICP can determine almost all the metals and metalloids of environmental significance (Cu, Zn, Pb, Cd, Ni, Hg, Cr, As). However, its application for the measurement of B, P, N, S and C is limited. In addition, ICP has very limited use for alkali metals such as Li, K, Rb and Cs. Explain why are there such limitations.

3. Explain why ICP rather than AA has recently been the instrument of choice in routine analysis of elements. What are the major advantages of ICP compared to FAA and GFAA?

REFERENCES

Chapter 9: Atomic Spectroscopy for Metal Analysis (this book)
US EPA (1986), Testing Methods for Evaluating Solid Wastes: Physical/Chemical Methods, EPA SW 846 on-line. (http://www.epa.gov/epaoswer/hazwaste).

EXPERIMENT 11

DETERMINATION OF EXPLOSIVES COMPOUNDS IN A CONTAMINATED SOIL BY HIGH PERFORMANCE LIQUID CHROMATOGRAPHY (HPLC)

Objectives—The objectives of this experiment are (1) to learn the principle and operation of high performance liquid chromatography (HPLC) for the analysis of semivolatile nitroaromatic compounds, (2) to gain an understanding of advantages and disadvantages of HPLC method over gas chromatography method, and (3) to become familiar with the procedure for the extraction of nitroaromatic compounds from contaminated soils.

Background—Nitroaromatic compounds have been used extensively as explosives since early this century. Three nitroaromatic compounds are of interest in this experiment: 2,4,6-trinitrotoluene (TNT), 2,4-dinitrotoluene (24DNT) and 2,6-dinitrotoluene (26DNT). TNT was once the world's most widely used explosive. 2,4-Dinitrotoluene (24DNT) and 2,6-dinitrotoluene (26DNT) are two major isomers of dinitrotoluene. They are the intermediates in the manufacture of TNT and the precursors of toluene diisocyanate used for the manufacture of polyurethane foams. Disposal practices associated with TNT manufacturing during and after World Wars I and II have resulted in an enormous contamination problem at ammunition production and handling facilities in the United States and worldwide. These three compounds are all listed as the priority pollutants by the U.S. Environmental Protection Agency.

Method Principle—HPLC is an analytical technique in which a liquid mobile phase transports a sample through a column containing a liquid stationary phase. The interaction of the sample with the stationary phase selectively retains individual compounds and permits separation of sample components. Detection of the separated sample compounds is achieved mainly through the use of absorbance detectors for organic compounds, and through conductivity and electrochemical detectors for metal and inorganic components.

A common detector in HPLC is photodiode array detector (PAD). The PAD measures the absorbance of a sample from an incident light source (UV-VIS). After passing through the sample cell, the light is directed through a grating device that separates the beam into its component wavelengths reflected on a linear array of photodiodes. This permits the complete absorbance spectrum to be obtained in 1 s or less and simultaneous multi-wavelength analysis.

The PAD is subject to the interference encountered with all absorbance detectors. Of special concern for HPLC is the masking of the absorbance region of the HPLC mobile phase and its additives. This may reduce the range and sensitivity of the detector to the sample components. Most interferences occur in monitoring the shorter wavelengths (200–230 nm). In this region, many organic compounds absorb light and can be sources of interference.

Time Required—Given a prior developed HPLC method, a 3-h session should be sufficient including sample extraction, instructor's demonstration on HPLC instrument, and running samples by students. Plan to reserve 15 min per sample. Hence, it would need a total of 2 h to run 8 samples (2 per group with a total of 4 student group). The instructor should also prerun the calibration standard solutions and make it available to students for their data analysis.

Apparatus and Reagents

Apparatus

(a) HPLC equipped with a UV detector and a auto-sampler (Hewlett-Packard series 1050 or equivalent), (b) Nova Pak ^{R} C_{18} column (4 μM, 2.1 mm × 150 mm), (c) analytical balance, (d) microcentrifuge, (e) micrcocentrifuge tube with 0.22 μm filter unit, (f) 200 μL Finnipipette pipettes, (g) 10 μL microsyringe if manual injection is used (Point type 3, no sharp tip syringe for HPLC sample injection!).

Reagents

(a) Mixed standard solutions containing 2,4,6-trinitrotoluene, 2,4-dinitrotoluene and 2,6-dinitrotoluene (0.1140 g TNT, 0.0911 g 24DNT and 0.0911 g 26DNT in 50 mL methanol; 20 mM each). TNT, 2,4-DNT and 2,6-DNT can be obtained from Chem Services (West Chester, PA), (b) Helium gas (for degassing mobile phases), (c) Mobile phase: 95% water (in 0.1% trichloroacetic acid; Bottle A) and 5% isopropanol (in 0.05% trichloroacetic acid; Bottle B), (d) HPLC grade acetonitrile and methanol, (e) A contaminated soil sample with unknown amount of TNT, 24DNT and 26DNT, or a clean soil spiked with TNT, 24DNT, and 26DNT.

Procedures

1. Preparation of soil extract
    (a) Using analytical balance, weigh approximately 0.05 g soil sample to the nearest 0.0001 g in a microcentrifuge tube. Place soil sample directly in the insert filter unit of the centrifuge tube. Label the centrifuge tube and record the weight of soil (w).
    (b) Pipet 200 μL HPLC grade acetonitrile into the filter unit and let it stand 5 minute for extraction.
    (c) Centrifuge for 2 min at approximately 10,000 rpm.
    (d) Repeat Step 1b and 1c twice. At the end, the final volume is 600 μL (record this volume v in mL). Discard the filter unit containing extracted soil. The soil extract is now ready for HPLC analysis.
2. HPLC operational procedures
    (a) Start-up procedure
        (i) Turn on the helium gas, perform degassing (not for the model with online vacuum degassing unit), purging, and priming of the pump if necessary. Make sure there are no bubbles in the solvent lines.
        (ii) Using predetermined ratio of mobile phase established in the method, continuously pump for at least 10 min until there is no significant noise on the baseline. The pressure should be stable but remain approximately above 300 bar.
        (iii) Load prior developed method file and edit the method parameters if needed. For the three compounds tested in this experiment, the method parameters include flow rate 0.7 mL/min, mobile phase ratio: water in Bottle A (95%) and isopropanol in Bottle B (5%), Select run time 15 minutes, Select wavelength $\lambda = 254$ nm, Select Initial Area Reject 300 (to reject small noise peaks), Save the method if changes have been made.

    (iv) Turn on the UV detector and continue pumping until stable pressure and good baseline are achieved. At the end, the instrument should show READY for both pump and UV detector.

(b) Running analyses

    (i) Make an injection using either a manual mode with a syringe or load sample vials in the autosample tray. In the operating program, input sample information and the location of sample vials.

    (ii) At the completion of the run, the system sends a printout to the printer, record the area from the printout.

    (iii) Repeat the injection of all other samples (soil extract). Use 10 μL sample for injection.

(c) Shut-down procedure: Follow the instrument manual to turn off both pump and UV detector, exit HPLC ChemStation, and turn off helium gas.

3. Data analysis: The concentrations reported in the printout will be either mg/L or mM depending on the unit used for the calibration curve. However, the concentration for the soil sample must be reported in mg/kg. Assume mM was used to express the concentration; the conversion are as follows:

(a) Convert mM into mg/L soil extract (Molecular weight of TNT, 24DNT and 26DNT are 228, 182, and 182, respectively)

(b) Convert mg/L soil extract into mg/kg soil (consider the volume of total soil extract $v = 600$ μL and the weight of soil w in grams).

Safety and Health Issues—The semivolatile nitroaromatic compounds are toxic but are not readily inhalable, avoid skin contact by always wearing glove. Minimize exposure to vapor of various solvents used during extraction and HPLC analysis.

Post Lab Assignment

1. The biodegradation of TNT and DNT under anaerobic conditions results in the formation of a series of hydroxylamino intermediates, which are more polar in nature than their parent compounds. In the reverse phase HPLC, what will be the retention time of these intermediates compared to their respective parent compounds?

2. The two isomers, 2,4-DNT and 2,6-DNT, have close retention time in C_8 or C_{18} columns. However, their separation can be readily achieved by using a Hypercarb column (Thermo Electron Inc.). Hypercarb consists of fully porous spherical carbon particles comprised of flat sheets of hexagonally arranged carbon atoms. The carbon atoms have a fully satisfied valence and offer completely different retention and selectivity to silica-based C_8 or C_{18} column. Explain.

3. The extraction method using acetonitrile is not a standard method for semivolatile compounds, but has been shown to be efficient for nitroaromatic compound in routine use. Suggest a verification method against which this method is reliable.

4. Give a list of HPLC-based U.S. EPA methods in the SW-846. Explain the possible reasons why there are only a few HPLC methods established by the U.S. EPA as compared to many available GC-based methods.

REFERENCES

Chapter 10: Chromatographic Methods for Environmental Analysis (this book)

ZHANG C, HUGHES JB, NISHINO SF SPAIN JC (2000), Slurry-phase biological treatment of 2,4- dinitrotoluene and 2,6-dinitrotoluene: Role of bioaugmentation and effects of high dinitrotoluene concentrations, *Environ. Sci. Technol.*, 34(13):2810–2816.

EXPERIMENT 12

MEASUREMENT OF HEADSPACE CHLOROETHYLENE BY GAS CHROMATOGRAPHY WITH FLAME IONIZATION DETECTOR (GC-FID)

Objectives—The objectives of this experiment are (1) to learn the use of gas chromatography with flame ionization detector (GC–FID) and its utility as a universal GC detector for carbon-containing chemicals, (2) to learn the use of static headspace extraction as an extraction method for volatile organic compounds, and (3) to learn the techniques in handling gaseous samples.

Background—The classical liquid–liquid extraction cannot be used for the extraction of any volatile organic compounds. Two common methods for VOCs are the static headspace extraction (SHE) and the dynamic headspace extraction or purge-and-trap. In this experiment, static headspace extraction of three chlorinated aliphatic hydrocarbons (CAHs) are demonstrated. They are tetrachloroethylene (PCE), trichloroethylene (TCE), and 1,2-*cis*-dichloroethylene (DCE). The chemical structure and properties of three test chemicals are given in Table E-11.

Because of the volatile nature of these three compounds, the vapor phase can be extracted and analyzed by FID. Flame ionization detector is one of the most widely used and generally applicable detectors for gas chromatography and, hence, is used for routine and general-purpose analysis. It is easy to use but is destructive of the sample. FIDs consist of a hydrogen/air flame and a collector plate. The construction of a typical ionization detector is shown in Chapter 10.

Table E-11 Properties of chlorinated solvents used in this experiment

Contaminant	Chemical structure	Molecular weight (g/mol)	Density (g/cm^3)	Solubility (mg/L)	log K_{ow}	Henry's constant (dimensionless)
1,2-*cis*-Dichloroethylene	ClCH=CHCl	96.94	1.28	3500	1.86	0.14
Trichloroethylene	Cl$_2$C=CHCl	131.39	1.46	1100	2.42	0.33
Tetrachloroethylene	Cl$_2$C=CCl$_2$	165.83	1.62	150	2.88	0.61

Densities measured 15.5–22 °C; Solubility, Henry's constants, and K_{ow} measured at 25 °C.

Method Principle—The static headspace extraction can be used to determine VOC concentration in the aqueous sample (C_o) by measuring only the gas phase VOC (C_g) in equilibrium with aqueous phase VOC. The underlying equation used to calculate C_o from C_g has been described in Chapter 7 (Eq. 7.21). For this calculation, the Henry's law constant (K) of the VOC at a specified temperature, and the phase ratio (β), that is, the ratio of the gas phase volume (V_g) to the liquid phase (sample) volume (V_s), must be known. The above equation also indicates that if the Henry's law constant is too small (nonvolatile compounds), the gas phase concentration (C_g) will likely be too low to be detectable.

Gas samples from the headspace are withdrawn and injected into GC–FID with a capillary column. The effluent from the column is mixed with hydrogen and air and then ignited electrically at a small metal jet. Most organic compounds produce ions and electrons that can conduct electricity through the flame. There is an electrode above the flame to collect the ions formed at a hydrogen/air flame. The number of ions hitting the collector is measured and a signal is generated.

In FID, the organic molecules undergo a series of reactions including thermal fragmentation, chemi-ionization, ion molecule, and free radical reactions to produce charged-species. The amount of ions produced is roughly proportional to the number of reduced carbon atoms present in the flame, and hence the number of molecules. Functional group, such as carbonyl, alcohol, halogen, and amine, yield fewer ions or none at all in a flame. In addition, the detector is insensitive toward noncombustible gases such as H_2O, CO_2, SO_2, and NO_x. FID is sensitive to compounds with C–H bonds. Some non-hydrogen containing organics (e.g., hexachlorobenzene) can result in a poor response.

Time Required—Considerable time will be needed to accurately prepare calibration standards used in this study and the static headspace extraction will typically take 24 h to reach equilibrium and stabilization. If these have been prepared in advance by the instructor, the demonstration and sample analysis with GC–FID can be accomplished within 3 h. Alternatively, this experiment can be combined with Experiment 13. The first 3-h session is used for preparation, and the second 3-h session for running samples simultaneously with FID and ECD.

Apparatus and Reagents

Apparatus

(a) GC with FID (HP 6890 series or equivalent), (b) HP-5 capillary column (30 m × 0.32 mm × 0.25 μm), (c) glass vials (volume accurately predetermined) with Teflon-lined septa and crimp aluminum caps, (d) gas-tight syringe, (e) shaker.

Reagents

(a) PCE (HPLC grade), (b) TCE (reagent-grade), (c) *cis*-DCE (reagent grade), (d) methanol.

Procedures

1. Preparation of calibration standard: A series of standards are prepared by completely filling vials with known concentrations of stock solutions of PCE, TCE, or DCE dissolved in methanol (available from AccuStandard). The vials are sealed with Teflon-lined septa and crimp aluminum caps allowing for zero headspace. An example is given below in Table E-12. The mass of each analyte is calculated by the known volume and the density given in Table E-11.

2. Static headspace extraction of volatile chlorinated solvents: Glass vials with a predetermined volume (e.g., 71.59 ± 0.48 mL, measured from 10 randomly selected vials) are used for static headspace extraction. Each vial is filled with 20 mL of aqueous solution. The vials are then sealed airtight with Teflon-lined septa and crimp aluminum caps. After the vials are sealed, a predetermined amount of stock solution (PCE, TCE, and DCE dissolved in methanol) is injected through a syringe. The amount of methanol in each vial was less than 0.5% (v/v). At such low concentrations, methanol does not affect vapor–liquid partitioning. The vials are then placed in a shaker to equilibrate for 24 h at room temperature. After equilibration, the vials are

Table E-12 Preparation of calibration standard for PCE, TCE, and DCE

Compound	Volume (μL)	Methanol (mL)	Mass (μg)	Conc. (mg/L)
PCE	10	26	16.2	623
	20	26	32.4	1246
	80	26	129.6	4985
TCE	5	26	7.3	281
	15	26	21.9	842
	45	26	65.7	2527
DCE	80	14	102.4	7314
	160	14	204.8	14629
	240	14	307.2	21943

removed from the shaker and further stabilized for several hours. To minimize potential volatilization losses, the vials are incubated in an inverted position, with the liquid phase in contact with septa.

3. GC–FID analysis of calibration standard and headspace samples

   (a) Turn on the hydrogen, air, and helium, and check the pressure of each gas. If the pressures are below 500 psi, you should have it changed. Turn on the GC instrument.

   (b) Turn on the computer and start the software. The default method then will be loaded automatically to GC. A "pop" sound should be heard, which means the detector has been ignited.

   (c) Load the prior developed GC–FID method file, and modify it if needed. Several important parameters suitable for the test compounds of this experiment include: helium pressure: 2.1; H_2 flow: 60; air flow: 450; front inlet temp: 200°C; detector temp: 275°C; initial oven temp: 40°C; hold time 3 min, program rate: 10°C/min; final temp: 100°C, hold time 0 min. Wait until the GC shows ready for sample injection.

   (d) For the liquid standard solutions contained in headspace-free vials, use 10 μL syringe to withdraw 4 μL samples for manual injection. The peak area and height of the signal will be integrated by the computer automatically and will be printed out. These will be used to determine a calibration curve.

   (e) For the unknown samples contained in headspace, use a gas-tight syringe to withdraw 400 μL of the head space gas sample. Inject the gas into the inlet as soon as possible. You should always inject each sample consistently and completely; if there is any gas left in the syringe, there will be a significant experimental error.

   (f) When you finish the experiment, just close the software, and turn off hydrogen and air. Leave helium gas flow at a smaller flow rate to preserve the column.

4. Data analysis

   (a) Plot separate calibration curve for PCE, TCE, and DCE (FID signals vs. μg)

   (b) Determine the gas phase concentration (C_g) of PCE, TCE, and DCE on the basis of calibration curve for each compound

   (c) Calculate the liquid phase concentrations of PCE, TCE, and DCE in the unknown sample using Eq. 7.20. The gas to liquid volume ratio: $\beta = (71.59\text{-}20)/20 = 2.58$, Henry's constant (K) is shown in Table E-11.

Safety and Health Issues—PCE, TCE, and DCE are very volatile toxic chemicals. Use with care and work under a ventilation hood. The hydrogen gas used for FID is flammable and can be explosive. Use leak detection device for potential leaks of hydrogen gas. Follow the safety guidelines regarding the use and transportation of high pressure gas cylinder.

Post Lab Assignment

1. The boiling points of PCE, TCE, and DCE are known to be 121°C, 87°C, and 60°C. Given this information, can you suggest the approximate temperature range for injection port, column, and detector? Can you predict the elution sequence of the three compounds?

2. Derive an equation from Eq. 7.20 so that the Henry's law constant of a volatile compound can be determined from a headspace analysis. Suggest an experimental procedure.

REFERENCES

Chapter 7: Fundamentals of Sample Preparation for Environmental Analysis (this book)

Chapter 10: Chromatographic Methods for Environmental Analysis (this book)

HEMOND HF, FECHNER-LEVY EJ (2000), Chemical Fate and Transport in the Environment. 2nd edition, Academic Press: San Diego, CA.

HOWARD PH, MEYLAR WM (1997), Handbook of Physical Properties of Organic Chemicals. CRC Lewis Publishers: Boca Raton, FL.

GOSSETT JM (1987), Measurement of Henry's law constant for C_1 and C_2 chlorinated hydrocarbons, *Environ. Sci. Technol.* 21(2):202–208.

ZHANG C, ZHENG G, NICHOLS CM (2006), Micellar partitioning and its effects on Henry's law constants of chlorinated solvents in anionic and nonionic surfactant solutions, *Environ. Sci. Technol.*, 40(1):208–214.

EXPERIMENT 13

DETERMINATION OF CHLOROETHYLENE BY GAS CHROMATOGRAPHY WITH ELECTRON CAPTURE DETECTOR (GC-ECD)

Objectives—The objectives of this experiment are (1) to learn the use of gas chromatography with an electron capture detector (GC–ECD), and (2) to understand the utility of GC–ECD as a selective and highly sensitive GC detector for halogenated compounds as compared to the universal but less sensitive FID detector.

Background—The electron capture detector was the first selective detector to be invented for gas chromatography since the 1950s. It still remains a very successful detector among many GC detectors commercially available today. Its use in the detection of DDT has revolutionized our understanding of the worldwide distribution of this once successfully used pesticide, which subsequently led to a regulation to ban the production of this pesticide worldwide.

The high sensitivity of DDT to ECD is because of the chlorinated nature of this pesticide. DDT and its metabolites are chlorinated aromatic compounds

(Appendix B). In general, the more halogenated the compound is, the more sensitive it will be by the ECD method. In this experiment, we use the same chlorinated aliphatic compounds that were prepared and tested by FID in Experiment 12. The extent of chlorination is in an increasing order of DCE, TCE and PCE. Comparisons can be made for the sensitivity between ECD and FID and between chlorinated hydrocarbons of various degrees of chlorination.

Method Principle—An electron capture detector (ECD) has a radioactive source (mostly ^{63}Ni) that emits electrons. The electrons from the emitter cause ionization of the carrier gas (often nitrogen) and the production of a burst of electrons. In the absence of organic species, a constant standing current between a pair of electrodes results from this ionization process. However, when organic molecules from the column effluent flow into the detector, the electronegative functional groups of such molecules (halogens, phosphorous, and nitro groups) will capture some of the electrons. As a result, the currents measured between the electrodes are reduced.

The ECD response is nonlinear unless the potential across the detector is pulsed. Although it is more sensitive than FID, it has a limited dynamic range and finds its greatest application only in analysis of halogenated compounds. The ECD is extremely sensitive to molecules containing highly electronegative functional groups such as halogens, peroxides, quinones, and nitro groups. It is therefore a popular detector for trace level determinations of chlorinated insecticides and halocarbon residues in environmental samples. But it is insensitive toward functional groups such as amines, alcohols, and hydrocarbons.

Time Required—This GC–ECD experiment can be combined with GC–FID experiment (refer to Experiment 12) either in a 3-h session or two 3-h sessions.

Apparatus and Reagents—GC equipped with ECD detector. Refer to Experiment 12 for other required apparatus and chemical reagents.

Procedures

1. Preparation of calibration standard: See Experiment 12.
2. Static headspace extraction of volatile chlorinated solvents: See Experiment 12.
3. GC–ECD analysis of calibration standard and headspace samples
   (a) Make sure helium gas is on and the remaining pressure is above 500 psi. Turn on the GC instrument, computer and start the software.
   (b) Load the GC–FID method and modify it if needed. Several important parameters suitable for the test compounds of this experiment include: oven temp: 40°C, hold 2 min; ramp: 15°C/min; final temp: 150°C, hold 2 min; ramp: 10°C/min, final temp: 250°C, hold 6 min; injector temp: 200°C; detector temp: 275°C. Wait until the GC shows ready for sample injection.

(c) Refer to Experiment 12 for the injection of liquid standard solutions and the gaseous headspace samples. Hold the syringe vertically and inject as quick as possible. When sample run is completed, the computer will automatically print out the peak area and peak height for your quantitation. *Unlike FID, you should NOT inject aqueous sample or any sample containing water! It may damage the column and ECD detector.*

(d) Reduce the helium gas flow rate but do not turn it off. Turn off the power only if you are not planning to use the GC for a long time.

Safety and Health Issues—PCE, TCE, and DCE are very volatile toxic chemicals. Use with care and work under a ventilation hood. The radioactive source is shielded inside the detector, which should not be of concern. When the detector service is required, however, only licensed service engineer can perform the work.

Post Assignment

1. Explain why ECD is in an increasing sensitivity of DCE < TCE < PCE.

2. From the calibration curve of PCE, TCE, and DCE, calculate the calibration sensitivity, which is defined as the slope of the calibration curve. Compare these with GC-FID.

REFERENCES

Chapter 7: Fundamentals of Sample Preparation for Environmental Analysis (this book).

Chapter 10: Chromatographic Methods for Environmental Analysis (this book).

ZHANG C, ZHENG G, NICHOLS CM (2006), Micellar partitioning and its effects on Henry's law constants of chlorinated solvents in anionic and nonionic surfactant solutions, *Environ. Sci. Technol.*, 40(1): 208–214.

EXPERIMENT 14

USE OF ION SELECTIVE ELECTRODE TO DETERMINE TRACE LEVEL OF FLUORIDE IN DRINKING AND NATURAL WATER

Objectives—The objectives of this experiment are: (1) to learn the principles and operation of electrochemical method using ion selective electrode (ISE), (2) to learn the standard addition method for quantitation, and (3) to understand the unique features of electrochemical methods and particularly the advantage of ISEs vs. other instrumental methods.

Background—Fluoride is an essential as well as a toxic element. Various forms of fluoride compounds in the environment have both natural and anthropogenic origins. Natural sources include the weathering of rocks and atmospheric emissions from volcanoes and seawater. Human activities related to the release of fluorides are mainly the mining and processing of phosphate rock, its use as agricultural fertilizer, and the manufacture of aluminum (Al). Other fluoride sources include the combustion of coal, other manufacturing processes of steel, copper, nickel, glass, brick, ceramic, use of fluoride-containing pesticides in agriculture, and addition of fluoride in drinking water supplies.

Fluorides can exist in various chemical forms. In atmosphere, fluorides may exist as gaseous compound or in particulates. Fluoride is usually transported through the water cycle in its complexed compound with Al. In soils, complexes of Al and Ca are the most predominant. Concentrations of fluorides in the environment also vary depending on the location and proximity to human sources. Reported concentrations in surface water are from 0.01 to 0.3 mg/L. Seawater contains more fluoride than fresh water, with concentrations ranging from 1.2 to 1.5 mg/L. In areas where the natural rock is rich in fluoride or where there is geothermal or volcanic activity, very high fluoride levels, up to 50 mg/L, may be found in groundwater or hot springs. In ambient air where emission sources are absent, the mean concentrations of fluoride are generally less than $0.1 \ \mu g/m^3$. In most soils, fluoride is present at concentrations ranging from 20 to 1000 mg/kg. A concentration level of several thousand mg/kg can be reached if mineral soils with natural phosphate or fluoride deposits are concerned.

The ISE method for the measurement of fluoride is one of the very successful applications of ISEs. Other electrodes have been commercially available for the in-situ measurement of a variety of environmental parameters. Operated under the similar electrochemical principles, they use relatively inexpensive devices, are fast to respond, and are unaffected by turbidity or color in the sample. Because the response is logarithmic in nature, the precision (up to the limit of detection) remains constant over the entire measuring range. This means that a fluoride measurement at 0.1 mg/L will have the same precision as a measurement at 1000 mg/L. Because of these advantages, standard U.S. EPA methods have been developed.

Method Principle—The measurement of fluoride is based on the linear relationship between potential (E) and the logarithmic form of the activity of fluoride ion in solution ($\log a_{F^-}$). Refer to Eq. 11.12 for a mathematical description of such relationship. The electrochemical device contains an internal reference electrode, an internal fluoride standard, the LaF_3 ion exchange crystal, and a potential measuring meter (almost all modern pH meters have the expended millivolt scale). An external reference electrode must be used to perform the measurement. Assuming that a saturated calomel electrode (SCE) is used as the external reference, the potentiometric cell may be represented as:

$$Ag|AgCl, KCl|Cl^- (0.3 \, M), F^- (0.001 \, M)|LaF_3|\text{test solution}||SCE$$

The fluoride selective electrode produces a potential across a LaF_3-containing solid ion exchange membrane. LaF_3 exhibits high and selective affinity toward F^-. With the exception of OH^-, it does not interact with other substances.

Keep in mind that the addition of a total ionic-strength adjustment buffer (TISAB) into all samples is critical for an accurate measurement of fluoride concentrations. This solution of mixed compounds serves several purposes. First, TISAB is to keep ionic strength constant to minimize the variations between standard and samples. The components of acetic acid buffer keep the solution pH between 5.3 and 5.5. At low pH, a weak acid HF can be formed which will affect the concentration of ionic fluoride (F^-) in the solution. This pH range will also minimize the interfering OH^-. Since fluoride forms complexes with many polyvalent cation species found in natural and processed waters (such as Si^{4+}, Fe^{3+}, and Al^{3+}) and the fluoride electrode only respond to ionic F^-, a strong complexing agent (cyclohexylenedinitrilotetraacetic acid) is added to prevent the formation of fluoride complexes with Si^{4+}, Fe^{3+}, and Al^{3+}.

Time Required—The time required for this experiment is 2.5 h. It can be expedited if each student group is provided a set of electrode and pH meter.

Apparatus and Reagents

Apparatus

(a) Fluoride electrode, (b) Calomel reference electrode or Ag/AgCl reference electrode, (c) digital pH meter, (d) magnetic stirrer with Teflon-coated stirring bar, (e) combination pH electrode, (f) stop watch (timer), (g) 50-mL plastic beakers.

Reagents

(a) 1000 mg/L F^- stock solution is prepared by adding 2.211 g NaF to water and diluting to 1.00 L in a volumetric flask, (b) Work solution containing 10 mg/L F^- (1 mL = 10 μg); dilute stock solution by 100 times, (c) TISAB solution: 57 mL glacial acetic acid, 58 g NaCl, 4 g CDTA (cyclohexylenedinitrilo-tetraacetic acid) in about 500 mL water, using combination electrode, adjusted pH to 5.0–5.5 with 5 M MaOH and diluted to a total volume of 1 L. This buffer is also available commercially, (d) pH standard solutions, (e) unknown samples (a tap water and a river water).

Procedures

1. Preparation of calibration standards according to Table E-13.
2. Measurement of calibration standards and samples
   (a) Pipet 25 mL standard solutions and add equal volume (25 mL) of TISAB solution (i.e., a 1:1 dilution for all standards and samples).

Table E-13 Preparation of fluoride calibration standards

No.	1	2	3	4	5	6	7	8
mL F⁻ (1 mL = 10 µg)	0	1	2	3	4	5	7.5	10
µg F⁻	0	10	20	30	40	50	75	100
mL DI H_2O	50	49	48	47	46	45	42.5	40
mg/L F⁻	0	0.2	0.4	0.6	0.8	1.0	1.5	2.0

(b) Set up the pH meter by connecting the fluoride selective electrode and reference electrode to the plugs at the rear of the instrument.

(c) Insert electrodes in solution, place beaker on magnetic stirrer and mix at medium speed. Obtain potential readings in millivolts (mV). Make sure that electrodes must remain in solution for at least 3 min until the potential readings have stopped drifting.

(d) Obtain the mV readings for all eight standard solutions. Repeat this for a tap water and a river water sample after 1:1 dilution with TISAB.

(e) Data analysis: Plot log µg of F⁻ against mV using Excel or other spreadsheet programs. This should be a linear calibration curve. Obtain the µg F⁻ for the unknown samples, and calculate the concentrations as: mg F⁻ = µg F⁻/V, where V is the sample volume (25 mL).

3. Measurement by standard addition method (an alternative to the standard calibration curve method described in Step 2.

(a) For the unknown sample, dilute with equal volume of TISAB solution. Measure the millivolts (E_1).

(b) Add a small volume of the F⁻ stock solution in the above measured sample. After mixing, obtain the millivolts (E_2). The change in the potentials before and after the standard addition is $\Delta E = E_2 - E_1$.

(c) Calculation of F⁻ concentration in the unknown sample (See Post Lab Assignment for the needed formula).

Safety and Health Issues—Sodium fluoride is very toxic and potentially carcinogenic. Care should also be exercised when handling solid sodium hydroxide and its solution with very high pH. Use these chemicals with care and avoid skin contact.

Post Lab Assignment

1. List the common water quality parameters that can be measured by ISEs in the field. List the pros and cons of the electrode probe methods in environmental analysis as compared to other instrumental methods.

2. What are the components in TISAB? Why must the buffer pH be kept 5.3–5.5, and why is CDTA present for fluoride measurement?

3. On the basis of the Nernst equation, derive an equation for the standard addition method described in the experimental procedure (Step 3). The derived formula for the calculation of F^- concentration in unknown samples should be as follows:

$$C = \frac{\Delta C}{anti \ \log \dfrac{\Delta E}{s} - 1} \tag{E-7}$$

where C is concentration of unknown sample, ΔC is the concentration of standard added to the sample, ΔE is the change in potential after the addition of standard, s is a temperature-dependent constant which is equal to 59 mV at 25°C.

4. For a 50-mL fluoride containing water sample, electrode potential was 95.3 mV after pH adjustment and TISAB. The same solution was added with 0.5 mg standard fluoride solution (neglect volume change), the potential was measured to be 55.3 mV, what is the concentration of F in mg/L?

5. Compare the results obtained by two methods for the concentrations of unknown samples, standard calibration method vs. standard addition method.

REFERENCES

Chapter 11: Electrochemical Methods for Environmental Analysis (this book)

ANDELMAN, JB (1968), Ion-selective electrodes - Theory and applications in water analysis, *JWPCF*, 40(11):1844.

FRANT MS (1974), Detecting pollutants with chemical-sensing electrodes, *Environ. Sci. Tech.*, 8(3): 224–228.

KISSA E (1983), Determination of fluoride at low concentrations with the ion-selective electrode, *Anal. Chem.*, 55(8):1445–1448.

RISEMAN JM (1969), Measurement of inorganic water pollutants by specific ion electrode, *Am. Lab.*, 1(7):32.

EXPERIMENT 15

IDENTIFICATION OF BTEX AND CHLOROBENZENE COMPOUNDS BY GAS CHROMATOGRAPHY-MASS SPECTROMETRY (GC–MS)

Objectives—The objectives of this experiment are to learn (1) the principle and operation of gas chromatography-mass spectrometry (GC–MS), and (2) the general GC–MS techniques for the qualitative (identification) and quantitative analysis of environmental contaminants.

Table E-14 Benzene-series hydrocarbons

Name	Formula	Mp (°C)	Bp (°C)	Sp. Gr.
Benzene	C_6H_6	5.51	80.09	0.879
Toluene	$C_6H_5 \cdot CH_3$	−95	110.8	0.866
o-Xylene	$C_6H_4 \cdot (CH_3)_2$	−29	144	0.875
m-Xylene	$C_6H_4 \cdot (CH_3)_2$	−53.6	138.8	0.864
p-Xylene	$C_6H_4 \cdot (CH_3)_2$	13.2	138.5	0.861
Ethylbenzene	$C_6H_5 \cdot C_2H_5$	−93.9	136.15	0.867

Background—Two series of homologous aromatic hydrocarbons are analyzed in this experiment: the benzene series and the chlorinated benzene series. The benzene series compounds are made up of alkyl substitution products of benzene. They are found along with benzene in coal tar and in many crude petroleums. Table E-14 lists the benzene-series hydrocarbons of commercial importance. Toluene, or methyl-benzene, is the simplest alkyl derivative of benzene. Xylene is a dimethyl derivative of benzene, which exists in three isomeric forms: ortho-xylene, meta-xylene, and para-xylene. All are isomers with ethylbenzene as they have the same general formula, C_8H_{10}. Together with benzene, these compounds are commonly referred to as the BTEX group.

The benzene-series hydrocarbons are used extensively as solvents and in chemical synthesis, and are common constituents of petroleum products (e.g., gasoline). Although they are relatively insoluble in water, wastewaters and leachates containing 10 to as high as 1000 mg/L for the BTEXs have been observed. These compounds are frequently detected in groundwaters, with a major source being leaking underground gasoline storage tanks.

The benzene-series compounds have been implicated in several human health effects, most notably cancer. Benzene is known to cause leukemia. The current drinking water MCLs are 5 µg/L for benzene, 700 µg/L for ethylbenzene, 1 mg/L for toluene, and 10 mg/L for the sum total of the xylenes.

Chlorinated benzenes are benzenes with one or more of the hydrogens replaced with chlorine. They are widely used industrial chemicals that have solvent and pesticide properties. Like the chlorinated aliphatics, they have been found at abandoned waste sites and in many wastewaters and leachates. They are fairly volatile, slightly to moderately soluble, and hydrophobic.

Chlorobenzene (monochlorobenzene) has a drinking water MCL of 100 µg/L. There are three dichlorobenzene isomers: 1,2-dichloribenzene (ortho isomer), 1,3-dichlorobenzene (meta isomer), and 1,4-dichlorobenzene (para isomer). The current MCLs are 75 µg/L for p-dichlorobenzene and 600 µg/L for o-dichlorobenzene; the meta isomer is not regulated at this time.

Method Principle—In gas chromatography a mobile phase (a carrier gas) and a stationary phase (column packing or capillary column coating) are used to separate individual compounds. The carrier gas is nitrogen, argon-methane, helium, or hydrogen. For packed columns, the stationary phase is a liquid that has been coated on an inert granular solid, called the column packing, which is held in borosilicate glass tubing. The column is installed in an oven with the inlet attached to a heated injector block and the outlet attached to a detector. Precise and constant temperature control of the injector block, oven, and detector is maintained. Stationary-phase material and concentration, column length and diameter, oven temperature, carrier-gas flow, and detector type are the controlled variables. When the sample solution is introduced into the column, the organic compounds are vaporized and moved through the column by the carrier gas. They travel through the column at different rates, depending on differences in partition coefficients between the mobile and stationary phases.

Gas chromatography is often coupled with the selective techniques of spectroscopy and electrochemistry, the so-called hyphenated methods. One such technique is the direct interface with rapid-scan mass spectrometers of various types. The flow rate from capillary column is generally low enough so that the column can be fed directly into the ionization chamber of the mass spectrometer. The most commonly used mass spectrometer is quadrupole-based detector, in which ions for mass analysis are produced by electron impact. Sample from GC is brought to a temperature high enough to produce a molecular vapor, which is then ionized by bombarding the resulting molecules with a beam of energetic electrons. The complex mass spectra that result from the electron-impact ionization are then used for quantification (compound identification) and quantitation.

Time Required—The instructor needs approximately an hour to introduce the GC–MS component, software, the operational parameters for a good separation, demonstration of calibrating (tuning) GC–MS, running sample, and performing the library search. The remaining 2.5 h will allow students to get on the software, familiarize the method parameters, load samples, run sample, and interpret fragmentation patterns from the acquired mass spectrum.

Apparatus and Reagents

Apparatus

(a) GC-MS (Hewlett-Packard Gas Chromatography 5890 Series II with Auto 7673 GC/SFC Injector with MS Selective Detector 5972 or equivalent), (b) Capillary column: HP-5MS 30 (m) $\times$ 0.25 (mm) $\times$ 0.25 (μm) film thickness, (c) 10 μL Microsyringe.

Reagents

(a) Certified mixed standard solution (2000 μg/mL in methanol) from AccuStandard Inc. (M-8020-10 X) containing the following 10 chemicals: benzene, chlorobenzene, 1,2-dichlorobenzene, 1,3-dichlorobenzene, 1,4-dichlorobenzene, ethylbenzene, toluene, o-xylene, m-xylene, and p-xylene, (b) Helium gas cylinder (ultra purity), (c) Methanol, HPLC grade.

Procedures

1. A checklist before the start-up of GC–MS: (a) Carrier gas (helium) outlet pressure adjusted to 60 psi. (b) Periodically change the septum on the column inlet. (c) Mount the auto injector and turn on the controller, (d) Turn on the GC power (the MS power should be on all the time). (e) Turn on the computer and start up instrument program (e.g., ChemStation software).

2. Load the Method file and modify it if needed. The default parameters for the BTEX series compounds are as follows. Your method parameters may be slightly different depending on the specific GC–MS system and the column used. Save the method with a new method name if any modification has been made.

    (a) Acquisition mode: "Scan" not "SIM"

    (b) Solvent delay: 2 min. This is the time to turn on the mass spectrometer after the start of the run. This delay time will allow the solvent to elute without being ionized for the protection of the filament in the MS detector.

    (c) Scan range: The stating and ending mass to be scanned.

    (d) Inlet (Injector) temperature: 180°C; Oven (GC-column) temperature: Initial 40°C for 3 minutes, then increase at a rate of 10°C/min to reach final temperature of 150°C. Hold this temperature for 1 min.

    (e) Injection mode: GC ALS (auto liquid sampler). If you want to do manual injection without the use of auto sampler, select Manual.

    (f) Report format: Detailed report type. Select "Total" for TIC plot not "Extracted"

    (g) Output destination: Printer. You will get a printout automatically once the analytical run is finished.

3. Calibration of GC–MS: Follow instruction to perform an autotune. At the completion of the autotune, the system sends a report to the printer. Review the Tune Report to make sure the GC–MS is in good operational condition.

4. Data acquisition by auto or manual injection of samples

    (a) Load sample vials in sample tray and set up a sequence file. Place sample vials in the corresponding positions in the sample tray. Make sure you have sufficient wash solvent (methanol) in the vials located in the injector.

    (b) Save the sequence with a proper name.

    (c) Start running samples by activating the sample sequence table in 2.

5. While it is running

    (a) Monitor a run in progress by examining the real time display of total ion chromatography, or by taking a snapshot by activating the data analysis icon.

(b) You can do library search of those eluted chemical(s) while the sample is still running.

(c) At the completion of data acquisition, the system sends a report to the printer. The report format can be customized in the Method file.

6. After data acquisition is completed: You can retrieve data from Data Analysis Window. Refer to the data file name you set up in the sample log table.

(a) Select the peak of interest and double click the right mouse to get the mass spectrum.

(b) Double click the mass spectrum to do a library search with the right mouse. The software will give you a list of compounds that match the spectrum in the library.

7. Make sure you get a copy of the printed report at the end of this experiment.

8. Exit the software and the GC-MS is now ready to shut down.

(a) Adjust the helium flow to 15 psi (the helium tank is always turned on at 15 psi even when the GC-MS is not being used).

(b) Turn off the GC oven and the temperature program for the detector and injector

(c) Exit the software and log off

(d) Do not turn off the GC power and MS power. The vacuum pump should be on all the time unless it is shut down for maintenance purpose.

Safety and Health Issues—(a) Be aware of the potential hazard when dealing with high pressure gas cylinders. (b) Be alert not to approach the moving arm of the autosampler. (c) Temperature in the oven and injector port can be high and can cause burn. (d) The BTEX series and the chlorinated benzene series compounds are volatile and toxic. Avoid inhalation.

Post Lab Assignment

1. In your lab report, include copies of your spectra and chromatography. Label each peak with the name of compound in the TIC chromatograms on the basis of the library search results. Include a table showing the compound's name, retention time, m/z of the major ion(s) for each compound. For each compound, compare the m/z of major ion(s) with the molecular weight of that compound (Table E-15). Discuss what you have found regarding the patterns of molecular fragmentation in the mass detector. Also compare the mass spectrum of the compound in the sample and the mass spectrum in the library. Are they similar to each other? Are there any differences?

2. From the chromatogram, the retention times and peak width information on the printout, calculate the following and comment on how good are the separation in this experiment:

Table E-15 Molecular formula and molecular weight of test chemicals

Name	CAS #	Formula	Molecular weight
Benzene	71-43-2	C_6H_6	78
Toluene	108-88-3	$C_6H_5 \cdot CH_3$	92
o-Xylene	95-47-6	$C_6H_4 \cdot (CH_3)_2$	106
m-Xylene	108-38-3	$C_6H_4 \cdot (CH_3)_2$	106
p-Xylene	106-42-3	$C_6H_4 \cdot (CH_3)_2$	106
Ethylbenzene	100-41-4	$C_6H_5 \cdot C_2H_5$	106
Chlorobenzene	108-90-7	$C_6H_5 \cdot Cl$	112
1,2-Dichlorobenzene	95-50-1	$C_6H_4 \cdot Cl_2$	146
1,3-Dichlorobenzene	541-73-1	$C_6H_4 \cdot Cl_2$	146
1,4-Dichlorobenzene	106-46-7	$C_6H_4 \cdot Cl_2$	146

    (a) The number of plates (N) from each peak (all chemicals)

    (b) The plate height (H) for the column based on the average number of plates

    (c) The retention factor (k) for each peak

    (d) The resolution between all neighboring peaks

    (e) The separation factor (α) between all neighboring peaks

3. The isomeric forms of xylenes are difficult to separate as demonstrated in this experiment. In order to have a better separation of these isomers, would you choose a column with a more or less polar stationary phase? Why?

4. The structural information obtained from the mass spectrum is complementary to what can be obtained from UV, IR, ^{1}H NMR and ^{13}C NMR. List the structural features of the molecules that each of these techniques can provide.

REFERENCES

Chapter 10: Chromatographic Methods for Environmental Analysis (this book)
Chapter 12: Other Instrumental Methods in Environmental Analysis (this book)

Appendices

APPENDIX A. COMMON ABBREVIATIONS AND ACRONYMS

^{13}H-NMR	Carbon thirteen NMR
^{1}H-NMR	Proton NMR
ACGIH	American Conference of Governmental Industrial Hygienists
ACS	American Chemical Society
AFM	Atomic force microscope
ANOVA	Analysis of variance
AOAC	Association of Official Analytical Chemists
APCI	Atmospheric pressure chemical ionization
APHA	American Public Health Association
APPI	Atmospheric pressure photoionization
ASA	American Society of Agronomy
ASE	Accelerated solvent extraction
ASTM	American Society for Testing and Materials
ASV	Anodic stripping voltammetry
AWWA	American Water Works Association
B/N	Base/neutral fraction
BHC	Hexachlorocyclohexane (HCH)
BOD	Biochemical oxygen demand

Fundamentals of Environmental Sampling and Analysis, by Chunlong Zhang
Copyright © 2007 John Wiley & Sons, Inc.

bp	Boiling point
Bq	Becquerel
BTEX	Benzene, toluene, ethylbenzene, xylene
CAA	Clean Air Act
CCL	Contaminant Candidate List
CCV	Continuing calibration verification
CERCLA	Comprehensive Environmental Response, Compensation, and Liability Act
CFR	Code of Federal Regulations
CI	Confidence Interval/Chemical Ionization
CLP-SOW	Contract Laboratory Program-Statement of Work
COD	Chemical oxygen demand
Coliwasa	Composite liquid waste sampler
CV	Coefficient of variation
CVAA	Cold vapor atomic absorption
CVS	Calibration verification standard
CWA	Clean Water Act
DCE	Dichloroethylene
DDT	1,1,1-Trichloro-2,2-bis(p-chlorophenyl)ethane and dichloro-diphenyl-trichloroethane
df	Degree of freedom
DIR	Dispersive infrared
DO	Dissolved oxygen
DOE	U.S. Department of Energy
DQIs	Data Quality Indicators
DQOs	Data quality objectives
ECD	Electron capture detector
EDTA	Ethylenediaminetetraacetic acid
EI	Electron Impact
EIA	Enzyme immunoassay
ELCD	Electrolytic conductivity detector
emf	Electromotive force (electrode potential)
ESI	Electrospray ionization
eV	Electron volts
EW	Equivalent weight
FAA	Flame atomic absorption
FAB	Fast atom bombardment
FAS	Ferrous ammonium sulfate
FDCA	Food, Drug, and Cosmetic Act
FID	Flame ionization detector
FIFRA	Federal Insecticide, Fungicide, and Rodenticide Act
FIR	Far-infrared
FTIR	Fourier transform infrared
GC	Gas chromatography
GC-MS	Gas chromatography–mass spectrometry
GFAA	Graphite furnace atomic absorption
GLC	Gas–liquid chromatography
GLP	Good laboratory practice
G–M	Geiger–Muller

GPC	Gel permeation chromatography
GPS	Global positioning system
GSC	Gas–solid chromatography
HAPs	Hazardous air pollutants
HAZWOPER	Hazardous Waste Operations and Emergency Response
HCL	Hollow cathode lamp
HDME	Hanging mercury drop electrode
HGAA	Hydride generation atomic absorption
HPLC	High performance liquid chromatography
IA	Immunoassay
IC	Ion chromatography
ICP–AES	Inductively coupled plasma–atomic emission spectroscopy
ICP–MS	Inductively coupled plasma–mass spectrometry
ICP–OES	Inductively coupled plasma–optical emission spectroscopy
ICS	Interference check standard
ICV	Initial calibration verification
IR	Infrared light
ISE	Ion selective electrode
K–D	Kuderna–Danish
LC–MS	Liquid chromatography–mass spectrometry
LCS	Laboratory control samples
LLC	Liquid–liquid chromatography
LLE	Liquid–liquid extraction
LSC	Liquid–solid chromatography
LUST	Leaking underground storage tank
m/z	Mass to charge ratio
MB	Method blank
MBAS	Methylene blue active substances
MCL	Maximum contaminant level
MDL	Method detection limit
MeV	Million electron volt
MFE	Mercury film electrode
MHTs	Maximum holding time
MHz	megahertz
MIR	Mid–infrared
MS/MSDs	Matrix spikes and matrix spike duplicates
MSD	Mass selective detector
MSDS	Material Safety Data Sheet
MS–MS	Tandem mass spectrometry
MTBE	Methyl t–butyl ether
MW	Molecular weight
NDIR	Nondispersive infrared
NEMI	National Environmental Methods Index
NIOSH	National Institute for Occupational Safety and Health
NIR	Near–infrared
NIST	National Institute for Standards and Technology
NMR	Nuclear magnetic resonance
NOAA	National Oceanic and Atmospheric Administration
NPD	Nitrogen–phosphorus detector

NPDES	National Pollutant Discharge Elimination System
NP–HPLC	Normal phase HPLC
NVOCs	Nonvolatile organic compounds
ORP	Oxidation–reduction potential
OSHA	Occupational Safety & Health Administration
P&T	Purge–and–trap
PAHs	Polycyclic aromatic compounds
PARCC	Precision, accuracy, recovery, comparability, and completeness
PBMS	Performance–based Measurement System
PC	Paper chromatography
PCBs	Polychlorinated biphenyls
PCE	Tetrachloroethylene (Perchloroethylene)
PDA	Photodiode array
PE	Performance Evaluation
PFE	Pressured fluid extraction
PID	Photoionization detector
PLOT	Porous layer open-tubular
PM	Particulate matter
PMT	Photomultiplier tube
POP	Persistent organic pollutants
ppb	Parts per billion
PPE	Personal protection equipment
ppm	Parts per million
ppt	Parts per trillion/Parts per thousands
PQL	Practical quantitation limit
PUF	Polyurethane foam
PVC	Polyvinyl chloride
QA	Quality Assurance
QAPP	Quality Assurance Project Plan
QC	Quality Control
RCRA	Resource Conservation and Recovery Act
Redox	Oxidation-reduction
RF	Radio frequency
RPD	Relative percent difference
RP–HPLC	Reverse phase HPLC
RSD	Relative standard deviation
RT	Regulatory threshold
SCE	Saturated calomel electrode
SCF	Supercritical fluid
SCOT	Support coated open tubular
SDWA	Safe Drinking Water Act
SEM	Standard error of mean
SEM	Scanning electron microscope
SFC	Supercritical fluid chromatography
SFE	Supercritical fluid extraction
SHE	Static headspace extraction/Standard hydrogen electrode
SIM	Selected ion monitoring
SOP	Standard operating procedures
SPE	Solid phase extraction

SPME	Solid phase microextraction
SRM	Standard reference materials
SS	Suspended solids
SSSA	Soil Science Society of American
STM	Scanning tunneling microscope
SVOCs	Semivolatile organic compounds
SW-846	Test Methods for Evaluating Solid Waste, Physical/Chemical Methods
TCD	Thermal conductivity detector
TCE	Trichloroethylene
TCLP	Toxicity characteristic leaching procedure
TDS	Total dissolved solids
TIC	Total ion chromatogram
TISB	Total ionic-strength adjustment buffer
TKN	Total Kjeldahl nitrogen
TLC	Thin layer chromatography
TMS	Tetramethylsilane
TOC	Total organic carbon
TOF	Time-of-Flight
TPHs	Total petroleum hydrocarbons
TS	Total solids
TSCA	Toxic Substances Control Act
TSP	Total suspended particle
TSS	Total suspended solids
TVS	Total volatile solids
U.S. EPA	U.S. Environmental Protection Agency
USGS	United States Geological Survey
USP	United States Pharmacopia
UV–VIS	Ultraviolet and visible light
VOCs	Volatile organic compounds
VOST	Volatile Organic Sampling Train
vp	Vapor pressure
WCOT	Wall-coated open-tubular
WEF	Water Environment Federation
XRF	X-ray fluorescence

APPENDIX B. STRUCTURES AND PROPERTIES OF IMPORTANT ORGANIC POLLUTANTS

Chemical Name CAS Registry Number	Structure	Water (PPI List)	Air (HAP List)	VOC/ SVOC	MW	BP (°C)	Solubility (mg/L)
1. Acenaphthene (83-32-9)		√		B/N	154.21	279	3.5–7.4
2. Acenaphthylene (208-96-8)		√		B/N	152.2	280	3.93–16
3. Acrolein (107-02-8)		√	√	VOC	56.1	52.5	20.8% (20°C)
4. Acrylonitrile (107-13-1)		√	√	VOC	53.06	77.4	NA
5. Aldrin (309-00-2)		√		P	364.93	NA	0.01
6. Anthracene (120-12-7)		√		B/N	178.23	340	1.29
7. Aroclor 1016 (12674-11-2)	Cl_{1-10} Cl_2-Cl_3	√		PCB	mixture	NA	0.22–0.25
8. Aroclor 1221 (11104-28-2)	Cl-Cl_2	√		PCB	mixture	NA	0.59 (24°C)
9. Aroclor 1232 (11141-16-5)	C_1-Cl_3	√		PCB	mixture	NA	NA
10. Aroclor 1242 (53469-21-9)	Cl_2-Cl_4	√		PCB	mixture	NA	0.10 (24°C)
11. Aroclor 1248 (12672-29-6)	Cl_3-Cl_5	√		PCB	mixture	NA	NA
12. Aroclor 1254 (11097-69-1)	Cl_4-Cl_6	√		PCB	mixture	NA	0.057 (24°C)

Name (CAS)	Structure			Class	MW	BP	Value
13. Aroclor 1260 (11096-82-5)	Cl_5-Cl_8	√		PCB	mixture	NA	0.080 (24°C)
14. Benz(a)anthracene (56-55-3)		√		B/N	228	437	0.014
15. Benzene (71-43-2)		√	√	VOC	78.11	80.1	1780 (20°C)
16. Benzidine (92-87-5)	H_2N—⟨ ⟩—⟨ ⟩—NH_2	√	√	B/N	184.23	402	400 (12°C)
17. Benzo(a)pyrene (50-32-8)		√		B/N	253.2	495	0.003
18. Benzo(b)fluoranthene (205-99-2)		√		B/N	252	NA	0.0012
19. Benzo(g, h, i)perylene (191-24-2)		√		B/N	276	>500	0.00026
20. Benzo(k)fluoranthene (207-08-9)		√		B/N	252	480	0.00055
21. Benzyl *butyl* phthalate (85-68-7)		√		B/N	312.4	370	2.9 ± 1.2
22. Bis(2-chloro-1-methylethyl) ether (108-60-1)		√		VOC	171.07	189	1700
23. Bis(2-chloroethoxy) methane (111-91-1)		√		B/N	NA	NA	NA

Compound	Structure			Class	MW		
24. Bis(2-chloroethyl) ether (111-44-4)	Cl–CH₂CH₂–O–CH₂CH₂–Cl	√		B/N	143.02	178	10200
25. Bis(2-chloroiso-propyl)ether (39638-32-9)	ClCH₂(CH₃)-CHOCH(CH₃)CH₂Cl	√		B/N	NA	NA	NA
26. Bis(2-ethylhexyl) phthalate (117-81-7)	(structure)	√	√	B/N	390.56	385	0.04–0.285
27. Bis(Chloromethyl) Ether (542-88-1)	Cl–CH₂–O–CH₂–Cl	√	√	VOC	114.96	104	NA
28. 4-Bromophenyl phenyl ether (101-55-3)	(structure)	√		B/N	249.11	304	NA
29. Camphechlor (Toxaphene) (8001-35-2)	(structure)	√		P	414 approx.	NA	3 (20°C)
30. Carbon tetrachloride (56-23-5)	CCl₄	√	√	VOC	153.82	76.7	1160
31. 4-Chlor-*m*-Cresol (59-50-7)	(structure)	√		A	142.6	235	4000
32. Chlordane (57-74-9)	(structure)	√		P	409.8	175	0.056
33. Chlorobenzene (108-90-7)	(structure)	√	√	VOC	112.56	132	500 (20°C)
34. Chlorodibromo-methane (124-48-1)	CHClBr₂	√		VOC	208.3	116–122	4400 (22°C)
35. Chloroethane (75-00-3)	CH₃CH₂Cl	√		VOC	64.5	12.4	5740 (20°C)
36. 2-Chloroethyl vinyl ether (110-75-8)	CH₂=CH–O–CH₂CH₂–Cl	√		VOC	106.55	109	3

Compound (CAS)	Structure			Class	MW	BP	Solubility
37. Chloroform (67-66-3)	(structure)	√	√	VOC	119.38	62	9300
38. Chloromethane (Methyl chloride) (74-87-3)	(structure)	√		VOC	51	−24	6500 (30°C, 1013 hPa)
39. 2-Chloronaphthalene (91-58-7)	(structure)	√		B/N	162.62	256	NA
40. 2-Chlorophenol (95-57-8)	(structure)	√		A	128.56	175.6	28500 (20°C)
41. 4-Chlorophenyl phenyl ether (7005-72-3)	(structure)	√		B/N	NA	NA	NA
42. Chrysene (218-01-9)	(structure)	√		B/N	228.2	488	0.006
43. 4,4'-DDD (72-54-8)	(structure)	√		P	320.1	112	0.160 (24°C)
44. 4,4'-DDE (72-55-9)	(structure)	√		P	NA	NA	0.065 (24°C)
45. 4,4'-DDT (50-29-3)	(structure)			P	354.5	NA	0.0031–0.0034
46. di-n-octyl phthalate (117-84-0)	(structure)	√		B/N	390.56	384	3
47. di-n-propylnitrosamine (621-64-7)	(structure)	√		B/N	130.22	77–78	10000 Approx
48. Dibenz(a,h)anthracene (53-70-3)	(structure)	√		B/N	278.35	524	0.0005

49. 1,2-Dibromoethane (106-93-4)		√		VOC	187.88	131.6	4310 (30°C)
50. Dibutyl phthalate (84-74-2)		√	√	B/N	278.34	340	400
51. 1,2-Dichloro-benzene (95-50-1)		√		B/N	147.01	179	145
52. 1,4-Dichloro-benzene(106-46-7)		√	√	B/N	147.01	173.4	79
53. 1,3-Dichloro-benzene (541-73-1)		√		B/N	147.01	172	123–143
54. 3,3'-Dichloro-benzidine (91-94-1)		√	√	B/N	253.13	>250	3.99 (22°C, pH 6.9)
55. Dichlorobromo-methane (75-27-4)		√		VOC	163.8	90	NA
56. 1,1-Dichloro-ethane (75-34-3)		√		VOC	98.96	57.3	5500 (20°C)
57. 1,2-Dichloro-ethane (107-06-2)		√		VOC	99	83.5	8690 (20°C)
58. 1,1-Dichloro-ethylene (75-35-4)		√		VOC	96.95	31.9	2500
59. Dichloromethane (Methylene chloride) (75-09-2)		√		VOC	84.93	40–42	16700
60. 2,4-Dichlorophenol (120-83-2)		√		A	163.01	210	4500
61. 1,2-Dichloro-propane (78-87-5)		√		VOC	112.99	96.8	2700 (20°C)
62. 1,3-Dichloropropene (Mixed Isomers) (542-75-6)		√	√	VOC	110.97	104 (cis) 112 (trans)	2770 (cis) 2800 (trans)

63. Dieldrin (60-57-1)		√			P	381	NA	0.1
64. Diethyl phthalate (84-66-2)		√			B/N	222.2	298	210
65. Dimethyl phthalate (131-11-3)		√	√		B/N	194.18	282	1744
66. 2,4-Dimethylphenol (105-67-9)		√			A	122.16	211.5	NA
67. 4,6-Dinitro-o-Cresol (534-52-1)		√	√		A	198.13	220	97 (20°C)
68. 2,4-Dinitrophenol (51-28-5)		√	√		A	184.11	NA	5600 (18°C)
69. 2,6-Dinitrotoluene (606-20-2)		√			B/N	182.14	285	208
70. 2,4-Dinitrotoluene (121-14-2)		√	√		B/N	182.13	300	270 (22°C)
71. 1,2-Diphenylhydrazine (122-66-7)		√	√		B/N	182.24	miscible	NA
72. α-Endosulfan (959-98-8)		√			P	NA	NA	NA

73. β-Endosulfan (33213-65-9)		√		P	NA	NA	NA
74. Endosulfan sulfate (1031-07-8)		√		P	NA	NA	NA
75. Endrin (72-20-8)		√		P	380.90	NA	NA
76. Endrin aldehyde (7421-93-4)		√		P	NA	NA	NA
77. Ethylbenzene (100-41-4)		√	√	VOC	106.17	136.2	206 (15°C)
78. Fluoranthene (206-44-0)		√		B/N	202	250	0.265
79. Fluorene (86-73-7)		√		B/N	NA	293	1.9
80. Heptachlor (76-44-8)		√	√	P	373.34	NA	NA
81. Heptachlor epoxide (1024-57-3)		√		P	389.30	NA	0.3
82. Hexachloro-1,3-butadiene (87-68-3)		√	√	B/N	261	210–220	2
83. Hexachlorobenzene (118-74-1)		√	√	B/N	284.80	326/322	0.004–0.006

84. Hexachlorocyclo-pentadiene(77-47-4)		√	√	B/N	273	234/239	0.13
85. Hexachloroethane (67-72-1)		√	√	B/N	236.76	NA	50 (22°C)
86. Indeno(1,2,3-c, d) pyrene (193-39-5)		√		B/N	276.34	536	0.062
87. Isophorone (78-59-1)		√	√	B/N	138.2	215	12000
88. α-Lindane (α-HCH) (319-84-6)		√	√	P	290.82	NA	1.4 (salt water)
89. β-Lindane (β-HCH (319-85-7)		√	√	P	NA	NA	NA
90. γ-Lindane (γ-HCH (58-89-9)		√	√	P	290.95	NA	17.0 (24°C)
91. δ-Lindane (δ-HCH) (319-86-8)		√	√	P	NA	NA	NA
92. Methanamine, N-methyl-N-nitroso (62-75-9)		√		VOC	74.08	151/153	NA
93. Methyl bromide (74-83-9)	—Br	√	√	VOC	94.95	4.6	900 (20°C)
94. N-Nitrosodiphenyl-amine (86-30-6)		√		B/N	198.24	NA	<1000 (19°C)

95. Naphthalene (91-20-3)		√	√	B/N	128.16	217.9	31–34
96. Nitrobenzene (98-95-3)		√	√	B/N	123.1	211	1900 (20°C)
97. 2-Nitrophenol (88-75-5)		√		A	139.11	214/217	2100 (20°C)
98. 4-Nitrophenol (100-02-7)		√	√	A	139.11	279	16000
99. Pentachlorophenol (87-86-5)		√	√	A	266.35	310	14 (20°C)
100. Phenanthrene (85-01-8)		√		B/N	178.22	340	0.82 (21°C)
101. Phenol (108-95-2)		√	√	A	94.11	182	82000 (15°C)
102. Pyrene (129-00-0)		√		B/N	202.26	360	0.16 (26°C)
103. 2,3,7,8-Tetra-chlorodibenzo-p-dioxin (TCDD) (1746-01-6)		√	√	TCDD	321.96	NA	NA
104. 1,1,2,2-Tetrachloroethane (79-34-5)		√	√	VOC	167.86	146.4	2900 (20°C)
105. Tetrachloroethylene (127-18-4)		√	√	VOC	165.83	121	150
106. 2,3,4,6-Tetrachlorophenol (58-90-2)		√		A	231.9	164	NA
107. Toluene (108-88-3)		√	√	VOC	92.1	110.8	515 (20°C)
108. 1,2-trans-Dichloroethylene (156-60-5)		√		VOC	96.95	48	6300 (20°C)

109. Tribromomethane (Bromoform) (75-25-2)	Br structure	√		VOC	252.77	149	3190 (30°C)
110. 1,2,4-Trichlorobenzene (120-82-1)	structure	√	√	B/N	181.46	213	19 (22°C)
111. 1,1,2-Trichloroethane (79-00-5)	structure	√	√	VOC	133.41	114	4500 (20°C)
112. 1,1,1-Trichloroethane (71-55-6)	structure	√		VOC	133.41	71-81	4400 (20°C)
113. Trichloroethylene (79-01-6)	structure	√	√	VOC	131.5	87	1100
114. 2,4,6-Trichlorophenol (88-06-2)	structure	√	√	A	197.46	244.5	800
115. Vinyl chloride (75-01-4)	structure	√	√	VOC	62.5	−14	1100

Note: (1) The selected chemicals are the priority organic pollutants listed by the U.S. EPA. They are arranged in an alphabetical order. A total of 15 inorganic pollutants are excluded in the table. Ref: Keith LH, Telliard WA (1979), Priority pollutants I – a perspective view. *Environ. Sci. Technol.*, 13(4):416–423. (2) PPI list = Organic compounds listed by the U.S. EPA as the priority pollutants. HAP list = Organic compounds listed by the U.S. EPA as the hazardous air pollutants. (3) Among 114 compounds, 31 are VOCs which are purgeable (VOC), the remaining 83 compounds are semi-volatile compounds (SVOCs), including 46 base / neutral extractable organic compounds (denoted as B/N), 11 acid extractable organic compounds (denoted as A), 18 pesticides (denoted as P), 7 PCBs, and a dioxin-series compound (TCDD). (4) Molecular weight (MW), boiling point, and solubility data are compiled from Verschueren K (2001), Handbook of Environmental Data on Organic Chemicals. 4th Edition. Vol. 1–2. John Wiley & Sons, NY. Solubility values are at 25°C unless otherwise indicated.

APPENDIX C1. STANDARD NORMAL CUMULATIVE PROBABILITIES

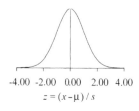

-4.00 -2.00 0.00 2.00 4.00

$z = (x - \mu) / s$

The entries in the table are the areas (probablities) under the standard normal curve from $z = 0$ to $z = z$. Examples are given below (see the shaded number in the table):

(1) $P\ (0 < z < 1.05) = 0.3531$
(2) $P\ (z > 1.05) = 0.5 - 0.3531 = 0.1469$
(3) $P\ (z < 1.05) = 0.5 + 0.3531 = 0.8531$
(4) $P\ (-1.05 < z < 0) = P(0 < z < 1.05) = 0.3531$
(5) $P\ (z > -1.05) = 0.5 + p(0 < z < 1.05) = 0.5 + 0.3531 = 0.8531$
(6) $P\ (z < -1.05) = P(z > 1.05) = 0.1469$

z	0	1	2	3	4	5	6	7	8	9
0.0	0.0000	0.0040	0.0080	0.0120	0.0160	0.0199	0.0239	0.0279	0.0319	0.0359
0.1	0.0398	0.0438	0.0478	0.0517	0.0557	0.0596	0.0636	0.0675	0.0714	0.0753
0.2	0.0793	0.0832	0.0871	0.0910	0.0948	0.0987	0.1026	0.1064	0.1103	0.1141
0.3	0.1179	0.1217	0.1255	0.1293	0.1331	0.1368	0.1406	0.1443	0.1480	0.1517
0.4	0.1554	0.1591	0.1628	0.1664	0.1700	0.1736	0.1772	0.1808	0.1844	0.1879
0.5	0.1915	0.1950	0.1985	0.2019	0.2054	0.2088	0.2123	0.2157	0.2190	0.2224
0.6	0.2257	0.2291	0.2324	0.2357	0.2389	0.2422	0.2454	0.2486	0.2517	0.2549
0.7	0.2580	0.2611	0.2642	0.2673	0.2704	0.2734	0.2764	0.2794	0.2823	0.2852
0.8	0.2881	0.2910	0.2939	0.2967	0.2995	0.3023	0.3051	0.3078	0.3106	0.3133
0.9	0.3159	0.3186	0.3212	0.3238	0.3264	0.3289	0.3315	0.3340	0.3365	0.3389
1.0	0.3413	0.3438	0.3461	0.3485	0.3508	0.3531	0.3554	0.3577	0.3599	0.3621
1.1	0.3643	0.3665	0.3686	0.3708	0.3729	0.3749	0.3770	0.3790	0.3810	0.3830
1.2	0.3849	0.3869	0.3888	0.3907	0.3925	0.3944	0.3962	0.3980	0.3997	0.4015
1.3	0.4032	0.4049	0.4066	0.4082	0.4099	0.4115	0.4131	0.4147	0.4162	0.4177
1.4	0.4192	0.4207	0.4222	0.4236	0.4251	0.4265	0.4279	0.4292	0.4306	0.4319
1.5	0.4332	0.4345	0.4357	0.4370	0.4382	0.4394	0.4406	0.4418	0.4429	0.4441
1.6	0.4452	0.4463	0.4474	0.4484	0.4495	0.4505	0.4515	0.4525	0.4535	0.4545
1.7	0.4554	0.4564	0.4573	0.4582	0.4591	0.4599	0.4608	0.4616	0.4625	0.4633
1.8	0.4641	0.4649	0.4656	0.4664	0.4671	0.4678	0.4686	0.4693	0.4699	0.4706
1.9	0.4713	0.4719	0.4726	0.4732	0.4738	0.4744	0.4750	0.4756	0.4761	0.4767
2.0	0.4772	0.4778	0.4783	0.4788	0.4793	0.4798	0.4803	0.4808	0.4812	0.4817
2.1	0.4821	0.4826	0.4830	0.4834	0.4838	0.4842	0.4846	0.4850	0.4854	0.4857
2.2	0.4861	0.4864	0.4868	0.4871	0.4875	0.4878	0.4881	0.4884	0.4887	0.4890
2.3	0.4893	0.4896	0.4898	0.4901	0.4904	0.4906	0.4909	0.4911	0.4913	0.4916
2.4	0.4918	0.4920	0.4922	0.4925	0.4927	0.4929	0.4931	0.4932	0.4934	0.4936
2.5	0.4938	0.4940	0.4941	0.4943	0.4945	0.4946	0.4948	0.4949	0.4951	0.4952
2.6	0.4953	0.4955	0.4956	0.4957	0.4959	0.4960	0.4961	0.4962	0.4963	0.4964
2.7	0.4965	0.4966	0.4967	0.4968	0.4969	0.4970	0.4971	0.4972	0.4973	0.4974
2.8	0.4974	0.4975	0.4976	0.4977	0.4977	0.4978	0.4979	0.4979	0.4980	0.4981
2.9	0.4981	0.4982	0.4982	0.4983	0.4984	0.4984	0.4985	0.4985	0.4986	0.4986
3.0	0.4987	0.4987	0.4987	0.4988	0.4988	0.4989	0.4989	0.4989	0.4990	0.4990
3.1	0.4990	0.4991	0.4991	0.4991	0.4992	0.4992	0.4992	0.4992	0.4993	0.4993
3.2	0.4993	0.4993	0.4994	0.4994	0.4994	0.4994	0.4994	0.4995	0.4995	0.4995
3.3	0.4995	0.4995	0.4995	0.4996	0.4996	0.4996	0.4996	0.4996	0.4996	0.4997
3.4	0.4997	0.4997	0.4997	0.4997	0.4997	0.4997	0.4997	0.4997	0.4997	0.4998
3.5	0.4998	0.4998	0.4998	0.4998	0.4998	0.4998	0.4998	0.4998	0.4998	0.4998
3.6	0.4998	0.4998	0.4999	0.4999	0.4999	0.4999	0.4999	0.4999	0.4999	0.4999
3.7	0.4999	0.4999	0.4999	0.4999	0.4999	0.4999	0.4999	0.4999	0.4999	0.4999
3.8	0.4999	0.4999	0.4999	0.4999	0.4999	0.4999	0.4999	0.4999	0.4999	0.4999
3.9	0.5000	0.5000	0.5000	0.5000	0.5000	0.5000	0.5000	0.5000	0.5000	0.5000

Note: Probablity values are calculated using a standard normal function @NORMSDIST(z) in Microsoft Ⓡ Excel

APPENDIX C2. PERCENTILES OF *t*-DISTRIBUTION

At a significant level of α, the 2-sided confidence level $= (1 - \alpha)^*100$, and the 1-sided confidence level $= (1-\alpha/2)^*100$
Example (see the shaded number in the table):
At a 90% 2-sided confidence level (or 95% 1-sided confidence level) and degree of freedom $df = 10, t = 1.812$

Significant level (α)	0.8	0.6	0.4	0.2	0.1	0.05	0.02	0.01
2-sided confidence level	20	40	60	80	90	95	98	99
1-sided confidence level	60	70	80	90	95	97.5	99	99.5
1	0.325	0.727	1.376	3.078	6.314	12.706	31.821	63.657
2	0.289	0.617	1.061	1.886	2.920	4.303	6.965	9.925
3	0.277	0.584	0.978	1.638	2.353	3.182	4.541	5.841
4	0.271	0.569	0.941	1.533	2.132	2.776	3.747	4.604
5	0.267	0.559	0.920	1.476	2.015	2.571	3.365	4.032
6	0.265	0.553	0.906	1.440	1.943	2.447	3.143	3.707
7	0.263	0.549	0.896	1.415	1.895	2.365	2.998	3.499
8	0.262	0.546	0.889	1.397	1.860	2.306	2.896	3.355
9	0.261	0.543	0.883	1.383	1.833	2.262	2.821	3.250
10	0.260	0.542	0.879	1.372	1.812	2.228	2.764	3.169
11	0.260	0.540	0.876	1.363	1.796	2.201	2.718	3.106
12	0.259	0.539	0.873	1.356	1.782	2.179	2.681	3.055
13	0.259	0.538	0.870	1.350	1.771	2.160	2.650	3.012
14	0.258	0.537	0.868	1.345	1.761	2.145	2.624	2.977
15	0.258	0.536	0.866	1.341	1.753	2.131	2.602	2.947
16	0.258	0.535	0.865	1.337	1.746	2.120	2.583	2.921
17	0.257	0.534	0.863	1.333	1.740	2.110	2.567	2.898
18	0.257	0.534	0.862	1.330	1.734	2.101	2.552	2.878
19	0.257	0.533	0.861	1.328	1.729	2.093	2.539	2.861
20	0.257	0.533	0.860	1.325	1.725	2.086	2.528	2.845
21	0.257	0.532	0.859	1.323	1.721	2.080	2.518	2.831
22	0.256	0.532	0.858	1.321	1.717	2.074	2.508	2.819
23	0.256	0.532	0.858	1.319	1.714	2.069	2.500	2.807
24	0.256	0.531	0.857	1.318	1.711	2.064	2.492	2.797
25	0.256	0.531	0.856	1.316	1.708	2.060	2.485	2.787
26	0.256	0.531	0.856	1.315	1.706	2.056	2.479	2.779
27	0.256	0.531	0.855	1.314	1.703	2.052	2.473	2.771
28	0.256	0.530	0.855	1.313	1.701	2.048	2.467	2.763
29	0.256	0.530	0.854	1.311	1.699	2.045	2.462	2.756
30	0.256	0.530	0.854	1.310	1.697	2.042	2.457	2.750
40	0.255	0.529	0.851	1.303	1.684	2.021	2.423	2.704
50	0.255	0.528	0.849	1.299	1.676	2.009	2.403	2.678
60	0.254	0.527	0.848	1.296	1.671	2.000	2.390	2.660
120	0.254	0.526	0.845	1.289	1.658	1.980	2.358	2.617
∞	0.253	0.524	0.842	1.282	1.645	1.960	2.327	2.576

Note: The critical *t* values are calculated using a function @TINV(α, df) in Microsoft ® Excel

APPENDIX C3. CRITICAL VALUES FOR THE *F*-DISTRIBUTION

The entries in the table give the critical F-values that are exceeded with probablity 0.05 by random values from F-distributions with degree of freedom df_1 in the numerator (population with a larger s^2) and df_2 in the denominator (population with a smaller s^2).

Example (see the shaded number in the table):
The critical F-value for $df_1 = 10$ and $df_2 = 15$ is 2.54. If F has $df_1 = 10$ and $df_2 = 15$, then Probablity $(F > 2.54) = 0.05$.

df_2 \ df_1	1	2	3	4	5	6	7	8	9	10	11	12	13	14	15	20	30	40	50	60
1	161.4	199.5	215.7	224.6	230.2	234.0	236.8	238.9	240.5	241.9	243.0	243.9	244.7	245.4	245.9	248.0	250.1	251.1	251.8	252.2
2	18.5	19.0	19.2	19.2	19.3	19.3	19.4	19.4	19.4	19.4	19.4	19.4	19.4	19.4	19.4	19.4	19.5	19.5	19.5	19.5
3	10.13	9.55	9.28	9.12	9.01	8.94	8.89	8.85	8.81	8.79	8.76	8.74	8.73	8.71	8.70	8.66	8.62	8.59	8.58	8.57
4	7.71	6.94	6.59	6.39	6.26	6.16	6.09	6.04	6.00	5.96	5.94	5.91	5.89	5.87	5.86	5.80	5.75	5.72	5.70	5.69
5	6.61	5.79	5.41	5.19	5.05	4.95	4.88	4.82	4.77	4.74	4.70	4.68	4.66	4.64	4.62	4.56	4.50	4.46	4.44	4.43
6	5.99	5.14	4.76	4.53	4.39	4.28	4.21	4.15	4.10	4.06	4.03	4.00	3.98	3.96	3.94	3.87	3.81	3.77	3.75	3.74
7	5.59	4.74	4.35	4.12	3.97	3.87	3.79	3.73	3.68	3.64	3.60	3.57	3.55	3.53	3.51	3.44	3.38	3.34	3.32	3.30
8	5.32	4.46	4.07	3.84	3.69	3.58	3.50	3.44	3.39	3.35	3.31	3.28	3.26	3.24	3.22	3.15	3.08	3.04	3.02	3.01
9	5.12	4.26	3.86	3.63	3.48	3.37	3.29	3.23	3.18	3.14	3.10	3.07	3.05	3.03	3.01	2.94	2.86	2.83	2.80	2.79
10	4.96	4.10	3.71	3.48	3.33	3.22	3.14	3.07	3.02	2.98	2.94	2.91	2.89	2.86	2.85	2.77	2.70	2.66	2.64	2.62
11	4.84	3.98	3.59	3.36	3.20	3.09	3.01	2.95	2.90	2.85	2.82	2.79	2.76	2.74	2.72	2.65	2.57	2.53	2.51	2.49
12	4.75	3.89	3.49	3.26	3.11	3.00	2.91	2.85	2.80	2.75	2.72	2.69	2.66	2.64	2.62	2.54	2.47	2.43	2.40	2.38
13	4.67	3.81	3.41	3.18	3.03	2.92	2.83	2.77	2.71	2.67	2.63	2.60	2.58	2.55	2.53	2.46	2.38	2.34	2.31	2.30
14	4.60	3.74	3.34	3.11	2.96	2.85	2.76	2.70	2.65	2.60	2.57	2.53	2.51	2.48	2.46	2.39	2.31	2.27	2.24	2.22
15	4.54	3.68	3.29	3.06	2.90	2.79	2.71	2.64	2.59	2.54	2.51	2.48	2.45	2.42	2.40	2.33	2.25	2.20	2.18	2.16
16	4.49	3.63	3.24	3.01	2.85	2.74	2.66	2.59	2.54	2.49	2.46	2.42	2.40	2.37	2.35	2.28	2.19	2.15	2.12	2.11
17	4.45	3.59	3.20	2.96	2.81	2.70	2.61	2.55	2.49	2.45	2.41	2.38	2.35	2.33	2.31	2.23	2.15	2.10	2.08	2.06
18	4.41	3.55	3.16	2.93	2.77	2.66	2.58	2.51	2.46	2.41	2.37	2.34	2.31	2.29	2.27	2.19	2.11	2.06	2.04	2.02
19	4.38	3.52	3.13	2.90	2.74	2.63	2.54	2.48	2.42	2.38	2.34	2.31	2.28	2.26	2.23	2.16	2.07	2.03	2.00	1.98
20	4.35	3.49	3.10	2.87	2.71	2.60	2.51	2.45	2.39	2.35	2.31	2.28	2.25	2.22	2.20	2.12	2.04	1.99	1.97	1.95
21	4.32	3.47	3.07	2.84	2.68	2.57	2.49	2.42	2.37	2.32	2.28	2.25	2.22	2.20	2.18	2.10	2.01	1.96	1.94	1.92
22	4.30	3.44	3.05	2.82	2.66	2.55	2.46	2.40	2.34	2.30	2.26	2.23	2.20	2.17	2.15	2.07	1.98	1.94	1.91	1.89
23	4.28	3.42	3.03	2.80	2.64	2.53	2.44	2.37	2.32	2.27	2.24	2.20	2.18	2.15	2.13	2.05	1.96	1.91	1.88	1.86
24	4.26	3.40	3.01	2.78	2.62	2.51	2.42	2.36	2.30	2.25	2.22	2.18	2.15	2.13	2.11	2.03	1.94	1.89	1.86	1.84
25	4.24	3.39	2.99	2.76	2.60	2.49	2.40	2.34	2.28	2.24	2.20	2.16	2.14	2.11	2.09	2.01	1.92	1.87	1.84	1.82
26	4.23	3.37	2.98	2.74	2.59	2.47	2.39	2.32	2.27	2.22	2.18	2.15	2.12	2.09	2.07	1.99	1.90	1.85	1.82	1.80
27	4.21	3.35	2.96	2.73	2.57	2.46	2.37	2.31	2.25	2.20	2.17	2.13	2.10	2.08	2.06	1.97	1.88	1.84	1.81	1.79
28	4.20	3.34	2.95	2.71	2.56	2.45	2.36	2.29	2.24	2.19	2.15	2.12	2.09	2.06	2.04	1.96	1.87	1.82	1.79	1.77
29	4.18	3.33	2.93	2.70	2.55	2.43	2.35	2.28	2.22	2.18	2.14	2.10	2.08	2.05	2.03	1.94	1.85	1.81	1.77	1.75
30	4.17	3.32	2.92	2.69	2.53	2.42	2.33	2.27	2.21	2.16	2.13	2.09	2.06	2.04	2.01	1.93	1.84	1.79	1.76	1.74
40	4.08	3.23	2.84	2.61	2.45	2.34	2.25	2.18	2.12	2.08	2.04	2.00	1.97	1.95	1.92	1.84	1.74	1.69	1.66	1.64
60	4.00	3.15	2.76	2.53	2.37	2.25	2.17	2.10	2.04	1.99	1.95	1.92	1.89	1.86	1.84	1.75	1.65	1.59	1.56	1.53
120	3.92	3.07	2.68	2.45	2.29	2.18	2.09	2.02	1.96	1.91	1.87	1.83	1.80	1.78	1.75	1.66	1.55	1.50	1.46	1.43

Note: The critical F values are calculated using a function @FINV(α, df_1, df_2) in Microsoft ® Excel

APPENDIX D. REQUIRED CONTAINERS, PRESERVATION TECHNIQUES, AND HOLDING TIMES

Parameter	Container[1]	Preservative[2]	Holding Time[3]
Physical & Aggregate			
Acidity / alkalinity	P, G	Cool, 4°C	14 days
Hardness	P, G	HNO_3 or H_2SO_4 to pH <2	6 months
Residual (total)	P, G	Cool, 4°C	7 days
Residual (filterable)	P, G	Cool, 4°C	7 days
Residual (nonfilterable)	P, G	Cool, 4°C	7 days
Residual (settleable)	P, G	Cool, 4°C	48 hours
Residual (volatile)	P, G	Cool, 4°C	7 days
Specific conductance	P, G	Cool, 4°C	28 days
Temperature	P, G	Not required	Analyze ASAP
Turbidity	P, G	Cool, 4°C	48 hours
Inorganic Non-metallic			
Ammonia	P, G	Cool, 4°C, H_2SO_4 to pH<2	28 days
Bromide / Chloride / Fluoride	P, G	Not required	28 days
Chlorine, total residual	P, G	Not required	Analyze ASAP
Cyanide, total & amenable to chlorination	P, G	Cool, 4°C, NaOH to pH >12; 0.6 g ascorbic acid if residual Cl_2 present	14 days (24 hours if S^{2-} is present
Dissolved oxygen (probe)	G with top	Not required	Analyze ASAP
Dissolved oxygen (Winkler)	G with top	Fix on site, store in dark	8 hours
Kjeldahl and organic nitrogen	P, G	Cool, 4°C, H_2SO_4 to pH<2	28 days
Nitrate	P, G	Cool, 4°C	48 hours
Nitrite	P, G	Cool, 4°C	48 hours
Nitrate plus nitrite	P, G	Cool, 4°C, H_2SO_4 to pH<2	28 days
pH (hydrogen ion)	P, G	Not required	Analyze ASAP
Orthophosphate	P, G	Cool, 4°C, filter ASAP	48 hours
Phosphorus (elemental)	G only	Cool, 4°C	48 hours
Phosphorus (total)	P, G	Cool, 4°C, H_2SO_4 to pH<2	28 days
Silica	P only	Cool, 4°C	28 days
Sulfate	P, G	Cool, 4°C	28 days
Sulfide	P, G	Cool, 4°C, add zinc acetate plus NaOH to pH<2	7 days
Sulfite	P, G	Not required	Analyze ASAP

Parameter	Container[1]	Preservative[2]	Holding Time[3]
Metals[4]			
Chromium (VI)	P, G	Cool, 4°C	24 hours
Mercury	P, G	HNO_3 to pH <2	28 days
Metals, except above	P, G	HNO_3 to pH <2	6 month
Aggregate Organic			
BOD	P, G	Cool, 4°C	48 hours
COD	P, G	Cool, 4°C, H_2SO_4 to pH<2	28 days
Oil & Grease	G only	Cool, 4°C, HCl or H_2SO_4 to pH<2	28 days
Organic carbon	P, G	Cool, 4°C, HCl or H_2SO_4 to pH<2	28 days
Surfactants	P, G	Cool, 4°C	48 hours
Individual Organic			
Purgeable halocarbons	G, Teflon-lined septum	Cool, 4°C, 0.008% $Na_2S_2O_3$	14 days
Purgeable aromatics	G, Teflon-lined septum	Cool, 4°C, 0.008% $Na_2S_2O_3$	14 days or 7 days without pH adjustment
Acrolein & acrylonitrile [5]	G, Teflon-lined septum	Cool, 4°C, 0.008% $Na_2S_2O_3$, adjust pH to 4–5	14 days
Phenols	G, Teflon-lined cap	Cool, 4°C, 0.008% $Na_2S_2O_3$	7 d until extraction, 40 d after extraction
Benzidines [6]	G, Teflon-lined cap	Cool, 4°C, 0.008% $Na_2S_2O_3$	7 d until extraction
Phthalate esters	G, Teflon-lined cap	Cool, 4°C	7 d until extraction, 40 d after extraction
Nitrosamines	G, Teflon-lined cap	Cool, 4°C, store in dark, 0.008% $Na_2S_2O_3$	7 d until extraction, 40 d after extraction
PCBs	G, Teflon-lined cap	Cool, 4°C	7 d until extraction, 40 d after extraction
Nitroaromatics & isophorone	G, Teflon-lined cap	Cool, 4°C, store in dark, 0.008% $Na_2S_2O_3$	7 d until extraction, 40 d after extraction
PAHs	G, Teflon-lined cap	Cool, 4°C, store in dark, 0.008% $Na_2S_2O_3$	7 d until extraction, 40 d after extraction
Haloethers	G, Teflon-lined cap	Cool, 4°C, 0.008% $Na_2S_2O_3$	7 d until extraction, 40 d after extraction

(*Continued*)

Parameter	Container[1]	Preservative[2]	Holding Time[3]
Chlorinated hydrocarbons	G, Teflon-lined cap	Cool, 4°C	7 d until extraction, 40 d after extraction
TCDD (dioxin)	G, Teflon-lined cap	Cool, 4°C	7 d until extraction, 40 d after extraction
Pesticides [7]	G, Teflon-lined cap	Cool, 4°C, pH 5–9	7 d until extraction, 40 d after extraction
Radiological Tests			
Alpha, beta, and radium	P, G	HNO_3 to pH < 2	6 months
Bacterial Tests			
Coliform, fecal & total	P, G sterile	Cool, 4°C, 0.008% $Na_2S_2O_3$	6 hours
Fecal streptococci	P, G sterile	Cool, 4°C, 0.008% $Na_2S_2O_3$	6 hours
Soil, sediment, sludge			
Metals	P	Cool, 4°C	6 months
Organic volatile	Widemouth G, Teflon-lined cap	Cool, 4°C	Analyze ASAP
Organic extractable	Widemouth G, Teflon-lined cap	Cool, 4°C	Analyze ASAP
Biological sample			
Fish samples	Wrap in Al foil	Freeze	Analyze ASAP

Source: 40 CFR 136.3. Notes: [1] P = Polyethylene, G = glass; [2] 0.008% $Na_2S_2O_3$ is required only if residual chlorine is present; [3] A longer holding time than listed in the table requires data validation to show that the specific types of samples under study are stable. ASAP = as soon as possible.
[4] For dissolved metals, samples should be filtered immediately on-site before adding preservatives.
[5] The pH adjustment is not required if acrolein will not be measured. Sample for acrolein receiving no pH adjustment must be analyzed within 3 days of sampling. [6] Extracts may be stored up to 7 days before analysis if storage is conducted under an inert (oxidant-free) atmosphere. [7] The pH adjustment may be performed upon receipt at the laboratory and may be omitted if the samples are extracted within 72 hours of collection. For the analysis of aldrin, add 0.008% $Na_2S_2O_3$.

Index

MPM 051018
Printed in Singapore